WILLIAM F. MAAG LIBRARY
YOUNGSTOWN STATE UNIVERSITY

The Mechanical Properties of Semiconductors

SEMICONDUCTORS
AND SEMIMETALS
Volume 37

Semiconductors and Semimetals

A Treatise

Edited by R. K. Willardson
CONSULTING PHYSICIST
SPOKANE, WASHINGTON

Albert C. Beer
CONSULTING PHYSICIST
COLUMBUS, OHIO

Eicke R. Weber
DEPARTMENT OF MATERIALS SCIENCE
AND MINERAL ENGINEERING
UNIVERSITY OF CALIFORNIA AT BERKELEY

The Mechanical Properties of Semiconductors

SEMICONDUCTORS
AND SEMIMETALS

Volume 37

Volume Editors

KATHERINE T. FABER
NORTHWESTERN UNIVERSITY
EVANSTON, ILLINOIS

KEVIN J. MALLOY
THE UNIVERSITY OF NEW MEXICO
ALBUQUERQUE, NEW MEXICO

ACADEMIC PRESS, INC.
Harcourt Brace Jovanovich, Publishers
Boston   San Diego   New York
London   Sydney   Tokyo   Toronto

This book is printed on acid-free paper. ∞

Copyright © 1992 by Academic Press, Inc.
All rights reserved.
No part of this publication may be reproduced or
transmitted in any form or by any means, electronic
or mechanical, including photocopy, recording, or
any information storage and retrieval system, without
permission in writing from the publisher.

ACADEMIC PRESS, INC.
1250 Sixth Avenue, San Diego, CA 92101-4311

United Kingdom Edition Published by
ACADEMIC PRESS LIMITED
24-28 Oval Road, London NW1 7DX

The Library of Congress has catalogued this serial title as follows:

Semiconductors and semimetals.—Vol. 1—New York: Academic Press. 1966-

    v.: ill.; 24 cm.

Irregular.
Each vol. has also a distinctive title.
Edited by R. K. Willardson, Albert C. Beer, and Eicke R. Weber
ISSN 0080-8784 = Semiconductors and semimetals

1. Semiconductors—Collected works. 2. Semimetals—Collected works.
I. Willardson, Robert K. II. Beer, Albert C. III. Weber, Eicke R.
QC610.9.S48   621.385'2—dc19   85-642319
                                                           AACR 2   MARC-S

Library of Congress   [8709]
ISBN 0-12-752137-2 (v. 37)

PRINTED IN THE UNITED STATES OF AMERICA
92 93 94 95 BC 9 8 7 6 5 4 3 2 1

Contents

CONTRIBUTORS viii
PREFACE ix

Chapter 1 **Elastic Constants and Related Properties of Semiconductor Compounds and Their Alloys**   1

A.-B. Chen, Arden Sher and W. T. Yost

I.	Introduction	2
II.	Measurement Methods	11
III.	Theoretical and Experimental Results	25
IV.	Dislocations and Hardness	63
V.	Concluding Remarks	70
	Acknowledgments	72
	References	73

Chapter 2 **Fracture of Silicon and Other Semiconductors**   79

David R. Clarke

I.	Introduction	80
II.	Crack Extension: The Mechanics of Fracture	83
III.	Fracture Mechanisms	90
IV.	Fracture of Semiconductors	106
V.	The Brittle-to-Ductile Transition	116
VI.	Environmental Effects	133
VII.	Closing Remarks	135
	Acknowledgments	136
	References	136
	Appendix: Fracture Resistance Measurements by Indentation	140

Chapter 3 The Plasticity of Elemental and Compound Semiconductors 143

Hans Siethoff

I.	Introduction	143
II.	General	145
III.	Lower Yield Point and Creep at Inflection Point	154
IV.	Dynamical Recovery	164
V.	Effect of High Doping	175
VI.	Addendum	182
	Acknowledgments	184
	References	184

Chapter 4 Mechanical Behavior of Compound Semiconductors 189

Sivaraman Guruswamy, Katherine T. Faber and John P. Hirth

I.	Introduction	190
II.	General Deformation Behavior of Elemental and Compound Semiconductors	190
III.	Hardness of III-V and II-VI Compounds	196
IV.	Compressive/Tensile Strength of Compound and Isovalent-Doped Compound Semiconductors	200
V.	Non-isovalent Group II and VI Dopants in III-V Compounds	218
VI.	The Role of Si Doping in GaAs	220
VII.	Crystal Growth	226
VIII.	Summary	226
	References	226

Chapter 5 Deformation Behavior of Compound Semiconductors 231

S. Mahajan

I.	Introduction	231
II.	Dislocations and Deformation-Induced Stacking Faults	233
III.	Deformation Characteristics of Binary Compound Semiconductors	237
IV.	Photoplastic Effects in Compound Semiconductors	252
V.	Atomic Ordering and Surface Phase Separation in Ternary and Quaternary III-V Semiconductors and Their Ramifications in Degradation Resistance of Light-Emitting Devices	257
VI.	Summary	264
	Acknowledgments	265
	References	265

Chapter 6 Injection of Dislocations into Strained Multilayer Structures — 267

John P. Hirth

I.	Introduction	267
II.	Dislocation Behavior	268
III.	Interface Equilibrium	271
IV.	Partial Equilibrium Arrays	280
V.	Dislocation Spreading	282
VI.	Dislocation Injection	284
VII.	Summary	290
	Acknowledgments	291
	References	291

Chapter 7 Critical Technologies for the Micromachining of Silicon — 293

Don L. Kendall, Charles B. Fleddermann, and Kevin J. Malloy

I.	Introduction	293
II.	Wet Chemical Etching of Single-Crystal Silicon	295
III.	Dry Etching Techniques for Micromachining	311
IV.	Bonding and Micromachining	317
V.	Conclusions	333
	Acknowledgments	333
	References	333

Chapter 8 Processing and Semiconductor Thermoelastic Behavior — 339

Ikuo Matsuba and Kinji Mokuya

I.	Introduction	339
II.	Thermoelastic Model of Dislocations	342
III.	Prediction of Defect Onset	355
IV.	Application of Thermoelastic Model	361
V.	Conclusions	365
	Acknowledgments	366
	References	366
	List of Symbols	367

INDEX — 369
CONTENTS OF VOLUMES IN THIS SERIES — 377

List of Contributors

Numbers in parentheses indicate the pages on which the authors' contributions begin.

A.-B. CHEN, *Department of Physics, Auburn University, Auburn, Alabama 36849* (1)

DAVID R. CLARKE, *Materials Department, College of Engineering, University of California, Santa Barbara, California 93106* (79)

KATHERINE T. FABER, *Department of Materials Science and Engineering, Northwestern University, 2225 Sheridan Road, Evanston, Illinois 60208–3108* (189)

CHARLES B. FLEDDERMANN, *The Center for High Technology Materials, The University of New Mexico, EECE Building, Room 125, Albuquerque, NM 87131* (293)

SIVARAMAN GURUSWAMY, *Department of Metallurgical Engineering, University of Utah, 412 William C. Browning Building, Salt Lake City, Utah 84112* (189)

JOHN P. HIRTH, *Department of Mechanical and Materials Engineering, Washington State University, Pullman, Washington 99164–2920* (189, 267)

DON KENDALL, *The Center for High Technology Materials, The University of New Mexico, EECE Building, Room 125, Albuquerque, NM 87131* (293)

S. MAHAJAN, *Department of Materials Science, Carnegie Mellon University, Pittsburgh, PA 15213* (231)

KEVIN J. MALLOY, *The Center for High Technology Materials, The University of New Mexico, EECE Building, Room 125, Albuquerque, NM 87131* (293)

IKUO MATSUBA, *Systems Development Laboratory, Hitachi, Ltd., 1099 Ohzenji, Asao-ku, Kawasaki-shi, 215 Japan* (339)

KINJI MOKUYA, *Takasaki Works, Hitachi, Ltd., 111 Nishiyokote-machi, Takasaki-shi, 370–11 Japan* (339)

HANS SIETHOFF, *Physical Institute, University of Würzburg, D-8700 Würzburg, Germany* (143)

ARDEN SHER, *Physical Electronics Laboratory, SRI International, 333 Ravenswood Avenue, Menlo Park, California, 94025* (1)

W. Y. YOST, *NASA Langley Research Center, Hampton, Virginia 23665* (1)

Preface

While the electronic and optical properties of semiconductors have been and are still the subject of extensive investigation, the mechanical properties of semiconductors often dictate fundamental limits on the fabrication and packaging of modern semiconductor devices. The goal of this volume is to describe the mechanical properties of semiconductors and the role they play in modern semiconductor technology. To that end, we have chosen to emphasize mechanical properties *per se* and refer the reader elsewhere for the influence of mechanical properties on electrical and optical properties. Mechanical properties define the regions of dislocation-free boule growth and epitaxial layer growth. They define processing limitations and often dominate issues related to packaging and failure of semiconductor devices. We have chosen to emphasize the fundamental mechanical phenomena instead of merely detailing examples of these phenomena.

The volume begins with a discussion of elastic properties of semiconductors, including elemental, compound and the pseudobinary alloy semiconductors. It is the elastic constant that dictates the first response of the material under mechanical or thermal loading. Experimental measurements of these constants are examined in Chapter 1 and are compared to three theoretical methods for estimating the elastic constants: *ab initio* calculations, valence force field methods, and tight binding theory. The emphasis of the chapter is on the comparison of these theoretical predictions of elastic constants with experimental measurements and a discussion of the accuracy of the approximations needed for timely theoretical calculations. Chapter 2 describes the conditions for failure when the elastic limit of brittle semiconductors is reached. Both the mechanics and the mechanisms by which failure occurs are discussed, along with ample examples of measured fracture energies and microscopic evidence for fracture mechanisms. Perhaps most revealing is recent *in situ* transmission electron microscopic evidence for the brittle-to-ductile transition that occurs at elevated temperature.

Three chapters (Chapters 3, 4 and 5) are devoted to deformation of semiconductor materials. Chapter 3 provides the necessary background for understanding plasticity in semiconducting materials and demonstrates how these phenomena are applied to both elemental and compound semiconductors. The stages of dynamical recovery during creep and constant strain-rate experiments are reviewed in this chapter. The role of dopants in deformation processes in elemental semiconductors is also explored. Chapter 4 is devoted to deformation of compound semiconductors and the influence of both isovalent and non-isovalent dopants. The implication of solid solution strengthening by dopants as it affects crystal growth is reviewed. In the final part of the sequence (Chapter 5), the deformation of ternary and quaternary semiconductor alloys is included. Photoplastic effects, whereby electron-hole recombinations occur near a dislocation and the released energy facilitates glide, are also discussed.

As a further treatise on deformation behavior, Chapter 6 examines conditions for the stability of strained layer superlattices. In order to avoid the nucleation of misfit dislocations at interfaces between the strained layers, an equilibrium argument is used to establish a critical layer thickness. These results have significant practical import for the growth of devices based on strained multilayers.

A further practical aspect of device fabrication involves the micromachining of semiconducting materials to provide three-dimensional structures. Wet chemical etching as well as dry etching techniques are reviewed in Chapter 7. The physics of bonding and its relevance to micromachining also is considered.

Finally, the thermoelastic behavior of semiconductors as it relates to annealing and growth of oxide films on semiconductors is described in Chapter 8. A model of dislocation generation is proposed by considering the temperature profile of the wafer during growth of the oxide film and the conditions whereby the temperature distribution is altered and the thermal stresses reduced. The model has been applied successfully to the fabrication of VLSIs.

It is intended that this volume be useful for both the physical or materials scientist interested in the science of fracture and deformation, as well as for crystal growers who desire a more fundamental understanding of the parameters that influence the growth process. It is further anticipated that electrical engineers or device manufacturers would be interested in this volume to provide models in which an understanding of the fundamental science can directly aid in the production of more reliable devices.

We are indebted to the contributors and editors for making this volume possible.

Katherine T. Faber
Kevin J. Malloy

CHAPTER 1

Elastic Constants and Related Properties of Semiconductor Compounds and Their Alloys

A.-B. Chen

PHYSICS DEPARTMENT
AUBURN UNIVERSITY
AUBURN, ALABAMA

and

Arden Sher

PHYSICAL ELECTRONICS LABORATORY
SRI INTERNATIONAL
MENLO PARK, CALIFORNIA

and

W.T. Yost

NASA LANGLEY RESEARCH CENTER
HAMPTON, VIRGINIA

I. INTRODUCTION	2
1. *Goals*	2
2. *Definition and Calculation of Elastic Constants*	2
3. *Elastic Constants and Sonic Wave Propagation*	6
II. MEASUREMENT METHODS	11
4. *Velocity Measurements*	11
5. *Ultrasonic Measurement Techniques*	15
6. *Corrections to Pulse Measurements: An Example*	18
7. *Optical Techniques*	22
III. THEORETICAL AND EXPERIMENTAL RESULTS	24
8. *Ab Initio Theory*	24
9. *Valence Force Field Model*	29
10. *Tight-Binding Theory*	33
11. *Semiconductor Alloys*	50
IV. DISLOCATIONS AND HARDNESS	63
12. *Slip Systems*	64
13. *Peierls Energy*	65
14. *Temperature Dependence*	70
V. CONCLUDING REMARKS	70
ACKNOWLEDGMENTS	72
REFERENCES	73

I. Introduction

1. Goals

The goals of this paper are to review the current state of knowledge of the elastic constants of elemental, compound, and pseudobinary alloy semiconductors. To accomplish this objective, we will review

- experimental methods currently used to measure elastic constants,
- experimental results,
- binding and elastic constant theory, and
- related mechanical properties.

Heavy emphasis is placed on comparisons between theory and experiment, and the accuracy of approximations currently in vogue. The theories discussed range from first-principles methods, requiring heavy comutations, to parametrized physical models. The intent is to identify a logical path between these extremes, and thereby provide insight into the connection between atom potentials and semiconductor mechanical properties. In the course of this presentation, there are a number of instances where improvements in the theory or additional experimental results, are needed. We have tried to highlight these situations and suggest possible remedies.

2. Definition and Calculation of Elastic Constants

The theory of elasticity of solids has been well formulated in many treatises (Love, 1944; Landau and Lifshitz, 1959), so we need not review the different formalisms and conventions. However, as we do wish to present a coherent account of the essence of the theory, it is necessary to define terms and describe general calculations to be used later.

In linear elasticity theory, deformation is assumed to be infinitesimal. The relative displacement vector $\tilde{x}'$ between two points in a deformed solid is related to the corresponding vector $\tilde{x}$ in the undeformed solid by the following equation, in component form, through a nine-component strain sensor ε:

$$x'_\alpha = x_\alpha + \Sigma_\beta \varepsilon_{\alpha\beta} x_\beta. \tag{1}$$

The change in the internal energy associated with ε is also small and will be denoted UV, where V is the equilibrium volume of a solid. Under the condition that the entropy and the electrostatic displacement field are constant, U is only a function of ε, and is quadratic in ε:

$$U = \tfrac{1}{2}\Sigma_{\alpha\beta\mu\nu}\varepsilon_{\alpha\beta}C_{\alpha\beta\mu\nu}\varepsilon_{\mu\nu}, \tag{2}$$

where $C_{\alpha\beta\mu\nu}$ are the elastic stiffness coefficients, which are characteristic properties of the solid. From this definition, $C_{\mu\nu\alpha\beta} = C_{\alpha\beta\mu\nu}$ is required for U to be an analytical function of ε. Moreover, $C_{\alpha\beta\mu\nu} = C_{\beta\alpha\mu\nu} = C_{\alpha\beta\nu\mu}$ is also required to ensure that U is zero under any infinitesimal rigid rotation. In light of these properties, the energy density can be expressed in terms of a symmetrical strain tensor η as

$$U = \tfrac{1}{2}\Sigma_{\alpha\beta\mu\nu}\eta_{\alpha\beta}C_{\alpha\beta\mu\nu}\eta_{\mu\nu}, \tag{3}$$

where $\eta_{\alpha\beta}$ is defined as

$$\eta_{\alpha\beta} = \tfrac{1}{2}(\varepsilon_{\alpha\beta} + \varepsilon_{\beta\alpha}). \tag{4}$$

The strain tensor η is a thermodynamic parameter with stress tensor σ as its conjugate variable (Brugger, 1964). The components of σ are given by

$$\sigma_{\alpha\beta} = \frac{\partial U}{\partial \eta_{\alpha\beta}} = \Sigma_{\mu\nu}C_{\alpha\beta\mu\nu}\eta_{\mu\nu}. \tag{5}$$

It is clear that σ is also a symmetrical tensor.

The most frequently used notation is the engineering convention, in which the strain tensor e is related to η of Eq. (4) by $e_{\alpha\alpha} = \eta_{\alpha\alpha}$ for the diagonal components, but $e_{\alpha\beta} = 2\eta_{\alpha\beta}$ for $\alpha \neq \beta$. Furthermore, since e has at most six independent components, it is treated as a six-component vector, with the vector components 1 to 6 corresponding respectively to the tensor components xx, yy, zz, yz, xz and xy. In terms of e the strain energy density is written as

$$U = \tfrac{1}{2}\Sigma_{ij}C_{ij}e_ie_j, \tag{6}$$

where C_{ij} can be identified as $C_{\alpha\beta\mu\nu}$ with $i = \alpha\beta$ and $j = \mu\nu$. Because C_{ij} is symmetrical, it has at most 21 independent components for any crystal. Crystal symmetries reduce this number further (Love, 1944; Ashcroft and Mermin, 1976). For a cubic lattice, to which class the zincblende and diamond semiconductors belong, there are only three independent components, namely C_{11}, C_{12}, and C_{44}.

The three independent elastic constants of the diamond and zincblende (zb) semiconductors can be calculated by considering the following three strains.

(i) Under a uniform expansion, which changes a displacement $\vec{x}$ into $\vec{x}' = (1 + e)\vec{x}$, then $e_1 = e_2 = e_3 = e$ and the other strain components are zero; the elastic energy density is then given by

$$U = 3(C_{11} + 2C_{12})e^2/2. \tag{7}$$

U can also be expressed in terms of the adiabatic bulk modulus B defined by $\delta P = -B(\delta V/V)$, where V is the crystal volume, δV is its change, and δP is the corresponding pressure change. The result is $U = B(\delta V/V)^2/2 = 9Be^2/2$, because the dilatation is $\delta V/V = 3e$ in the present case. This establishes the relationship

$$B = (C_{11} + 2C_{12})/3. \tag{8}$$

(ii) The next case to consider is a tetragonal shear strain e which changes a displacement according to

$$(x, y, z) \rightarrow (x + ex, y - ey, z). \tag{9}$$

The only nonzero strain components are $e_1 = -e_2 = e$. Then U becomes

$$U = (C_{11} - C_{12})e^2. \tag{10}$$

(iii) To calculate C_{44} we consider a shear strain e that changes a displacement according to

$$(x, y, z) \rightarrow (x + ey/2, y + ex/2, z). \tag{11}$$

This strain contains e_6 as the only nonzero component. Because C_{44} equals C_{66}, the energy density is simply

$$U = C_{44} e^2/2. \tag{12}$$

Although the macroscopic crystal distortion of the Bravais lattice caused by this strain is described by Eq. (11), microscopically there is a relative displacement $\vec{u} = (0, 0, u)$, the so-called Kleiman (1962) internal displacement, between two successive atomic planes perpendicular to the z axis. In other words, the relative displacements between the atoms on the same fcc sublattice are governed by Eq. (11), but there is an additional induced relative

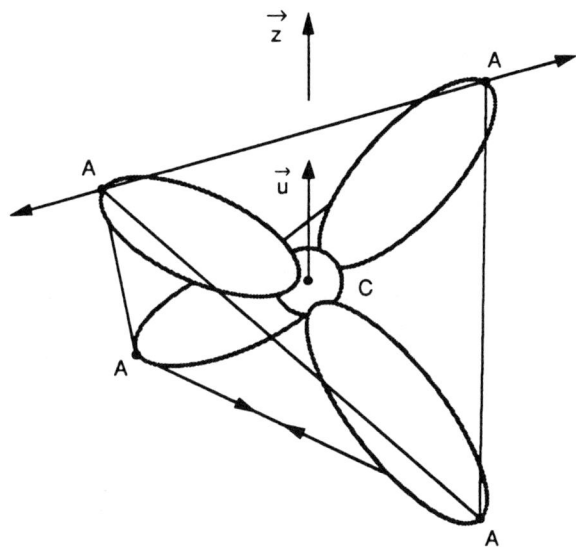

FIG. 1. Distortion of a tetrahedron corresponding to the C_{44} elastic constant; u is the internal dispacement between the anion and cation sublattices.

displacement $\tilde{u}$ between the two sublattices. The directions of the displacements of atoms in a tetrahedral cell are shown in Fig. 1. In calculations one can use an arbitrary infinitesimal pair of e and u to obtain the coefficients in the following quadratic expansion of the strain energy density:

$$U = \Phi u^2/2 + Deu + C_{44}^{(0)}e^2/2, \tag{13}$$

where the force function Φ is related to the transverse optical (TO) phonon frequency ω at Γ (the center of Brillouin zone) by $\Phi = \mu\omega^2$, with μ being the reduced mass. In Eq. (13), D is a constant, and $C_{44}^{(0)}$ would be the shear stiffness coefficient if the internal displacement were not allowed. Kleiman (1962) defined an internal displacement parameter that is related to the equilibrium value of u by $u = \zeta ae/4$ for a fixed e, where a is the lattice constant. Taking the first derivative of U with respect to u in Eq. (13) and setting it equal to zero, one finds that the ζ value is given by

$$\zeta = -4D/(a\Phi). \tag{14}$$

Finally, the sought-after C_{44} is given by

$$C_{44} = C_{44}^{(0)} - \zeta^2 a^2 \Phi/16. \tag{15}$$

Thus, the internal displacement is an essential part of C_{44}. These procedures will be used in the theoretical calculations to be presented in subsequent sections.

3. Elastic Constants and Sonic Wave Propagation

The purpose of this section is to give the reader sufficient information to measure ultrasonic wave velocities and to deduce from them the elastic coefficients of materials of interest to the electronics industry. We start from the definition of strain and develop equations that relate the elastic coefficients with wave propagation velocities in various crystallographic directions. The cases taken as examples, cubic and isotopic, were chosen because of their prevalent role in electronic materials.

A number of techniques used to measure wave propagation velocities in materials are discussed. The researcher's decision on technique depends upon a number of factors, such as accuracy needed, size and crystallographic orientation of available samples, equipment, etc. Sample preparation is significant, since a carelessly prepared sample yields useless velocity data. Because the equipment needs vary widely among the techniques that can be used, references containing information on each technique are given. Key papers chosen for citation were picked for their readability and their direct application to the measurement technique. Primary emphasis has been given to techniques that are generally accepted and that give results that are understood. Throughout this section a general theory of elasticity due to Murnaghan (1951) is used.

a. Elastic-to-Wave Relationship

Consider, as before, a point in a lossless elastic medium at rest, whose displacement vector, $\vec{x} = (x_\alpha)$, $\alpha = 1, 2, 3$, are the cartesian coordinates of a point in the medium. In the presence of a stress wave disturbance, the point moves to a new location, $\vec{x}' = (x'_\alpha)$ at time t. The Lagrangian strains, η, defined consistently with Eqs. (1) and (4) but with higher-order terms retained (Landau and Lifshitz, 1986; Murnaghan, 1951), are as follows:

$$\eta_{\eta\nu} = \frac{1}{2}\left(\frac{\partial x'_\alpha}{\partial x_\mu}\frac{\partial x'_\alpha}{\partial x_\nu} - \delta_{\mu\nu}\right) \tag{16}$$

where $\delta_{\mu\nu}$ is the Kronecker delta. Einstein summation convention over repeated indices is used in this section.

By considering the internal energy per unit mass of the medium $E(x_\alpha, \eta_{\alpha\beta}, S)$, where S is the entropy, we obtain a relationship between wave propagation and the internal energy of the medium. If we restrict this discussion to a medium that is initially unstressed, then the internal energy per unit volume of the medium is $\rho_0 E(r_\alpha, \eta_{\alpha\beta}, S)$, where ρ_0 is the mass density of the medium in the unstrained state. By expanding this in a power series of the Lagrangian strains, we obtain

$$\rho_0 E(x_\alpha, \eta_{\alpha\beta}, S) = \rho_0 E(x_\alpha, 0, S) + \frac{1}{2!} C_{\mu\nu\alpha\beta} \eta_{\mu\nu} \eta_{\alpha\beta} + \frac{1}{3!} C_{\mu\nu\alpha\beta\gamma\delta} \eta_{\mu\nu} \eta_{\alpha\beta} \eta_{\gamma\delta} + \cdots \quad (17)$$

The coefficients $C_{\mu\nu\alpha\beta}$ and $C_{\mu\nu\alpha\beta\gamma\delta}$ are the second-order and third-order elastic coefficients (adiabatic) as defined by Brugger (1964):

$$C_{\mu\nu\alpha\beta} = \rho_0 \left(\frac{\partial^2 E}{\partial \eta_{\mu\nu} \partial \eta_{\alpha\beta}} \right), \quad (18)$$

$$C_{\mu\nu\alpha\beta\gamma\delta} = \rho_0 \left(\frac{\partial^3 E}{\partial \eta_{\mu\nu} \partial \eta_{\alpha\beta} \partial \eta_{\gamma\delta}} \right). \quad (19)$$

The wave equation can be derived (Goldstein, 1965, p. 347) from Lagrange's equation

$$\frac{d}{dt}\left(\frac{\partial L}{\partial \dot{x}_\mu}\right) + \frac{d}{dx_\alpha}\left(\frac{\partial L}{\partial\left(\frac{\partial x'_\mu}{\partial x_\alpha}\right)}\right) - \frac{\partial L}{\partial x'_\mu} = 0, \quad (20)$$

where $\dot{x}_\mu = \partial x_\mu / \partial t$ and L is given by

$$L = \tfrac{1}{2}\rho_0 \dot{u}_\mu \dot{u}_\mu - \rho_0 E(x_\alpha, \eta_{\alpha\beta}, S). \quad (21)$$

Defining the particle displacement vector, **u**, as

$$\mathbf{u}_\mu(\vec{r}, t) = x'_\mu - x_\mu, \quad (22)$$

and differentiating it gives

$$\frac{\partial u_\mu}{\partial x_\nu} = \frac{\partial x'_\mu}{\partial x_\nu} - \delta_{\mu\nu}. \quad (23)$$

Combining Eqs. (22), (23), and (16) gives the strain tensor in terms of the particle displacements,

$$\eta_{\mu\nu} = \frac{1}{2}\left(\frac{\partial u_\mu}{\partial x_\nu} + \frac{\partial u_\nu}{\partial x_\mu} + \frac{\partial u_\beta}{\partial x_\mu}\frac{\partial u_\beta}{\partial x_\nu}\right). \tag{24}$$

Using Eqs. (21) to (24) in Eq. (20), and retaining first-order terms in $\partial u_\mu/\partial x_\nu$, gives

$$\rho_0 \ddot{u}_\mu = \frac{d}{dx_\nu}\left(C_{\mu\nu\alpha\beta}\frac{\partial u_\alpha}{\partial x_\beta}\right). \tag{25}$$

We assume a plane wave solution to Eq. (25) of the form

$$u_\mu = \hat{e}_\mu \cdot \hat{\mathbf{u}} \cos(k_\nu x_\nu - \omega t), \tag{26}$$

where

$\hat{e}_\mu$ = unit displacement vector in the μth direction,
$\hat{\mathbf{u}} = (u_\mu)$ = particle displacement vector,
$\vec{k} = (k_\nu)$ = wave propagation vector, and
ω = angular frequency.

Substituting Eq. (26) into Eq. (25), we obtain the eigenvalue–eigenvector equation

$$|C_{\mu\nu\alpha\beta}\kappa_\nu\kappa_\beta - \rho_0 v^2 \delta_{\mu\alpha}| = 0, \tag{27}$$

where

(κ_μ) = the set of direction cosines of $\vec{\kappa}$,
$v = \omega/|\vec{k}|$ is the phase speed, and
$\vec{\kappa} = \vec{k}/|k|$ indicates the direction of wave propagation in the medium.

For any given wave propagation direction, there are three eigenvalue solutions to Eq. (27); these correspond to one quasilongitudinal and two quasitransverse polarization modes. Relationships have also been derived for wave propagation in the presence of residual stresses and for retention of the higher-order elastic coefficients (Cantrell, 1982; Breazeale and Ford, 1965).

Because both the stress and strain tensors are symmetric, one can reduce the number of independent second-order elastic coefficients to 21 (Landau and Lifshitz, 1986, p. 32). It is also convenient to use the Voight (1928) contraction of the indices as in Eq. (6), which is used in the remainder of this section. The number of independent elastic coefficients varies with crystal structure[1] and is listed in the accompanying table.

Crystal Class	Number of Independent Elastic Coefficients
Triclinic	21
Monoclinic	13
Orthorhombic	9
Tetragonal (C_4, S_4, C_{4h})	7
Tetragonal (C_{4v}, D_{2d}, D_4, D_{4h})	6
Rhombohedral (C_3, S_6)	7
Rhombohedral (C_{3v}, D_3, D_{3d})	6
Hexagonal	5
Cubic	3
(Isotropic)	2

Green (1973) and others have solved Eq. (27) for isotropic materials and for cubic crystals in the [100], [110], and [111] directions. Using the symmetry properties of the crystals, one obtains the results presented in Section I.3.b for cubic crystals, and in Section I.3.c for isotropic solids.

b. Cubic Crystals

For plane waves propagating along the [100] direction:

$$(\kappa_1 = 1, \kappa_2 = 0, \kappa_3 = 0);$$

longitudinal (compressional) waves (pure mode with particle displacements u_1 in direction of propagation, u_2 and $u_3 = 0$) have a phase speed

$$v_1 = \sqrt{\frac{C_{11}}{\rho_0}}; \tag{28}$$

and transverse (shear) waves (pure mode with particle displacements $u_1 = 0$, u_2 and u_3 perpendicular to the direction of propagation) have values given by

$$v_2 = v_3 = \sqrt{\frac{C_{44}}{\rho_0}}. \tag{29}$$

[1] We assume that the angles defining the orientation of axes in the crystal are not specified. For a further discussion, see Landau and Lifshitz (1986).

For plane waves propagating along the [110] direction:

$$(\kappa_1 = 1/\sqrt{2},\ \kappa_2 = 1/\sqrt{2},\ \kappa_3 = 0);$$

for longitudinal (compressional) waves (pure mode with particle displacements $u_1 = u_2$, $u_3 = 0$), one obtains

$$v_1 = \sqrt{\frac{C_{11} + C_{12} + 2C_{44}}{2\rho_0}};\qquad(30)$$

and for transverse (shear) waves (pure mode with $u_1 = -u_2$, $u_3 = 0$), one obtains

$$v_2 = \sqrt{\frac{C_{11} - C_{12}}{2\rho_0}};\qquad(31)$$

or for transverse (shear) waves (pure mode with $u_1 = u_2 = 0$, $u_3 \neq 0$), one obtains

$$v_3 = \sqrt{\frac{C_{44}}{2\rho_0}}.\qquad(32)$$

For plane waves propagating along the [111] direction:

$$(\kappa_1 = 1/\sqrt{3},\ \kappa_2 = 1/\sqrt{3},\ \kappa_3 = 1/\sqrt{3});\qquad(33)$$

the speed for longitudinal (compressional) waves (pure mode with $u_1 = u_2 = u_3$) is given by

$$v_1 = \sqrt{\frac{C_{11} + 2C_{12} + 4C_{44}}{3\rho_0}};\qquad(34)$$

and that for transverse (shear) waves (pure mode with $u_1\kappa_1 + u_2\kappa_2 + u_3\kappa_3 = 0$ or particle displacement perpendicular to wave propagation) is given by

$$v_2 = v_3 = \sqrt{\frac{C_{11} - C_{12} + C_{44}}{3\rho_0}}\qquad(35)$$

c. Isotropic Solids

For the case of isotropy, all directions are equivalent, we are left with two independent constants (Lamé constants). The elastic coefficients can be expressed in terms of the Lame' constants (Green, 1973) as follows:

$$\begin{aligned} C_{11} &= C_{22} = C_{33} = \lambda + 2\mu, \\ C_{12} &= C_{13} = C_{23} = \lambda, \\ C_{21} &= C_{31} = C_{32} = \lambda, \\ C_{44} &= C_{55} = C_{66} = \mu. \end{aligned} \qquad (36)$$

For plane waves propagating in any direction (e.g., the x-direction):

$$(\kappa_1 = 1, \kappa_2 = 0, \kappa_3 = 0).$$

the speed for longitudinal (compressional) wave (pure mode with particle displacements u_1 in direction of propagation, u_2 and $u_3 = 0$) is given by

$$v_1 = \sqrt{\frac{\lambda + 2\mu}{\rho_0}}, \qquad (37)$$

and that for transverse (shear) waves (pure mode with particle displacements $u_1 = 0$, and u_2 and u_3 perpendicular to the direction of propagation) is given by

$$v_2 = v_3 = \sqrt{\frac{\mu}{\rho_0}}. \qquad (38)$$

II. Measurement Methods

4. Velocity Measurements

To measure the second-order elastic coefficients (SOEC), one can determine the sound velocity and the density of the sample, and calculate the combination of SOEC that the direction of propagation requires. For cubic systems, sound velocities measured in the pure mode directions equivalent to [100], [110], and [111] make it possible to determine the three independent elastic constants, C_{11}, C_{12}, and C_{44}. Using care in sample preparation, and appropriate corrections for bond thickness and diffraction, the sound velocities can be determined to parts in 10^5. The accuracy of a velocity determination is usually limited by the accuracy of the path length measurement in pulsed and continuous wave techniques. With optical techniques, deflection angles, wavelength measurements, and frequencies are determinant factors.

In general, there is little problem in obtaining a sound velocity measurement on the order of 1% uncertainty. But to improve on this, one must exercise additional care in the preparation of surfaces, the control of temperature, the determination of acoustic path length, and the determination of travel time of the acoustic wave.[2] For single transducer configurations, where a transducer is used to send and receive the acoustic wave, the base equation for determining the velocity using pulsed techniques or pulse echo without corrections is as follows:

$$v = \frac{2L}{T}. \tag{39}$$

where v is the wave velocity, L is the sample length, and T is the round-trip time. Corrections[3] are made in the evaluation of T and vary with the technique used in the measurement. Corrections for bond thickness (McSkimin, 1961), and for diffraction effects (Papadakis, 1967) can be measured and/or calculated. Accuracies of some of the more frequently used pulsed techniques are given in Table I. A comparison of the accuracies of various techniques are discussed by Papadakis (1972, 1976). Using continuous wave techniques, the propagating plane wave model (Bolef and Miller, 1971), the basic equation without corrections is

$$v = 2L\Delta f, \tag{40}$$

where Δf is the frequency difference between two adjacent mechanical resonance modes. Typically, velocity measurement accuracy is good to about 1% to 10%. A correction factor (Chern et al., 1981) can be applied to the right side of Eq. (39) that corrects for the effects of the transducer and bond on the mechanical resonances of the sample. When applied, the inaccuracy can be as low as five parts per 100,000, neglecting inaccuracies in path length measurements.

If the sample is transparent, optical techniques offer a convenient method to measure sound velocity. For some of these techniques, the accuracy is on order of parts per thousand. Still, some offer accuracies of parts per ten thousand and can be used with small samples, and very high-frequency ultrasonic waves (Breazeale et al., 1981). In general, the calculation for

---

[2]The accuracy depends upon the correct choice for the resonant frequency of the transducer, which can be obtained from the fact that it is half a wavelength in thickness. Using a micrometer, and the wave velocity of the transducer material, one can calculate the resonant frequency of the transducer. Transducer off-resonant conditions can have a relatively large influence on the measurement of round-trip time for pulsed studies.

[3]Other influences on the measurement of round-trip time are bond thickness and diffraction. These are discussed in various papers, including McSkimin (1961) and Papadakis (1967, 1972).

TABLE I

ACCURACIES OF MEASUREMENT TECHNIQUES

Method	Features	Accuracy	Corrections
Pulsed			
Pulse echo	· cm or longer samples · one transducer · commercial units available	a few parts in 1,000	none
Gated double pulse superposition[a]	· cm or longer samples · matched transducers	1 part in 10,000	bond
Pulse superposition[b,c]	· cm or longer samples · one transducer	1 part in 5,000[a]	bond, diffraction estimate (b)
Pulse echo overlap[d,e]	· cm or longer samples · one transducer · commercial units available	(delay time accuracy several parts per 1,000,000)	bond, diffraction
Long pulse buffer rod[f]	· thin lossy samples (as thin as 2 mm)	1 or 2 parts in 10,000	bond
Continuous Wave	· bonds must be thin · cm or longer length · sample preparation more critical		requires corrections for all results. Otherwise, errors can be very large.[p] Difficult to correct for diffraction. All these techniques require compound resonator corrections.
Continuous wave transmission[g,h,i]	· can be automated · electromagnetic crosstalk can give problems	parts in 10,000,000 under ideal conditions; considerably less with thick bonds and/or thin samples[k]	
Sampled continuous wave method[j,k]	· single transducer · eliminates electromagnetic crosstalk · can be automated		
Spectrometer methods[l,m]	· can track small changes · can require specialized transducers		
Optical Methods (for Use with Optically Transparent Media)			
Measurement of wavelength	· uses traveling microscope to measure wavelength	0.1 to 0.01 %	

TABLE I (*Continued*)

ACCURACIES OF MEASUREMENT TECHNIQUES

Method	Features	Accuracy	Corrections
Raman–Nath diffraction	• measurement of angles of incidence and diffraction to measure wavelength of sound	1%	
Bragg diffraction	• useful at frequencies between ≈ 100 MHz and several GHz • useful for measurement of local variations in velocity	0.01 to 0.1%	
Brillouin scattering	• measurement of optical frequency shift and angle • useful near phase transitions • frequencies in the GHz range	1 to 3% (0.1% precision)	
Stimulated Brillouin scattering[n,o]	• needs high power electromagnetic source • measurement can result in fracture of sample • frequencies to 60 GHz reported	3%	

[a] Williams and Lamb (1985).
[b] McSkimin (1961).
[c] Papadakis (1972).
[d] Papadakis (1967).
[e] Papadakis (1972).
[f] McSkimin (1950).
[g] Bolef et al. (1962).
[h] Bolef and Miller (1971).
[i] Chern et al. (1981).
[j] Miller (1973).
[k] Chern et al. (1981).
[l] Conradi et al. (1974).
[m] Heyman (1976).
[n] Chiao et al. (1964).
[o] Brewer and Rieckhoff (1964).
[p] Chern et al. (1981).

velocity involved the determination of acoustic wavelength in the medium, by using the equation

$$v = f\lambda, \tag{41}$$

where f is the frequency of the sound wave, determined by the drive frequency of the transducer, and λ is the wavelength. Brillouin scattering also makes it possible to measure the sound velocity (Beyer and Letcher, 1969). In addition to measurement of an angle, the accuracy also depends on determination of the wavelength of light in the medium; this requires a determination of the index of refraction of the medium.

5. ULTRASONIC MEASUREMENT TECHNIQUES

Surveys of techniques and details for measuring ultrasonic velocities are available from various sources (Breazeale et al., 1981; Papadakis, 1976; Truell et al., 1969). Consideration in this work will be given to three classes of techniques for absolute velocity measurements. A compilation of some of the features of each technique is given in Table I.

a. Sample Preparation

The tolerance selected for preparing the sample depends largely upon the accuracy needed in the determination. If a 1% measurement is in order, surface preparation and parallelism are less critical than if an accuracy of 0.01% is desired. Also important is the consideration of correction factors needed to compensate for bond thickness and diffraction effects. At present, diffraction corrections exist only for compressional waves. Therefore, under equivalent experimental conditions, the most accurate determination of combinations of elastic coefficients would be those that are calculated from compressional wave velocities.

It is assumed that a surface on the sample has been ground to optical tolerances, and that the crystallographic direction of the axis of this surface has been determined. Typically, the crystallographic directions are measured to within minutes of arc (McSkimin and Andreatch, 1964). The remaining critical issues are the parallelism of the reflecting surface to the reference surface, and the flatness of these surfaces. Parallelism of the surfaces can be measured, for example, with an autocolliminator or a He–Ne gas laser and some mirrors. By placing the sample's reference surface on a stationary flat, dust-free surface, rotating the sample, and measuring the diameter of the circle traced by the beam reflected from the sample's top surface, and measuring the total path length of the beam from the sample to the image screen, one can determine the parallelism of the surfaces. The surfaces should be parallel to an angle better than 0.01 $\lambda_{\text{acoustic}}$/transducer diameter. For a typical half-inch diameter, 10 MHz transducer and a typical solid, the surfaces must be parallel to better than 4×10^{-4} radians (1 degree, 23 minutes). Higher frequencies, multiple reflections, and continuous-wave techniques require proportionally smaller tolerances.

Typically, one also tries for a flatness of better than 1/100 of an acoustic wavelength for accurate determination of transit times. If the resonant frequency of the transducer is 10 MHz, then the wavelength in a solid is in the neighborhood of 5×10^{-4} meters, which puts the flatness requirement near 5 micrometers. Using an optical flat and Newton's rings analysis, one can determine the flatness of the sample. For higher-frequency transducers and continuous-wave techniques, the number of fringes are appropriately decreased. At 100 MHz, pulsed mode, for example, the tolerance is down to less than several fringes. By using flat lapping surfaces, this is easily achieved. When multiple reflections are involved in the measurement, one must consider that for each reflection, changes occur in the wavefront direction; the phase change across the surface of the transducer at each reflection is due to lack of parallelism between the faces. For a case where a 20 MHz transducer was used, and multiple reflections employed for the measurement, Papadakis (1967) quotes a sample surface parallelism and flatness of better than 10^{-4} inches per inch.

b. Piezoelectric Transducers

Selection of the piezoelectric transducer depends largely upon the wave mode, the electronic equipment to be used for the measurement, and the personal preference of the researcher. In general, piezoelectric crystals[4] such as quartz (high electric impedance) and lithium niobate, or poled ceramics such as PZT (lead zirconate titanate), are chosen. Their physical and electrical properties are covered elsewhere (O'Donnell *et al.*, 1981). Other piezoelectric transducers, including polymeric materials,[5] are also available. The transducer diameter should be smaller than the sample to assure that propagation modes are not affected by the location of lateral boundaries (Tu

---

[4]Information on piezoelectric transducers is available from Valpey-Fisher Corp., 75 South Street, Hopkinton, Massachusetts 01748, and Crystal Technology Inc., 1060 E. Meadow Circle, Palo Alto, California 94303.

[5]A film sold under the tradename KYNAR is an example of this. Information on this material can be obtained from Pennwalt Corp., Box C, King of Prussia, Pennsylvania 19406-0018. Some of the properties are covered in Bloomfield *et al.* (1978). Lead attachment to KYNAR films is covered in Scott and Bloomfield (1981).

et al., 1955). This is especially important where measurements are taken on small samples that may have their lateral sides close to each other. One also must consider the problem as it relates to wave propagation, since wave modes other than those considered here can be excited in materials of small dimensions. From studies on cylindrical specimens, the minimum sample diameter can be no less than approximately 2.5λ, where λ is the wavelength of the ultrasound in the medium. The transducer diameter is chosen after the frequency is selected. Generally, the transducer diameter should be no longer than half the diameter of the sample, to prevent any interference in the measurement from reflections off the lateral boundaries caused by diffraction effects.

Selection of a bonding material depends upon the type of wave (compressional or shear), and the temperature range through which the measurements are to be taken. At room temperature, a good choice for both types of waves in phenyl salicylate or phenyl benzoate (Papadakis, 1964), while Dow-Corning DC-200 silicone is good for longitudinal waves. At other temperatures (McSkimin, 1957), other bonding materials, such as Nonaq stopcock grease, are used for compressional and shear waves. Various resins can be used with shear wave transducers; commercially prepared bonding materials are also available.[6]

Application of the bond requires care to keep surfaces clean and free of dust. The bond should be as thin as practical, taking care not to break the transducer. Often, applying some heat helps with viscous bonding materials. When using phenyl salicylate or phenyl benzoate for bonds, certain procedures can be followed to assure uniform bonds (Papadakis, 1964a).

c. Noncontacting Transducers

There are three types of non-contacting excitation transducers: capacitive transducers (compressional wave excitation only) (Cantrell and Breazeale, 1977), electromagnetic transducers (EMAT) (Vasile and Thompson, 1977; Johnson and Mase, 1984), and optically stimulated acoustic transducers (Prosser and Green, 1985). With the exception of the electrostatic transducer (compressional with circular piston geometry), diffraction correction data have not been developed. However, bond corrections are not necessary as these methods generate the wave directly on the sample surface. Generally, the signal levels are quite small, and require high gain and specialized circuits or devices to bring the signals to usable levels.

---

[6]For example, suppliers of damped ultrasonic transducers can supply material suitable for high temperature and shear measurements. Two companies are Panametrics, Inc., 221 Crescent St., Waltham, Massachusetts 02254, and Harisonics, Inc., 7 Hyde St., Stamford, Connecticut 06907.

6. Corrections to Pulse Measurements: An Example

With some pulsed systems, and with care in preparing the sample and in taking the measurements, one can expect round-trip time determinations to have standard deviations in the neighborhood of 100 picoseconds. As corrections for bond thickness and diffraction are often larger than this, it is necessary to correct for these sources of systemic error where possible. As an example, consider, for a properly prepared sample, that the pulse-echo overlap technique is to be used to measure the round-trip time for a tone burst. The time between the first-received and the second-received echos will be determined. For simplicity, we will treat the case of only one round trip in the sample. The measured time T is composed of several terms: the true travel time δ, the bond thickness contribution Δ^B, and the diffraction contribution Δ^D.

a. Bond Thickness

The model for the sample-bonded transducer system has been developed by Williams and Lamb (1958) and has been used extensively by McSkimin (1961) and by Papadakis (1967). They showed that if the transducer bond sample system is treated as an acoustic transmission line, it is possible to calculate the effects of bond thickness on the measurement of transit time. McSkimin (1961) showed that by measuring the change in the transit time when the system is detuned by a known amount, it is possible to determine the bond thickness in terms of measured quantities. This permits one to correct for the transit time through the bond material. Papadakis (1967) discussed the correction in some detail, and applied it to his measurements on fused quartz and silicon.

Consider an undamped transducer bonded to one end of a sample with some bonding material of known velocity and density; a relatively long wave is reflected from the interface between the sample and the transducer bond system. The reflected wave from this interface experiences a phase shift from the impedance mismatch, which can be calculated from the model. The calculation uses the real and imaginary parts of the effective impedance of the components to obtain the contribution to the phase shift. By detuning the frequency of the applied tone burst, we can measure a corresponding change in transit time. Then, a comparison with the measured values permits the calculation of the bond thickness, so that its effect on the measurement can be determined. The approach is outlined below.

The reflection coefficient for the system described above is given by

$$\frac{E_b}{E_i} = \frac{Z_d - Z_s}{Z_d + Z_s}, \qquad (42)$$

where E_b is the reflected pressure wave, E_i is the incident pressure wave, Z_s is the specific acoustic impedance (Elmore and Heald, 1969) of the sample ($Z_s = \rho c$, where ρ is the mass density, and c is the wave propagation speed), and Z_d is the effective specific acoustic impedance of the transducer bond system, given by McSkimin (1961),

$$Z_d = jZ_1 \left[\frac{\left(\dfrac{Z_1}{Z_2}\right)\tan k_1 \Delta_1 + \tan k_2 \Delta_2}{\left(\dfrac{Z_1}{Z_2}\right) - \tan k_2 \Delta_1 \tan k_1 \Delta_1} \right], \qquad (43)$$

where Z_1 and Z_2 are the specific acoustic impedances of the bond material and transducer material, respectively. Using similar conventions, k_1 and k_2 are propagation constants in the respective materials, as Δ_1 and Δ_2 are the respective bond and transducer thicknesses, and j is $\sqrt{-1}$.

The phase angle (McSkimin, 1961) is calculated by writing Eq. (42) in complex polar form and gives

$$\gamma = \arctan\left(\frac{2|Z_d|Z_s}{|Z_d|^2 - Z_s^2}\right), \qquad (44)$$

where $|Z_d|$ is the modulus of Z_d, and $(\pi + \gamma)$ is the phase shift between the reflected and incident pressure waves impinging on the interface of the transducer bond system with the sample.

When the system is driven at the transducer resonant frequency, f_r can be calculated by using the fact that the transducer at fundamental resonance has a thickness of one-half of a wavelength; then $|Z_d|$ becomes $Z_1 \tan k_1 \Delta_1$, which is generally quite small for a thin bond. Frequencies of off-transducer resonance, generally chosen as $0.9 f_r$, give a larger phase angle, since the dependence of Z_d upon off-resonance frequency excitation is large. As outlined in McSkimin (1961) and Papadakis (1967), we can calculate the time difference between phase-matched echo trains, ΔT, caused by the change of drive frequency, and compare it with experimental results to determine the correct condition for overlap.

Assume an alignment of, for example, the first and second echoes in the sample. The change in the measured time between echos is given by

$$\Delta T = \frac{1}{f_L}\left(n - \frac{p\gamma_L}{2\pi}\right) - \frac{1}{f_R}\left(n - \frac{p\gamma_R}{2\pi}\right), \qquad (45)$$

where p is the number of round trips (for our case, $p = 1$) of the ultrasonic tone burst in the sample, n is the number of cycles of mismatch in the tone burst, f_L is the off-transducer resonance drive frequency, f_R is the drive frequency at transducer resonance, γ_L is the phase angle at the off-resonance frequency (radians), and γ_R is the phase angle (radians) at the transducer resonance. One can calculate ΔT for the case of no cycle mismatch ($n = 0$) and no bond thickness ($\gamma_R = 0$) from Eqs. (43) through (45). The measured value closest to the calculated value determines the correct experimental cycle for cycle match. Following the experimental identification of the correct match, the measured ΔT can be used to determine the correction for bond thickness by adjusting Δ_1 in Eq. (43) and solving Eq. (44) to bring Eq. (45) into agreement with the measured ΔT values. The phase angle at resonance can be determined, and the travel time correction for bond thickness, Δ^B, is given by

$$\Delta^B = \frac{\gamma_R}{2\pi f_R}, \qquad (46)$$

where γ_R and f_R are the phase angle, and drive frequency respectively, both taken at transducer resonance.

b. Diffraction Corrections

Diffraction effects for compressional waves in various crystalline symmetries have been treated by several investigators (Seki et al., 1956; Papadakis, 1963, 1964b, 1966; Williams, 1970), while others have treated isotropic media (Benson and Kiyohara, 1974; Khimunin, 1972; Rogers and Van Buren, 1974). Papadakis (1972) has shown that without diffraction corrections, one can expect errors in travel times as large as $0.25/f_R$. He also discussed (1972) diffraction corrections for the technique of pulse-echo overlap technique. Using the dimensionless quantity

$$S = \frac{z\lambda}{a^2}, \qquad (47)$$

where S is the Seki parameter, a is the transducer radius, λ is the wavelength in the sample, and z is the distance of propagation, he wrote the phase correction due to diffraction, Δ^D, as

$$\Delta^D = \frac{[\phi(S_n) - \phi(S_m)]}{2\pi f}, \tag{48}$$

where ϕ is the phase shift in radians of the received wave front that is due to diffraction effects. For the example, the sample is of length L, which gives $S_n = 2L\lambda/a^2$, and $S_m = 4L\lambda/a^2$.

The true travel time, δ, can be written in terms of the measured time, T, and the corrections:

$$\delta = T + \Delta^B + \Delta^D. \tag{49}$$

The thickness of the sample, L, is measured by conventional means, such as a high-precision micrometer, and the velocity of sound is given by

$$v = \frac{2L}{\delta}. \tag{50}$$

7. Optical Techniques

Optical techniques have been used with success in the measurement of sound velocities in transparent media. The methods include diffraction and scattering of light by sound waves. A direct method of optically measuring an acoustic wavelength in the material can also be used.

The diffraction techniques considered here permit the determination of the wavelength of sound. The wavelength, λ^*, is combined with the ultrasonic frequency, f, of the sound beam to calculate its velocity, according to the expression

$$f\lambda^* = v, \tag{51}$$

where v is the speed of sound in the sample. In any optical technique, however, one must specify the type of diffraction experienced by the light.

a. Fraunhofer Diffraction of Light by Sound

Consider an ultrasonic transducer, bonded to a transparent solid, generating ultrasonic compressional waves that propagate into the solid. Further suppose that the sound wave encounters light traversing the same medium so that the light is diffracted. There are two different physical regimes that can produce the diffraction effects. One involves the formation of a corrugation in the phase fronts of the light that is due to the spatial variation in the index of refraction of the solid, as caused by the sound waves. This is called Raman–Nath diffraction (Raman and Nath, 1935a, 1935b, 1936a, 1936b, 1936c; Born and Wolf, 1970). The second regime involves the reflection of light from the evenly spaced crests of the sound waves. These reflections occur under some conditions that are similar to x-ray diffraction by a crystal lattice; this is called Bragg diffraction (Bhatia and Noble, 1953). Both types of diffraction can be used to determine the sound velocity of the material.

In order to determine which type of diffraction effect predominates (Nomoto, 1942), a dimensionless parameter (Klein et al., 1965), Q, is defined as

$$Q = \frac{K^{*2}L}{\mu_0 K}, \tag{52}$$

where K^* is the ultrasonic propagation constant, L is the width of the ultrasonic beam, μ_0 is the index of refraction, and K is the propagation constant of light in vacuum. If $Q > 9$, one has Bragg diffraction. If $Q < 1$, Raman–Nath diffraction occurs. For $1 < Q < 9$, the diffraction is mixed. For illustrative purposes, consider a typical transparent solid ($c_{\text{sound}} \approx 5 \times 10^3$ m/s, index of refraction ≈ 1.5, and ultrasonic beam width $\approx 1.27 \times 10^{-2}$ m), illuminated with light at wavelength 632.8 nm. If the ultrasonic frequency is approximately 27 MHz or less, the interaction satisfied Raman–Nath diffraction conditions. If the frequency is greater than approximately 81 MHz, the interaction is governed by Bragg diffraction conditions.

b. Raman–Nath Diffraction

For the case of Raman–Nath diffraction, we consider the light beam impinging on the medium at an angle ϕ, with the light normal to the sound beam. The location of the diffraction orders are given by the expression (Breazeale et al., 1981)

$$\sin(\theta_n + \phi) - \sin(\phi) = \frac{n\lambda}{\lambda^*}, \tag{53}$$

where ϕ is the angle of incidence, θ is the angle of diffraction, λ is the wavelength of light in the medium, λ^* is the ultrasonic wavelength, and n is an integer. One can experimentally set $\phi = 0$, which reduces Eq. (53) to

$$\sin(\theta_n) = \frac{n\lambda}{\lambda^*}. \tag{54}$$

The ultrasonic wavelength can be determined from Eq. (53). With the measurement of the ultrasonic frequency, one can determine the sound velocity of the compressional wave (Barnes and Hiedemann, 1957). In determining the wavelength measurement and its uncertainty, it is necessary to make an analysis of the optical setup, including the effects of Snell's Law at interfaces.

c. Bragg Diffraction

For the case of Bragg diffraction, the angle of incidence, ϕ, is set to the angle of diffraction, and

$$n\lambda = 2\lambda^* \sin \phi_B, \tag{55}$$

where ϕ_B is the Bragg angle. Bragg diffraction is used to measure wave velocities in the frequency range from approximately 100 MHz to the low end of the gigahertz scale. For example (Krischer, 1968), the technique has been used to measure wave velocity to an estimated accuracy of better than 0.1%. It is also useful in measuring the local velocities within a sample (Simondet et al., 1976; Michard and Perrin, 1978). Measurements with precision of better than 0.01% in homogenous samples have been reported (Simondet et al., 1976).

d. Direct Measurements (Fresnel Diffraction)

Consider a standing ultrasonic wave in a sample through which collimated light is passed so that the collimated light beam is perpendicular to the sound beam. A measuring microscope or similar optical device is focused so that the image of the wave can be viewed at the instrument focal plane. Because the images of the wave fronts are $\lambda/2$ apart, it is possible to determine the ultrasonic velocity from measurements of ultrasonic frequency and measurements of the wavelength in the medium. The technique is sensitive enough to detect local variations in velocity greater than 0.01% (Hiedemann and Hoesch, 1934, 1937; Mayer and Hiedemann, 1958, 1959).

e. Brillouin Scattering

Consider the scattering of a photon by a high-frequency phonon traveling in a specific direction within a crystal. Application of conservation of energy and momentum to the scattering, coupled with the approximation that any photon frequency shift is small (Benedek and Fritsch, 1966; Beyer and Letcher, 1969, pp. 47–50), gives

$$v^* = \pm 2v\left(\frac{v}{C}\right)\sin\left(\frac{\theta}{2}\right), \qquad (56)$$

where v is the photon frequency, v^* is the phonon frequency as well as the difference in frequency of the scattered photons, C is the speed of the photon in the medium, v is the speed of the phonon, and θ is the scattering angle. If desirable, the Bragg condition can be used for constructive reinforcement by adjusting θ (Pollard, 1977). The technique gives three lines in the scattered photon spectra, which can be separated and measured with appropriate optical devices, such as Fabry–Perot interferometers, to obtain v^*. Equation (56) can be solved for v in terms of the other quantities. Brillouin scattering is useful in the investigation of the sound velocity of a material near a phase transition (Fleury, 1970, pp. 37–42). The technique has been used for both longitudinal and mixed modes. However, in one study on cubic crystals, no Brillouin scattering was observed from an acoustic branch that consisted of pure transverse waves (Benedek and Fritsch, 1966). Uncertainties in the index of refraction, the scattering angle, and the width of the Stokes and anti-Stokes lines influence the accuracy of the determination. The precision is considerably better, however, with values of 0.1% as mentioned.

In stimulated Brillouin scattering, the photon scattering process is dependent upon the intensity of the radiation striking the surface. With high enough light intensity, nonlinear effects occur, which result in scattering by frequencies and harmonics created by harmonic generation (Brewer, 1965). A large buildup of acoustic intensity, both compressional and shear waves, accompanies a threshold in optical intensity (Chiao *et al.*, 1964). Other effects include the possibility of sample destruction from the intense radiation, and the line pulling effects of the laser cavity on the scattered light, which affects accuracy (Fleury, 1970, pp. 57–58). Amplified acoustic frequencies have been reported as high as 60 GHz.

III. Theoretical and Experimental Results

8. AB INITIO THEORY

It is clear from Section 2 that the calculation of elastic constants requires an accurate computation of the variation in the total energy of a solid, from equilibrium to distorted configurations. Thanks to the availability of powerful computers, great advances in *ab initio* total-energy calculation have been made in recent years (Anderson et al., 1985). In this section we briefly discuss the approximations used in *ab initio* theory and summarize the calculated results.

The Hamiltonian of a solid consists of five parts: the two kinetic energies of the ions and electrons, K_I and K_e respectively, and the potential energies U_{II} among ions, U_{ee} among electrons, and U_{el} between electrons and ions. The Born–Oppenheimer (1927) adiabatic approximation is a simple way to separate the electronic from the ionic variables. In this approximation, because the ion speed is at least two orders smaller than the electron speed, one freezes the ionic motion in a configuration specified by a set of ion position vectors $\vec{R}_n$, then solves the Schrodinger equation for that part of the Hamiltonian that involves the set of electronic coordinates $\{\vec{r}_i\}$:

$$H\psi_\gamma(\{\vec{r}_i\}, \{\vec{R}_n\}) = E_\gamma(\{\vec{R}_n\})\psi_\gamma(\{\vec{r}_i\}, \{\vec{R}_n\}), \tag{57}$$

where $H = K_e + U_{ee} + U_{el}$. Thus the energy E_γ is a function of the ionic configuration. The lowest energy curve of the sum of $E_g = E_\gamma + U_{II}$ as a function of $\{R_n\}$ then serves as the potential energy for the ionic motion. A Taylor expansion of E_g about its minimum value E_o, takes the form

$$E_g\{\vec{R}_n\} = E_o + \tfrac{1}{2}\Sigma_{nm\alpha\beta}\Phi^{\alpha\beta}_{nm}u_{n\alpha}u_{m\beta} + \cdots, \tag{58}$$

where $u_{n\alpha}$ is the α-component of the small displacement $\vec{R}_n - \vec{R}_n(0)$ with $\{\vec{R}_n(0)\}$ the equilibrium ionic positions at the minimum energy E_0. The force functions $\Phi^{\alpha\beta}_{nm}$ are the second derivatives of E_g with respect to the displacements evaluated at the ion equilibrium positions. The force functions are directly related to the elastic constants. For example, for a Bravais lattice they are given by Aschcroft and Mermin (1976) as

$$c_{\alpha\beta\mu\nu} = -\tfrac{1}{2}\Sigma_L L_\alpha \Phi^{\beta\mu}(\vec{L}) L_\nu, \tag{59}$$

where the sum is over all the lattice vectors $\vec{L}$.

Equation (57) is computationally the most difficult part of the problem, because it deals with about 10^{23} electrons that are interacting with each other and with the ions; the wave functions have to be the properly antisymmetrized many-body functions. Self-consistent density-functional theory (SCDFT) (Hohenberg and Kohn, 1964; Kohn and Sham, 1965; Callaway and March, 1984), which casts the Hamiltonian into a density functional, reduces the many-body problem to an effective one-electron problem. This theory has been tested in many crystalline solids, and the resulting elastic constants have been excellent, especially for semiconductors.

In SCDFT, the ground-state energy of a solid is completely specified by single-particle wave functions of the occupied states $\{\phi_v\}$. First, the electron density is given by

$$\rho(\vec{r}) = \Sigma_v |\phi_v(\vec{r})|^2. \tag{60}$$

Then, the ground state energy is computed as follows:

$$E_g = K_e + U_{ee} + U_{el} + U_{ll} + U_{xc}, \tag{61}$$

where the different terms are given by the expressions

$$K_e = \Sigma_v \int \phi_v^*(\vec{r}) \frac{p^2}{2m} \phi_v(\vec{r}) \, d^3r, \tag{62}$$

$$U_{ee} = \frac{e^2}{2} \int\int \frac{\rho(\vec{r})\rho(\vec{r}')}{|\vec{r} - \vec{r}'|} d^3r \, d^3r'. \tag{63}$$

$$U_{el} = -e^2 \Sigma_n Z_n \int \frac{\rho(\vec{r})}{|\vec{r} - \vec{R}_n|} d^3r, \tag{64}$$

$$U_{ll} = \frac{e^2}{2} \Sigma'_{mn} Z_m Z_n / |\vec{R}_m - \vec{R}_n|, \tag{65}$$

$$U_{xc} = \int \rho(\vec{r}) \varepsilon_{xc}[\rho(\vec{r})] d^3r. \tag{66}$$

Note that $n = m$ is excluded in the sum in Eq. (65). The meaning and the notations of the above equations are mostly self-evident, except the $\varepsilon_{xc}[\rho]$ in Eq. (66); this is the correction term arising from the many-body exchange and correlation effects. The square bracket means that ε_{xc} is a functional of ρ. Several different expressions for ε_{xc} as a function of ρ, available in the literature (for example, Wigner, 1934; Hedin and Lunquist, 1971; Ceperley and Alder, 1980; Perdew and Zunger, 1981), have yielded similar results for the structural properties of semiconductors. A minimization of E_g with respect to ϕ_v^*, with the constraint that the total number of electrons is a constant, leads to the familiar Schroedinger single-particle equation:

$$\left[\frac{p^2}{2m} + V(\vec{r})\right] \phi_v(\vec{r}) = \varepsilon_v \phi_v(\vec{r}), \tag{67}$$

where ε_v is a Lagrange multiplier and V is an effective one-electron potential containing three parts:

$$V = V_{ee} + V_{el} + V_{xc}, \tag{68}$$

$$V_{ee} = e^2 \int \frac{\rho(\vec{r}')}{|\vec{r} - \vec{r}'|} d^3r'. \tag{69}$$

V_{el} is the Coulomb potential due to ionic charges Z_n:

$$V_{el} = -e^2 \Sigma_n Z_n / |\vec{r} - \vec{R}_n|. \tag{70}$$

Finally, the exchange-correlation potential V_{xc} is given by

$$V_{xc} = \varepsilon_{xc} + \rho \frac{\partial \varepsilon_{xc}}{\partial \rho}. \tag{71}$$

Thus, Eqs. (60) and (67)–(71) form a repeated loop, $\rho \to V \to \phi \to \rho$, and the calculation must be iterated until self-consistency is achieved. Following this recipe for calculating the total energy, one calculates E_g as a function of ionic positions, finds the equilibrium configuration, then imposes a strain and calculates the strain energy to deduce the elastic constants, following the prescription of Section 2. The problem then becomes strictly computational. The most challenging task is an accurate solution for the single-particle eigen states of Eq. (67). For a crystalline solid, lattice translational symmetry simplifies the problem, and band-structure techniques can be applied to obtain the solution. Because the strain energy is many orders of magnitude smaller than the total energy, very precise computation is required if one hopes to obtain meaningful elastic constants. So far, at least two band-structure methods have been demonstrated as reliable for all three elastic constants: the plane-wave method using pseudopotentials (PP-PW) (Nielsen and Martin, 1983, 1985a), and the full-potential linearized-muffin-tin-orbital method (FP-LMTO) (Methfessel et al., 1989). Although the full-potential argumented plane-wave method (FP-APW) (Krahauer et al., 1979; Wimmer et al., 1981; Wei and Krahauer, 1985; Ferreira et al.1989) has produced excellent lattice constants, structural energies, and bulk moduli, the complete semiconductor elastic constants based on this method are not yet available.

Even if the total energies at different distortions can be calculated accurately, there is still the problem of searching for the equilibrium atomic positions in a distorted crystal, and the numerical determination of elastic constants from energy differences. If the strain energy can be calculated directly without taking the difference between two large energies, or if the derivatives can be calculated directly, not only can the computation time be shortened, but the numerical errors will also be reduced. The quantum mechanical theory of forces and stresses of Nielsen and Martin (1985b) and the closely related direct calculation of elastic constants from linear response theory of Baroni et al. (1987) represent the status of efforts in this direction. The former has been carried out for all three elastic constants for Si, Ge and GaAs (Nielsen and Martin, 1985a), while the latter has been done only on the bulk modulus for Si; both are based on the PW-PP method. Table II shows a

TABLE II

COMPARISON BETWEEN CALCULATED AND EXPERIMENTAL CONSTANTS[a]

		Expt[b]	FP-LMTO[c]	PP-PW[d]	TB
SI					
	a	5.431	5.41	5.45	5.431
	B	9.923	9.9	9.3	9.923
	$C_{11} - C_{12}$	10.274	10.2	9.8	10.274
	C_{44}	8.036	8.3	8.5	8.013
	$C_{44}^{(0)}$			11.1	11.30
	ζ	0.54[e]	0.51	0.53	0.51
	ω	523	518	521	572
Ge					
	a	5.65		5.59	5.65
	B	7.653		7.2	7.653
	$C_{11} - C_{12}$	8.189		8.5	8.189
	C_{44}	6.816		6.3	6.84
	$C_{4}^{(0)}$			7.7	9.46
	ζ			0.44	0.49
	ω	303		302	342
GaAs					
	a	5.642		5.55	5.642
	B	7.69		7.3	7.69
	$C_{11} - C_{12}$	6.63		7.0	6.63
	C_{44}	6.04		6.2	5.79
	$C_{44}^{(0)}$			7.5	7.83
	ζ			0.48	0.50
	ω	273		268	292

[a]Comparison between calculated and experimental lattice constant a, elastic constants B, $C_{11} - C_{12}$ and C_{44}, Kleiman (1962) internal distortion parameter ζ, and the TO optical phonon ω in wave numbers 1/cm. Also listed are $C_{44}^{(0)}$ defined in Eq. (13). The FP-LMTO and PP-PW are the *ab initio* theories described in Part 1, and TB is the tight-bonding theory in Part III, Section 10. All elastic constants are in units of 10^{11} dynes/cm^2.

[b]All the experimental lattice constants are those tabulated by Zallen (1982). The experimental elastic constants are taken from Table III, and the phonon frequencies are from Table V.

[c]Methfessel et al. (1989).

[d]Nielsen and Martin (1985a).

[e]Cousins et al. (1987).

comparison between theoretical calculations and experimental results. From this comparison, it is fair to say that we have a very reliable *ab initio* theory for the elastic constants for crystalline semiconductors based on the self-consistent local density-functional theory. Note that Table II also lists the results from an empirical tight-binding (TB) theory to be discussed in Section 10.

9. VALENCE FORCE FIELD MODEL

The preceding section showed that *ab initio* theory for the elastic constants requires complicated computations. Accurate *ab initio* calculations for semiconductors have been obtained only recently and only for several systems. On the other hand, phenomenological microscopic models of elastic constants for all semiconductors have been available for some time. Of these, the valence force-field model (VFF) is perhaps the simplest and the most useful. This topic has been reviewed and well analyzed in a paper by Martin (1970), and its conclusions constitute the main body of this section.

a. Diamond Structure

The original VFF model by Musgrave and Pople (1962) was for the diamond structure, in which the elastic energy is a quadratic form, in terms of the changes in each bond length Δr_i, in bond angles $\Delta \theta_{ij}$, and in the products $\Delta r_i \Delta r_j$ and $\Delta \theta_{ij}$, between nearest-neighbor bonds. For elastic constants, Keating (1966) showed that the VFF can be simplified by the following approximation to the elastic energy of the crystal:

$$\Delta E = \frac{3\alpha}{8d^2} \sum_i [\Delta(\vec{r}_i \cdot \vec{r}_i)]^2 + \frac{3\beta}{8d^2} \sum_{i>j} [\Delta(\vec{r}_i \cdot \vec{r}_j)]^2, \qquad (72)$$

where the bond index i runs over all the bonds, but the i and j sum only over those pairs of bonds that are connected to a common atom. In Eq. (72), d is the equilibrium bond length and $\Delta(\vec{r}_i \cdot \vec{r}_j)$ is the change in the dot product of the two bond vectors that start at the common atom, point along the bond directions, and end at the first neighbor atoms. Following the calculational procedure described in Section 2 for uniform expansion, the U in the VFF under a uniform expansion can be shown to be $U = 2(3\alpha + \beta)e^2d^2/\Omega$, where $\Omega = a^3/4$ is the equilibrium volume per unit cell. Thus,

$$B = (C_{11} + 2C_{12})/3 = (\alpha + \beta/3)/a. \qquad (73)$$

For the shear strain described by Eq. (9), Eq. (72) yields $U = 4\beta e^2/a$. Thus, according to Eq. (10),

$$C_{11} - C_{12} = 4\beta/a. \tag{74}$$

For the shear strain described by Eq. (11) and with an internal displacement $\tilde{u} = (0, 0, u)$, Eq. (72) yields the following expression:

$$U = [\alpha(e - \eta)^2 + \beta(e + \eta)^2]/(8a), \tag{75}$$

where $u = \eta a/4$. A comparison between Eqs. (13) and (75) shows that $\phi = 16(\alpha + \beta)/a^3$, $D = -4(\alpha - \beta)/a^2$, and $C_{44}^{(0)} = (\alpha + \beta)/a$. Using these results in Eq. (14), we find that the Kleiman internal displacement parameter in the present model is given by

$$\zeta = (\alpha - \beta)/(\alpha + \beta) = 2C_{12}/(C_{11} + C_{12}). \tag{76}$$

Equation (15) then produces

$$C_{44} = 2\alpha\beta/[(\alpha + \beta)a]. \tag{77}$$

The three elastic constants given above are not independent and can be shown to related to each other by the Keating identity (1966), or

$$I_K = 2C_{44}(C_{11} + C_{12})/[(C_{11} - C_{12})(C_{11} + 3C_{12})] = 1. \tag{78}$$

b. Zincblende Structure and Coulomb Force

The Keating identity (1966) holds very well for systems with the diamond structure but not so well for the zincblende compounds (see Table III). One obvious difference between the two structures is the presence of Coulomb interactions arising from charge shifts between the cation and anion sublattices in zb semiconductors. Martin (1970) incorporated Blackman's (1959) treatment of the Coulomb forces in the Keating VFF in the following manner. First, the Coulomb energy was treated as a screened Madelung energy E_M. For example, in a uniformly expanded crystal with a bond length r, the Coulomb energy was taken to be $E_M = -N\alpha_M Z^{*2} e^2/(\varepsilon r)$, where N is the total number of unit cells, $\alpha_M = 1.6381$ is the Madelung constant, and Z^{*2}/ε is the effective charge defined by the optic-mode splitting:

$$S = Z^{*2}/\varepsilon = \mu(\omega_l^2 - \omega_t^2)/(4\pi e^2). \tag{79}$$

TABLE III
EXPERIMENTAL VALUES FOR CUBIC SEMICONDUCTORS[a]

	C_{11}	C_{12}	C_{44}	s	α	β	I_K	I_M	I_{BOM}
C[b]	107.640	12.520	57.740	0.0	129.100	84.573	1.00	1.00	1.02
Si[b]	16.772	6.498	8.036	0.0	49.247	13.951	1.00	1.00	1.13
Ge[b]	13.112	4.923	6.816	0.0	39.438	11.583	1.08	1.08	1.05
AlSb[b]	8.769	4.341	4.076	1.684	33.768	6.653	1.11	1.05	1.08
GaP[b]	14.390	6.520	7.143	3.815	46.965	10.448	1.12	1.05	1.08
GaAs[b]	12.110	5.480	6.040	2.827	40.895	9.159	1.12	1.06	1.05
GaSb[b]	9.089	4.143	4.440	1.569	33.123	7.412	1.10	1.06	1.06
InP[b]	10.220	5.760	4.600	3.766	41.095	6.250	1.20	1.07	1.03
InAs[b]	8.329	4.526	3.959	2.820	33.744	5.531	1.22	1.11	1.00
InSb[b]	6.918	3.788	3.132	1.372	29.909	4.951	1.17	1.11	1.05
ZnS[c]	9.420	5.680	4.360	6.788	37.026	4.571	1.33	1.07	0.95
ZnS[c]	10.790	7.220	4.120	6.788	45.126	4.341	1.28	1.02	1.01
ZnS[c]	9.810	6.270	4.483	6.788	39.947	4.300	1.42	1.13	0.90
ZnS[c]	10.460	6.530	4.630	6.788	41.880	4.828	1.33	1.08	0.95
ZnSe[b]	8.95	5.39	3.984	4.368	34.432	4.716	1.28	1.09	0.98
ZnSe[c]	8.59	5.06	4.06	4.368	34.519	4.673	1.32	1.13	0.95
ZnSe[c]	8.720	5.240	3.920	4.368	35.469	4.603	1.29	1.10	0.98
ZnTe[c]	7.130	4.070	3.120	2.566	29.976	4.452	1.18	1.06	1.05
ZnTe[c]	7.220	4.090	3.080	2.566	30.204	4.558	1.14	1.03	1.08
CdTe[b]	5.33	3.65	2.04	3.105	27.058	2.455	1.34	1.07	0.98
CdTe[c]	6.150	4.300	1.960	3.105	31.546	2.731	1.16	0.94	1.13
HgTe[b]	5.971	4.154	2.259	2.381	30.300	2.542	1.37	1.16	0.96
HgTe[c]	5.63	3.66	2.11	2.381	26.919	2.542	1.37	1.15	0.95

[a]Experimental elastic constants for some cubic semiconductors and the parameters of Eq. (81) taken from Martin (1970), with the force constants α and β obtained from Eqs. (82) and (83) and the identity relations I_K, I_M, and I_{BOM} given by Eqs. (78), (88), and (119), respectively. The elastic constants are in units of 10^{11} dynes/cm^2, and the force constants are in 10^3 dynes/cm.

[b]Data quoted from "Landolt–Bornstein Numerical Data and Functional Relationships in Science and Technology," New Series, Vols. 17 and 22.

[c]Listed in the review by Mitra and Massa (1982).

In Eq. (79), ω_1 and ω_t are, respectively, the longitudinal and transverse phonon frequencies in the long-wave-length limit. Then, to counterbalance the Coulomb forces, a repulsive force term was added and assumed to contribute to the bond-stretching energy in the form

$$\Delta E_R = -\sum_i \alpha_M Z^{*2} e^2 \, \Delta r_i / (4\varepsilon d^2). \tag{80}$$

With the above two contributions added, the total strain energy is $\Delta E_T = \Delta E + \Delta E_M + \Delta E_R$, where ΔE is the VFF contribution in Eq. (72) and ΔE_M is the change in the Madelung energy. The energies ΔE_T are expanded in a power series, and only terms up to the second power in the strain are kept. The ΔE_M contributions arising from fixed values of the charge shift $S = Z^{*2}/\varepsilon$ on the atomic sites under different strains were worked out by Blackman (1959). Using these results and defining

$$s = Se^2/d^4 = e^2 Z^{*2}/(d^4\varepsilon), \tag{81}$$

Martin (1970) obtained the following modified expressions for the elastic constants:

$$C_{11} + 2C_{12} = (3\alpha + \beta)/a - 0.355s, \tag{82}$$

$$C_{11} - 2C_{12} = 4\beta/a + 0.053s, \tag{83}$$

$$\zeta = [(\alpha - \beta)/a - 0.294s]/C_M, \tag{84}$$

$$C_{44} = (\alpha + \beta)/a - 0.136s - C_M\zeta^2, \tag{85}$$

where C_M is defined as

$$C_M = (\alpha + \beta)/a - 0.266s. \tag{86}$$

The above equations can be combined to yield

$$\zeta = (2C_{12} - C')/(C_{11} + C_{12} - C'), \tag{87}$$

where $C' = 0.314s$. Since the extra parameter s is fixed by the optic modes and the bond length, the above results combine into a new identity, the Martin identity (1970):

$$I_M = \frac{2C_{44}(C_{11} + C_{12} - C')}{(C_{11} - C_{12})(C_{11} + 3C_{12} - 2C') + 0.831C'(C_{11} + C_{12} - C')}. \tag{88}$$

Table III lists a set of experimental values of the elastic constants and the s values for a number of diamond and zincblende semiconductors. These values are used to compute the force constants α and β and the identity expressions I_k and I_M given in Eqs. (78) and (88), respectively. Several sets of data are quoted for some of the systems to show the uncertainties in the experiments for these systems. The table results clearly show that the inclusion of the Coulomb energies improves the identity relation; the deviations of I_M from unity are 15% or less. Also listed are the values for another identity relation, I_{BOM} from Eq. (119), based on a tight-binding model to be discussed in Section 10. Martin further studied trends as functions of the bond lengths d and the ionicity scale f_i of Phillips (1973) and Van Vechten (1969). He found that α scales roughly as $1/d^3$, i.e.,

$$\alpha d^3/e^2 = \text{constant}, \qquad (89)$$

where e is the electron charge. He also found the ratio between the bond-angle and bond-stretching forces tend to decrease as f_i increases and scales roughly as

$$\beta/\alpha \propto 1 - f_i. \qquad (90)$$

He further observed that if S of Eq. (79) is set equal to f_i and if the α and β values are extrapolated, using Eqs. (89) and (90), from those fitted to the average values of the B and $C_{11} - C_{12}$ for Si and Ge, then all the elastic constants can be predicted from Eqs. (82) through (86) to an accuracy of 10%.

It is interesting to compare Eq. (82) using the results of Eqs. (89) and (90), with Cohen's (1985) empirical formula for the bulk modulus

$$B = (1971 - 22\lambda)/d^{3.5}, \qquad (91)$$

where B is in GPa, d in Å, and $\lambda = 0, 1,$ and 2, respectively, for the group IV, II-V, and II-VI semiconductors. Both Martin and Cohen's formulas give B values to better than 10% for all materials tabulated in Table III. The B in Eq. (91) scales as $1/d^{3.5}$, while in VFF it scales as $1/d^4$.

10. Tight-Binding Theory

In the semi-empirical tight-binding (TB) approach, the total energy of a semiconductor crystal is assumed to be the sum of the electron energies $\varepsilon_v(\vec{k})$ in the valence bands plus repulsive pair energies u_{ij} between the nearest-neighbor atoms (Chadi, 1978):

$$E_T = E_{bs} + U_r = \sum_v \sum_k \varepsilon_v(\vec{k}) + \sum_{i>} \sum_j u_{ij}. \qquad (92)$$

Furthermore, the band energies are interpolated by using a TB Hamiltonian that contains term values of the atoms, and a handful of interaction parameters between orbitals of the neighboring atoms. Despite the simplicity of Eq. (92), recent first-principles theories have given some support to this approximation.

One virtue of the TB approach over valence force-field models is that it is a quantum theory without much complication. As compared to first-principles theory, the TB approach is easier to execute, particularly when applied to complicated systems such as alloys and superlatices. In actual applications, the TB calculation either is carried out using the full band-structure calculation (BS), or is approximated by simpler local theory such as Harrison's (1980, 1983a, and 1983b) bond orbital model (BOM).

a. Band Structure Calculations

The simplest TB Hamiltonian contains the s and p atomic term values ε_s and ε_p for both cations and anions, and the nearest-neighbor two-center interactions $V_{ss\sigma}$, $V_{sp\sigma}$, $V_{pp\sigma}$, and $V_{pp\pi}$. To be more explicit, the 8×8 k-dependent Hamiltonian contains the term values as the diagonal matrix elements, while the off-diagonal matrix elements between the cation and anion orbitals are given by

$$H_{\gamma\gamma'}(\vec{k}) = \sum_d e^{i\vec{k}\cdot\vec{d}} h_{\gamma\gamma'}(\vec{d}), \tag{93}$$

where the sum runs over the four first neighbor atoms specified by the bond displacements $\vec{d}$. The γs are the orbital indices for s, p_x, p_y, and p_z. The $h_{\gamma\gamma'}$ values are related to the two-center V's by the Slater–Koster (1954) relations:

$$h_{ss} = V_{ss\sigma}, \tag{94}$$

$$h_{sx} = \alpha_1 V_{sp\sigma}, \tag{95}$$

$$h_{xx} = \alpha_1^2 V_{pp\sigma} + (1 - \alpha_1^2) V_{pp\pi}, \tag{96}$$

$$h_{xy} = \alpha_1 \alpha_2 (V_{pp\sigma} - V_{pp\pi}), \tag{97}$$

where $\alpha_i = x_i/d$ are the direction cosines of $\vec{d}$ and the V's depend only on the length d.

Once the values of these TB parameters and their dependences on the bond length are known, the Hamiltonian at each k inside the Brillouin zone (BZ) can be evaluated, and the summation of k carried out to obtain the band-structure energy, which, when added to the repulsive energy, gives the total energy of any specified geometry. All the elastic constants, associated internal displacements, and transverse optical phonon frequencies, are easily calculable following the procedure of Section 2. The only point to note is the k-

sum, which without strain can be calculated accurately by using the 10 special k points (Chadi and Cohen, 1973) in the irreducible wedge of the BZ. Under strain, the crystal symmetry changes; one needs to extend these special k points to other nonequivalent wedges. However, since the sum of the valence-band energies as a function of k is a rather smooth function, a uniform sampling over the whole BZ converges very quickly. A $5 \times 5 \times 5$ grid is sufficiently accurate for the required calculations. To avoid the numerical inaccuracy inherent in direct energy subtractions, one can also calculate the second derivatives directly by using perturbation theory.

Perturbation theory starts with the expansion of the k-dependent Hamiltonian H in powers of the infinitesimal strain parameter e, keeping terms up to second-order:

$$H(k) = H_0 + H_1 e + \tfrac{1}{2} H_2 e^2, \tag{98}$$

where H_0 is the strain-free Hamiltonian, and H_1 and H_2 are, respectively, the first and second derivatives with respect to e evaluated at $e = 0$. The band energy contribution to the strain coefficient then comes from the second derivative of E_{bs} with respect to e, denoted by

$$\frac{\partial^2 E_{bs}}{\partial e^2} = \sum_v \sum_k \langle v\vec{k}|H_2|v\vec{k}\rangle + 2 \sum_v \sum_c \sum_k \frac{|\langle v\vec{k}|H_1|c\vec{k}\rangle|^2}{\varepsilon_v(\vec{k}) - \varepsilon_c(\vec{k})}, \tag{99}$$

where $\varepsilon_c(\vec{k})$ and $|c\vec{k}\rangle$ are, respectively, the eigen energies and eigenvectors of H_o for the conduction bands, and $v\vec{k}$ stands for the valence bands. Note that the inter-valence-band contributions in the second-order perturbation sum cancel exactly so they are not needed in Eq. (99). To evaluate these matrix elements one needs to have the first and the second strain derivatives of the two-center interactions and the direction cosines α_i. For the strain parameters e defined in Section 2, and for the two center interactions V that scale as $1/d^n$, the following results are useful.

(i) For the bulk modulus, the direction cosines do not change, and we have

$$\partial V/\partial e = -nV \quad \text{and} \quad \partial^2 V/\partial e^2 = n(n+1)V.$$

(ii) For $C_{11} - C_{12}$ with e specified in Eq. (9), we get

$$\partial \alpha_i/\partial e = \alpha_i(\delta_{i1} - \delta_{i2}), \; \partial^2 \alpha_i/\partial e^2 = \alpha_i[3(\delta_{i1} + \delta_{i2}) - 4]/3, \quad \partial V/\partial e = 0,$$

and

$$\partial^2 V \partial e^2 = 4nV/3.$$

(iii) For C_{44} with the strain e given in Eq. (11) and an internal displacement u, we find

$$\partial V/\partial e = -n\alpha_1\alpha_2 V, \quad \partial V/\partial u = n\alpha_3 v/d,$$
$$\partial \alpha_i/\partial e = (\delta_{i1}\alpha_2 + \delta_{i2}\alpha_1)/2 - \alpha_i\alpha_1\alpha_2,$$
$$\partial \alpha_i/\partial u = -\delta_{i3}/d + \alpha_i\alpha_3/d, \quad \partial^2 V/\partial e^2 = n(n-1)V/9,$$
$$\partial^2 V/\partial u^2 = n(n-1)V/(3d^2),$$
$$\partial^2 \alpha_i/\partial e^2 = -\alpha_i(\delta_{i1} + \delta_{i2})/12, \quad \partial^2 \alpha_i/\partial u^2 = -2\delta_{i3}\alpha_3/d^2,$$
$$\partial^2 V/(\partial e\partial u) = -n(n+2)\alpha_1\alpha_2\alpha_3/d,$$

and finally

$$\partial^2 \alpha_i/(\partial e\partial u) = (\delta_{i1}\alpha_2\alpha_3 + \delta_{i2}\alpha_1\alpha_3 + 2\delta_{i3}\alpha_1\alpha_2 - 6\alpha_i\alpha_1\alpha_2\alpha_3)/(2d).$$

Also note that three second derivatives of the band-structure energy are needed, namely $\partial^2 E_{bs}/\partial e^2$, $\partial^2 E_{bs}/(\partial e\partial u)$, and $\partial^2 E_{bs}/\partial u^2$ for the evaluation of $C_{44}^{(0)}$, D, and ϕ of Eq. (13), respectively.

b. Bond Orbital Model

Harrison's (1980, 1983a, and 1983b) bond orbital model (BOM) emphasizes calculations of the TB total energy in terms of local energies. One special feature of the BOM is its universality. Another feature is that its simple and often analytical forms provide direct insight into the essential physics. Although BOM aims at predicting trends, it is reasonably accurate in many cases.

The band-structure energy, or the center of gravity of the valence band, in BOM is computed in the following steps. The terms involved are indicated in Fig. 2.

(1) Construct the sp^3 hybrid orbitals $|h\rangle$ for each atom; these hybrid orbitals are directed toward the neighboring atoms. For example, the hybrid in the [111] direction is given by

$$|h\rangle = (|s\rangle + |p_x\rangle + |p_y\rangle + |p_z\rangle)/2. \tag{100}$$

The hybrid energy is then given by $\varepsilon_h = \langle h|H|h\rangle = (\varepsilon_s + 3\varepsilon_p)/4$. In the zincblende structure, the cation hybrid energy ε_h^C is in general different from the anion hybrid energy ε_h^A. Note that the two hybrid orbitals of the same atom but in two different directions are now coupled by the so-called metallic energy $V_1 \equiv \langle h|H|h'\rangle = (\varepsilon_s - \varepsilon_p)/4$.

(2) Construct the bonding and antibonding molecular orbitals, $|b\rangle$ and $|a\rangle$, from the two hybrid orbitals $|h^C\rangle$ and $|h^A\rangle$ directed toward each other

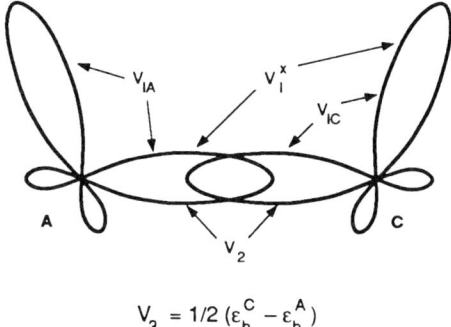

$V_3 = 1/2\,(\varepsilon_h^C - \varepsilon_h^A)$

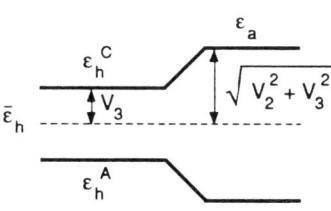

FIG. 2. Labels of the interactions between hybrids associated with an adjacent anion–cation pair. $V_{1A}(V_{1C})$ is the interaction between two hybrids on the same anion (cation); the "covalent energy" V_2 is the interaction between anion and cation hybrids that point toward one another along the bond direction; V_1^x is the interaction between an anion hybrid in one direction and an adjacent cation hybrid pointing in a different direction. The "ionic energy" V_3 is half the difference betwen cation and anion term values. The lower segment of the figure depicts the splitting of the hybrid energy levels by the V_2 and V_3 interactions.

along the same bond by diagonalizing a 2×2 matrix with ε_h^C and ε_h^A on the diagonal, and $V_2 = \langle h^c | H | h^A \rangle$ as the off-diagonal matrix elements. The resulting energies for $|b\rangle$ and $|a\rangle$ are $\varepsilon_b = \bar{\varepsilon}_h - \sqrt{V_2^2 + V_3^2}$ and $\varepsilon_a = \bar{\varepsilon}_h + \sqrt{V_2^2 + V_3^2}$, respectively, where $\bar{\varepsilon}_h = (\varepsilon_h^c + \varepsilon_h^A)/2$ is the mean hybrid energy, V_2 is the covalent energy, and $V_3 = (\varepsilon_h^c - \varepsilon_h^A)/2$ is the polar energy. The eigen states can also be written explicitly in terms of these energies:

$$|b\rangle = \sqrt{(1 + \alpha_p)/2}\,|h^A\rangle + \sqrt{(1 - \alpha_p)/2}\,|h^c\rangle, \qquad (101)$$

$$|a\rangle = -\sqrt{(1 - \alpha_p)/2}\,|h^A\rangle + \sqrt{(1 + \alpha_p)/2}\,|h^c\rangle, \qquad (102)$$

where α_p is called the polarity and is defined as

$$\alpha_p = V_3/(V_2^2 + V_3^2)^{1/2}. \tag{103}$$

(3) The quantity ε_b would be the center of gravity of the valence bands, if interactions between states on different bonds were neglected. Harrison (1983b) incorporated these interactions in a perturbation theory in which the change of the bonding energy is given by

$$\Delta\varepsilon_b = \sum_{a'} |\langle b|H|a'\rangle|^2/(\varepsilon_b - \varepsilon_{a'}), \tag{104}$$

where the sum runs over the antibonding states of the surrounding bonds. Note that the interactions among the bonding states lead to the formation of the valence bands, but do not shift their center of gravity; therefore, these interactions need not to be considered in the total energy calculation. Including the energy correction $\Delta\varepsilon_b$, the final band-structure energy per bond (which contains two electrons) is given by

$$E_b = 2\varepsilon_b + 2\Delta\varepsilon_b. \tag{105}$$

Harrison (1980) referred to the second term as the metallization energy. Besides providing a simple means for evaluating the center of gravity of the valence electron, Harrison's BOM also provides a set of universal TB parameters. Based on comparison with the free-electron band width (Froyen and Harrison, 1979) and with empirical TB parameters, Harrison (1983b) deduced the following set of universal two-center interactions:

$$V_{\alpha\alpha'} = \eta_{\alpha\alpha'}\hbar^2/(md^2), \tag{106}$$

with $\eta_{ss\sigma} = -1.32$, $\eta_{sp\sigma} = 1.42$, $\eta_{pp\sigma} = 2.22$, and $\eta_{pp\pi} = -0.63$, where m is the free-electron mass and d the bond length. In units where d is in Å and V in eV, $V_{\alpha\alpha'} = 7.62\, \eta_{\alpha\alpha'}/d^2$. The pair-repulsive energy u in the BOM is taken as resulting from the overlap of wavefunctions of the orbitals on the two centers and was shown to have the form

$$u = u_0(d_0/d)^4, \tag{107}$$

where d_0 is the equilibrium bond length. The value of u_0 is determined by requiring that d_0 is the experimental value. Note that the d dependences of both $V_{\alpha\alpha'}$ and u are taken to be the proper scaling not only among different systems, but also within the same system, as the bond length varies under distortions.

Under these assumptions, the bond energy E_{bond}, which is defined as the difference between the energy per bond in a semiconductor and the average energy per two electrons in the free atoms, i.e., $E_{\text{bond}} = E_b + u_0 - 2\bar{\varepsilon}$, takes the following simple form for a nonpolar semiconductor (Harrison 1983a):

$$E_{\text{bond}} = V_2(1 - \alpha_m + 9\alpha_m^2/16), \tag{108}$$

where α_m is called the metallicity and is defined as $2V_1/V_2$. For a polar semiconductor, E_{bond} becomes slightly more complicated:

$$E_{\text{bond}} = 2\bar{\varepsilon}_h - 2\bar{\varepsilon} - 2(V_2^2 + V_3^2)^{1/2}$$

$$\times \left[1 - \frac{1}{2}\alpha_c^2 + \frac{9}{16}\alpha_c^4(V_{1C}^2 + V_{1A}^2)/(V_2^2 + V_3^2)\right], \tag{109}$$

where $\alpha_c = \sqrt{1 - \alpha_p^2}$ is called the covalency, and V_{1C} and V_{1A} are the metallic energies for the cation and the anion, respectively. The bulk modulus also takes a very simple form; for a group-IV semiconductor, it reads

$$B = -2V_2(1 - 9\alpha_m^2/16)/(\sqrt{3}\,d^3), \tag{110}$$

and for a polar semiconductor, it becomes

$$B = -2V_2\left[\alpha_c^2 - \frac{9}{8}\alpha_c^3(5\alpha_c^2 - 4)(V_{1C}^2 + V_{1A}^2)/(V_2^2 + V_3^2)\right]/(\sqrt{3}\,d^3). \tag{111}$$

These expressions show that the bulk modulus varies as $1/d^5$ in the pure covalent case, and as $1/d^9$ in the extreme ionic limit $V_3 \gg V_2$. Note that this result is different from the $1/d^{3.5}$ dependence in Cohen's (1985) formula and the $1/d^4$ scale in VFF.

Shear strains cause a semiconductor to shift away from perfect tetrahedral symmetry. To deal with the shear elastic coefficient, the BOM has to be modified. A simple approximation is the rigid hybrid model (Harrison, 1983b; van Schilfgaarde and Sher, 1987a and b), in which the hybrid orbitals of each atom are assumed to remain in their original tetrahedral directions despite the lattice distortion. Then the hybrids of two nearest-neighbor atoms making up the bonding and antibonding states no longer are directed toward each other, as shown in Fig. 3. There is a misalignment angle θ between each hybrid and the line connecting the two atoms, and the convalent energy V_2 is given by

$$V_2(\theta) = \frac{1}{4}\left[V_{ss\sigma} - 2\sqrt{3}\cos\theta\, V_{sp\sigma} - \cos^2\theta\, V_{pp\sigma} + 3(1 - \cos^2\theta)V_{pp\pi}\right]. \quad (112)$$

The lowest-order change δV_2, caused by an infinitesimal angular misalignment $\delta\theta$, is then given by

$$\delta V_2 = \frac{1}{4}(\sqrt{3}V_{sp\sigma} + 3V_{pp\sigma} - 3V_{pp\pi})(\delta\theta)^2. \quad (113)$$

Under the strain e described in Eq. (9) for $C_{11} - C_{12}$, there is no bond length change, and $(\delta\theta)^2 = 2e^2/3$. If one assumes that the metallization coupling is only through the metallic energies V_{1C} and V_{1A}, as has been assumed so far, then the change of the crystal energy is the change of the band-structure energy due to δV_2. Then, according to Eqs. (10) and (105),

$$\begin{aligned}C_{11} - C_{12} &= \frac{\sqrt{3}}{2d^3}\frac{\partial E_b}{\partial V_2}\delta V_2/e^2 \\ &= \frac{\sqrt{3}}{4d^3}\alpha_c(\sqrt{3}V_{sp\sigma} + 3V_{pp\sigma} - 3V_{pp\pi}) \quad (114) \\ &\quad \times \left[1 + \left(\frac{3}{4} - \frac{9}{8}\alpha_c^2\right)(V_{1C}^2 + V_{1A}^2)/(V_2^2 + V_3^2)\right].\end{aligned}$$

Harrison (1983b) has pointed out, however, that in addition to V_1, other interactions such as V_1^x shown in Fig. 2 produce important contributions to the shear elastic constant. By arguing that these other contributions must cancel those associated with the change δV_2 arising from the metallization energy in a rigid rotation, Harrison deduced the following expression:

$$C_{11} - C_{12} = \frac{\sqrt{3}}{4d^3}\alpha_c^3(\sqrt{3}V_{sp\sigma} + 3V_{pp\sigma} - 3V_{pp\pi}). \quad (115)$$

Under the strain e for the C_{44} given in Eq. (11) and with an internal displacement given by $u = \eta d/\sqrt{3}$ as described in Section 2, the bond misalignment angles for the four bonds have the same magnitude with $(\delta\theta)^2 = 2(\eta + e/2)^2/9$. The four bond lengths also change, with the change for

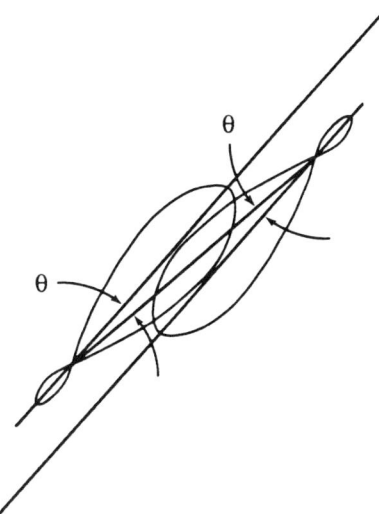

FIG. 3. A schematic picture of the hybrids, treated as rigid, in a shear distortion leading to $C_{11} - C_{12}$.

one pair given by $\delta r_1 = \delta r_2 = \delta + \varepsilon$ and by $\delta r_3 = \delta r_4 = -\delta + \varepsilon$ for the other pair, where $\delta = (e - \eta)d/3$ and $\varepsilon = (\eta + e/2)^2 d/9$. If again one assumes that the metallization is only through V_1, then the strain energy density can be shown to be given by

$$U = 9B\delta^2/(2d^2) + 3(C_{11} - C_{12})(\delta\theta)^2 \\ = B(e - \eta)^2/2 + (C_{11} - C_{12})(\eta + e/2)^2/3. \quad (116)$$

For a given strain e, U can be minimized with respect to η, which yields the Kleiman displacement parameter $\zeta = \eta/\varepsilon$ with ζ given by

$$\zeta = (B - C/3)/(B + 2C/3), \quad (117)$$

where $C = C_{11} - C_{12}$. Finally, from $U = C_{44} e^2/2$, the following relationship[7] is established:

$$9/C_{44} = 6/C + 4/B, \quad (118)$$

or

$$I_{BOM} = 9BC/[C_{44}(6B + 4C)] = 1. \quad (119)$$

---

[7] This is an expression corrected from one published previously (van Schilfgaarde and Sher, 1987a and b).

If one includes the effect of V_1^x, the energy density will involve an additional term that couples δr and $\delta \theta$. Then the analysis is no longer simple.

The preceding explicit formulas for the elastic constants in BOM are not much more complicated than the valence force-field model. However, they relate macroscopic forces to intrinsic atomic interactions. It is interesting to note that the simple identity relation of Eq. (119) holds very well. As can be seen in Table III, this result is certainly better than the I_k of Eq. (78), and is very competitive to Martin's identity I_M, which requires the inclusion of his particular treatment of Coulomb forces. Because the Coulomb energy is not included explicitly in Eq. (119), its contribution to the elastic constants is probably small.

c. Numerical Results and Quantitative Applications

To study the quantitative aspect of the theory, one first needs to establish the TB parameters. Table IV lists the term values we will use. The values of the outermost valence levels are taken to be minus the experimental first ionization energies listed in Kittel's (1986) book, and the other term values are deduced from calculated extraction or promotion energies using norm-conserved atomic pseudopotentials (Bachelet et al., 1982). These term values are very similar to Mann's (1967) Hartree–Fock calculations used by Harrison (1980), which are also given in Table IV. The major difference in the two sets occurs in heavy elements, where relativistic s-shifts are important, but were not included in Mann's results.

Table V lists the values of bond lengths, bond energies E_{bond}, the elastic coefficients B, $C \equiv C_{11} - C_{12}$, and C_{44}, and the zone-center TO phonon frequencies ω for a selected group of systems to be examined in the remainder of this section. Experimental values are also presented, with the exception of extrapolations for the elastic constants of AlP and AlAs. Table VI compares results between the BOM and the full band-structure (BS) calculation using Harrison's universal TB parameters. Except for the Ge and HgTe values, the agreement between the two calculations for E_{bond} is within 10% or better. The calculated E_{bond} values are also in fair agreement with the experimental values except for diamond. Since diamond has a much smaller bond length than the rest of the systems, this discrepancy is an indication of a limit to the scaling rules for both $V_{\alpha\alpha'}$ and u. Although the trends for the bulk moduli from both calculations are similar, the calculational errors in the BOM can be as large as 50%. Also note that the calculated values of B for most systems are only about one-half of the experimental values. The tabulated values of C and C_{44} for BOM are based on Eqs. (115) and (118), respectively. Considering the simplicity of these formulas, the agreement with the band structure calcula-

TABLE IV

TERM VALUES USED IN THE PRESENT CALCULATIONS AND MANN'S (1967) HARTREE-FOCK VALUES AS USED BY HARRISON (1980).

Element	Present		Mann	
	ε_s	ε_p	ε_s	ε_p
Cu	−7.72	−2.96	−7.72	−2.37
Ag	−7.57	−3.54	−7.06	−2.61
Au	−9.22	−3.91	−6.98	−2.67
Be	−9.32	−5.41	−8.41	−5.79
Mg	−7.62	−2.97	−6.88	−3.84
Zn	−9.39	−4.09	−7.96	−4.02
Cd	−8.99	−4.17	−7.21	−3.99
Hg	−10.43	−4.35	−7.10	−3.95
B	−14.00	−8.30	−13.46	−8.43
Al	−11.78	−5.98	−10.70	−5.71
Ga	−13.23	−5.90	−11.55	−5.67
In	−12.03	−5.56	−10.14	−5.37
C	−19.81	−11.26	−19.37	−11.07
Si	−15.03	−8.15	−14.79	−7.58
Ge	−16.40	−7.75	−15.15	−7.33
Sn	−14.53	−7.03	−13.04	−6.76
Pb	−15.25	−6.45	−12.48	−6.53
N	−26.08	−14.54	−26.22	−13.84
P	−19.62	−10.57	−19.22	−9.54
As	−20.02	−9.93	−18.91	−8.98
Sb	−17.56	−8.77	−16.02	−8.14
O	−28.55	−13.561	−34.02	−16.72
S	−21.16	−10.39	−24.01	−11.60
Se	−21.41	−9.90	−22.86	−10.68
Te	−19.12	−9.32	−19.12	−9.54
F	−36.23	−17.44	−42.78	−19.86
Cl	−25.81	−13.05	−29.19	−13.78
Br	−24.95	−12.01	−27.00	−12.43
I	−21.95	−10.79	−22.34	−10.97

tions is remarkable. However, the overall calculated values for these two shear coefficients are also consistently smaller than the experimental values.

The above comparisons show that the BOM and BS calculations predict similar qualitative trends for the binding energies and the elastic constants. In this regard, the BOM has the advantage of providing explicit forms to show the dependences on bond lengths and polarities. However, the merit of the TB theory over the valence-force model is in its ability to incorporate atomic quantities to mimic quantum mechanical effects. To be useful for specific material science applications (for example in alloy surface segregation, see Patrick et al., 1987 and 1988; Sher et al., 1988), the theory has to be more

TABLE V
EXPERIMENTAL PROPERTIES OF SEMICONDUCTORS[a]

	d	E_{bond}	B	C	C_{44}	ω
C	1.540	−3.68	44.227	95.120	57.740	1332
Si	2.352	−2.32	9.923	10.274	8.036	520
Ge	2.450	−1.94	7.653	8.189	6.816	301
AlP	2.367	−2.13	8.600	6.900	6.150	440
AlAs	2.451	−1.89	7.727	7.160	5.420	361
AlSb	2.656	−1.76	5.817	4.428	4.076	366
GaP	2.360	−1.78	9.143	7.780	7.143	367
GaAs	2.448	−1.63	7.690	6.630	6.040	269
GaSb	2.640	−1.48	5.792	4.946	4,440	231
InP	2.541	−1.74	7.247	4.460	4.600	304
InAs	2.622	−1.55	5.794	3.803	3.959	219
InSb	2.805	−1.40	4.831	3.130	3.132	185
ZnS	2.342	−1.59	7.637	3.990	4.558	279
ZnSe	2.454	−1.29	6.457	3.560	3.984	213
ZnTe	2.637	−1.20	5.090	3.060	3.120	177
CdTe	2.806	−1.10	4.210	1.680	2.040	141
HgTe	2.798	−0.81	4.759	1.817	2.259	116

[a]Values of bond length d, bond energy E_{bond}, bulk modulus B, and shear coefficient $C = C_{11} - C_{12}$ used to determine the parameters in Tables VII through XI. Also listed are the experimental values of C_{44} and the TO optical phonon mode ω at Γ to be compared with the calculations. All the elastic constants are in units of 10^{11} dynes/cm^2, d in Å, E_{bond} in eV, and ω in terms of wave numbers in 1/cm. All bond lengths are deduced from the lattice constants quoted by Zallen (1982). The values of E_{bond} are taken from Harrison (1980), Table 7-3, except for AlSb, ZnTe, CdTe, and HgTe, which are deduced from the Phillips (1973) Table 8.2. The elastic constants are taken from Table III, and the phonon frequencies are taken from values compiled in "Landolt–Borstein Numerical Data and Functional Relationships in Science and Technology," New Series, edited by K.-H. Helllwidge, Vols. 17 and 22.

quantitative. Successful quantitative application of the TB theory has been made by Chadi (1978, 1979, and 1984) in his study of semiconductor surfaces. The comparisons in Table VI indicate that the BOM should be treated differently from the BS calculation when considered for quantitative applications. If one wishes to calculate the properties using the local picture, one should use the BOM. If one wants to carry out the TB Hamiltonian precisely, one needs to adopt a different set of parameters based on the BS calculation. For this reason, we shall next consider quantitative applications of BOM and BS calculations separately.

There are many ways to parametrize the TB theory. Chadi (1979, 1984) used a simple form for the repulsive energy $u = a + b(d - d_0) + c(d - d_0)^2$ and the same $1/d^2$ scaling for the TB parameters $V_{\alpha\alpha'}$. To keep the theory as close to Harrison's (1983a and 1983b) form as possible but free it from the

TABLE VI

COMPARISON OF BOM AND BS MODELS' PREDICTIONS.[a]

	E_{bonds}		B		C		C_{44}	
	BS	BOM	BS	BOM	BS	BOM	BS	BOM
C	−7.17	−7.08	49.31	47.39	67.15	67.10	44.45	51.78
Si	−2.39	−2.50	4.65	4.18	7.19	8.07	5.10	5.29
Ge	−2.03	−2.34	2.89	1.91	4.72	6.58	3.69	3.00
AlP	−2.59	−2.54	3.81	4.21	4.45	4.83	3.45	4.10
AlAs	−2.22	−2.20	3.12	3.66	3.85	4.07	2.91	3.50
AlSb	−1.64	−1.62	1.99	2.47	2.75	3.12	2.11	2.54
GaP	−2.32	−2.16	3.63	4.40	4.48	5.15	3.69	4.34
GaAs	−1.99	−1.85	2.83	3.76	3.75	4.30	3.02	3.66
GaSb	−1.53	−1.43	1.76	2.45	2.63	3.40	2.17	2.65
InP	−2.21	−2.06	2.35	2.91	2.69	2.93	2.21	2.63
InAs	−1.89	−1.74	1.93	2.67	2.38	2.52	1.90	2.32
InSb	−1.43	−1.28	1.28	1.97	1.81	2.17	1.48	1.87
ZnS	−1.83	−1.79	3.46	3.74	3.53	3.62	2.71	3.30
ZnSe	−1.52	−1.48	2.67	3.06	2.75	2.67	2.08	2.53
ZnTe	−1.15	−1.05	1.82	2.16	1.88	1.87	1.50	1.78
CdTe	−1.06	−0.97	1.24	1.47	1.22	1.14	0.96	1.13
HgTe	−0.72	−0.49	1.23	1.72	1.33	1.31	1.11	1.30

[a] Comparison of the tight-binding theory using the full band structures (BS) and the bond orbital model (BOM for bond energyes E_{bond}, bulk moduli B, and shear coefficients $C = C_{11} - C_{12}$ and C_{44}. All energies are in eV and elastic constants in 10^{11} dynbes/cm^2.

$1/d^2$ and $1/d^4$ scaling rules for $V_{\alpha\alpha'}$ and u, respectively, we assume the following forms:

$$V_{\alpha\alpha'} = V^0_{\alpha\alpha'}(d_0/d)^n \tag{120}$$

and

$$u = u_0(d_0/d)^m, \tag{121}$$

where the superscript and subscript 0 indicate the values evaluated at the equilibrium bond length d_0. For simplicity, the values of $V^{(0)}_{\alpha\alpha'}$ are taken to be Harrison's values given in Eq. (106) scaled by a factor f:

$$V^{(0)}_{\alpha\alpha'} = fV^{\text{Harr}}_{\alpha\alpha'}. \tag{122}$$

Thus, there are four parameters for each system: the scaling parameter f, the powers n and m, and the value u_0. These parameters can be determined by requiring that the model produce the correct experimental values for E_{bond}, d_0, $C_{11} - C_{12}$, and B. Since $C_{11} - C_{12}$ is only governed by $V_{\alpha\alpha'}^{(0)}$ in both BOM and band calculation, it alone determines the scaling factor f. Then the bond energy E_{bond} can be used to determine u_0. The requirement that the first derivative of E_T be zero at d_0 then determines the ratio of the powers n/m, which couples with the equation for the bulk modulus to yield the values for n and m. One can then use these sets of parameters to check the validity of the model by calculating other quantities not employed in the fitting, e.g., C_{44}, the internal displacement parameter ζ, and the optical phonon frequencies ω at the zone center. If the results are acceptable, the model can be extended to more complicated systems such as alloys and superlattices with local environments similar to the bulk crystals.

Table VII shows the results for f, n, m, and u_0 obtained from the preceding fitting procedure by using the full band-structure calculations, and the corresponding values of C_{44}, ζ, and ω calculated for consistency checks. The scaling factor f ranges from 1 to 1.4 and tends to decrease with an increase in

TABLE VII

FULL BAND STRUCTURE CALCULATION[a]

	f	n	m	u_0	C_{44}	ζ	ω
C	1.390	2.840	3.767	21.924	48.393	0.121	1,459
Si	1.326	3.040	5.001	6.938	8.013	0.511	572
Ge	1.388	3.204	5.278	6.415	6.841	0.487	342
AlP	1.294	3.530	5.598	6.435	5.827	0.516	447
AlAs	1.464	3.524	5.430	7.089	5.598	0.459	384
AlSb	1.337	3.268	5.668	4.838	3.944	0.564	354
GaP	1.395	3.705	5.683	7.285	6.857	0.501	382
GaAs	1.397	3.633	5.716	6.530	5.791	0.500	292
GaSb	1.431	3.471	5.717	5.519	4.515	0.536	256
InP	1.323	4.240	6.633	5.603	4.260	0.584	304
InAs	1.300	3.997	6.427	4.962	3.564	0.552	220
InSb	1.353	3.773	6.399	4.350	3.092	0.602	200
ZnS	1.062	3.308	5.996	4.225	3.727	0.632	325
ZnSe	1.134	3.420	5.994	4.260	3.164	0.576	233
ZnTe	1.284	3.306	5.828	4.285	2.813	0.590	205
CdTe	1.171	3.656	6.761	3.092	1.701	0.694	156
HgTe	1.173	3.760	7.074	3.080	2.040	0.716	152

[a]The results for the parameters f, n, m, and u_0 obtained from the fitting of the bond energy, bond length, bulk modulus, and shear coefficient $C_{11} - C_{12}$ of Table V using the full band structure calculation. Also listed are the calculated C_{44}, internal displacement parameter ζ, and the TO optical phonon mode ω at Γ. All the elastic constants are in units of 10^{11} dynes/cm^2, u_0 is in eV, and ω are wave numbers in 1/cm.

polarity. In the power dependence of $V_{\alpha\alpha'} \propto (d_0/d)^n$, n ranges from 2.8 to 4.3, which is larger than the $n = 2$ used in Harrison's universal TB parameters. For the repulsive pair energy $u = u_0(d_0/d)^m$, the power m ranges from 3.8 to 6.8. The ratio m/n falls in the range from 1.3 to 1.9, which is smaller than the $m/n = 2$ used by Harrison. The calculated values of C_{44} for most systems agree with the experimental data to 10% or better, except for diamond and ZnS. Note that the experimental data for ZnS are rather dispersed. The calculated TO optical phonon modes at Γ in 1/cm for most group IV and III-V systems also agree with experiments to 10% or better. The discrepancies for the II-VI systems are larger, about 15%. Reliable results for ζ from both experiments and first-principles calculations are available only for a limited number of systems. The calculated ζ in the TB model agrees very well with those results, as shown in Table I. The overall results for C_{44}, ζ, and ω in the TB calculations are equivalent to those based on the valence force model, including Martin's Coulomb force corrections. By construction, the TB model also produces the correct cohesive energies, bond lengths, bulk moduli and shear coefficient $C_{11} - C_{12}$, because these quantities are used to fit the parameters.

The results in Table VII are based on the term values given in Table IV and the TB parameters scaled from Harrison's universal parameters. It is useful to know how the predictions are influenced by these parameters and the fitting procedure. Table VIII shows the results based on Chadi procedure in which the TB matrix elements $V_{\alpha\alpha'}$ are scaled as $1/d^2$, and the repulsive pair energy is taken to be $u = u_0 + u_1(d - d_0) + u_2(d - d_0)^2$. The parameter u_0 is set to produce the correct bond energy, u_1 is determined by requiring the correct equilibrium bond length, and u_2 is fixed by the bulk modulus. Two sets of TB parameters are tabulated for each system: One is the set used by Chadi (1978, 1979, and 1984), and the other is the set obtained by multiplying Harrison's $V_{\alpha\alpha'}$ by scaling factor f listed in Table VIII. For convenient comparison, the zero of the term values is set equal to the anion s energy. Despite considerable differences in these two sets of TB parameters, the results of the predictions from both sets are very similar and also very similar to those predicted from the other procedure used for the results in Table VII. The only noticeable difference between the predictions in Table VIII and Table VII is that the present procedure produces larger phonon frequencies and slightly smaller C_{44} values. We also note that the fitted parameters u_0, u_1, and u_2 and the predicted values for Chadi's elastic constant set in Table VIII are not the same as Chadi's published (1979) values; these give bulk moduli about 20% smaller than the experimental values, but give considerably better phonon frequencies.

To parametrize the BOM, several different stages of approximations can be made. However, for a general application, the full BOM steps presented in

TABLE VIII

Two Sets of TB Parameters.[a]

	\multicolumn{8}{c}{SI}							
	ε_s^A	ε_p^A	ε_s^C	ε_p^C	$V_{ss\sigma}$	$V_{sp\sigma}$	$V_{pp\sigma}$	$V_{pp\pi}$
Chadi	0.0	7.20	0.0	7.20	−2.03	2.55	4.55	−1.09
Present	0.0	6.88	0.0	6.88	−2.41	2.59	4.05	−1.15
	u_0	u_1	u_2	C	C_{44}	$C_{44}^{(0)}$	ζ	ω
Chadi	7.29	−9.98	23.90	10.66	7.89	11.38	0.49	620
Present	6.93	−9.70	23.42	10.27	7.83	11.39	0.51	592

	\multicolumn{9}{c}{GaAs}								
	ε_s^A	ε_p^A	ε_s^C	ε_p^C	$V_{ss\sigma}$	$V_{sp\sigma}^{AC}$	$V_{pp\sigma}^{CA}$	$V_{pp\pi}$	$V_{pp\pi}$
Chadi	0.0	9.64	5.12	11.56	−1.70	2.40	1.90	3.44	−0.89
Present	0.0	10.09	6.79	14.12	−2.34	2.52	2.52	3.94	−1.12
	u_0	u_1	u_2	C	C_{44}	$C_{44}^{(0)}$	ζ	ω	
Chadi	5.12	−7.12	18.22	6.36	5.60	8.77	0.54	339	
Present	6.53	−8.39	19.90	6.63	5.70	8.53	0.54	322	

[a] Comparison between the two different sets of TB parameters described in the text, the resultant expression coefficients u_0, u_1, u_2 of the repulsive pair energy u, and the predicted elastic constants, Kleinmann internal displacement parameters ζ, and phonon frequency ω from Chadi fitting scheme.

Eqs. (100) to (105) for calculating E_b should be followed, regardless of approximations. The simplest model, referred to as BOM(1), is to include only V_{1A} and V_{1C} in the matrix element $\langle b|H|a'\rangle$ for the calculation of the metallization energy in Eq. (104). The next approximation, BOM(2), is to include V_1^x as well. Finally, one can include all the first-neighbor interatomic TB parameters in $\langle b|H|a'\rangle$; this will extend the $|a'\rangle$ to those belonging to the second-neighbor bonds. This last approximation will be referred to as BOM(3).

Table IX shows the results for BOM(1) following the parameterization procedure described in Eqs. (120) to (122). The fittted parameters f, n, m, and u_0 are substantially different from those based on the BS calculations. The predicted C_{44} for the group IV and III-V systems are slightly larger than the experimental values, but good for the II-VI systems. The calculated ζ values, although not all smaller than those in Table VII, are smaller on the average. The predicted phonon frequencies are too high.

The parameters and the predicted results from BOM(2) are listed in

TABLE IX
BOM(1) CALCULATIONS[a]

	f	n	m	u_0	C_{44}	ζ	ω
C	1.440	2.896	3.809	22.345	61.098	0.142	2,103
Si	1.356	3.208	5.166	7.023	9.729	0.447	695
Ge	1.395	3.666	5.854	6.383	8.240	0.506	416
AlP	1.047	3.334	5.842	4.418	6.627	0.460	571
AlAs	1.264	3.530	5.685	5.538	6.668	0.431	491
AlSb	1.179	3.170	5.831	3.818	4.482	0.511	411
GaP	1.154	3.439	5.744	5.239	7.500	0.453	496
GaAs	1.179	3.337	5.647	4.863	6.340	0.456	358
GaSb	1.274	3.354	5.797	4.456	4.950	0.500	302
InP	0.999	3.579	6.490	3.413	4.558	0.522	388
InAs	0.995	3.167	5.960	3.081	3.787	0.496	260
InSb	1.126	3.228	6.105	3.091	3.267	0.543	224
ZnS	0.692	2.750	7.262	1.642	4.312	0.580	401
ZnSe	0.750	2.823	6.839	1.821	3.685	0.544	287
ZnTe	0.888	2.715	6.484	1.987	3.156	0.531	248
CdTe	0.734	2.720	8.394	1.018	1.949	0.660	189
HgTe	0.732	2.202	8.810	0.842	2.140	0.677	177

[a]The results for the parameters f, n, m, and u_0 obtained from the fitting of the bond energy, bond length, bulk modulus, and shear coefficient $C_{11} - C_{12}$ of Table V using the BOM(1) described in the text. Also listed are the calculated C_{44}, internal displacement parameter ζ, and the TO optical phonon mode ω at Γ. All the elastic constants are in units of 10^{11} dynes/cm², u_0 is in eV, and ω is given in terms of wave number in 1/cm.

Table X. These parameters more closely resemble those in Table VIII than do the values from BOM(1). However, the predicted C_{44} values are still too small, and the ζ values are too large, but the ω values are better than those from BOM(1).

Table XI contains the results from BOM(3); these approach those of the BS calculations. In comparison with Table VII and the experimental values in Table V, the BOM(3) does well for ω, produces slightly smaller C_{44}, and probably slightly larger ζ values.

In conclusion, the TB method is a reasonable approach to the static elastic properties of semiconductors. If carried out rigorously, the TB parameters in Table VII will provide quantitative results for superlattices, alloys, and possibly surfaces in which the local environments are similar to those in the bulk. The quantitative predications of BOM are not as good as the BS calculations, but are still reasonable. The fitted parameters given in Tables IX to XI allow different stages of approximations to be made using the BOM. This is especially useful for more complicated systems, because computationally the BOM is about two orders faster than the band structure calculations.

TABLE X

BOM(2) CALCULATIONS[a]

	f	n	m	u_0	C_{44}	ζ	ω_{TO}
C	1.440	2.896	3.809	22.345	55.161	0.319	1672
Si	1.356	3.208	5.166	7.023	7.520	0.652	597
Ge	1.395	3.666	5.854	6.383	6.106	0.720	358
AlP	1.283	3.440	5.493	6.171	5.094	0.677	511
AlAs	1.472	3.543	5.419	6.979	5.103	0.640	440
AlSb	1.342	3.249	5.580	4.771	3.256	0.706	373
GaP	1.371	3.540	5.490	6.901	5.916	0.688	428
GaAs	1.387	3.455	5.452	6.306	4.864	0.681	317
GaSb	1.414	3.398	5.593	5.305	3.681	0.717	264
InP	1.307	3.846	6.115	5.339	3.446	0.744	354
InAs	1.298	3.528	5.731	4.834	2.806	0.706	243
InSb	1.342	3.464	5.910	4.204	2.351	0.747	207
ZnS	1.062	3.102	5.728	4.075	3.016	0.726	373
ZnSe	1.141	3.150	5.587	4.181	2.577	0.688	271
ZnTe	1.283	3.067	5.498	4.140	2.270	0.704	231
CdTe	1.178	3.271	6.216	3.011	1.319	0.764	177
HgTe	1.169	3.046	6.116	2.916	1.457	0.793	162

[a]The results for the parameters f, n, m, and u_0 obtained from the fitting of the bond energy, bond length, bulk modulus, and shear coefficient $C_{11} - C_{12}$ of Table V using the BOM(2) described in the text. Also listed are the calculated C_{44}, internal displacement parameter ζ, and the TO optical phonon model ω at Γ. All the elastic constants are in units of 10^{11} dynes/cm^2, u_0 is in eV, and ω is wave number in 1/cm.

11. SEMICONDUCTOR ALLOYS

The systems to be considered in this section are alloys of the diamond and zincblende semiconductors, both the ordered and disordered alloys. The ordered alloys include binary compounds such as SiC, and ternary compounds, such as GaInAs$_2$, in three crystal structures of the types CuAuI, chalcopyrite, and CuPt as shown in Fig. 4 (Bernard et al., 1988). The disordered alloys include binary solutions such as Si$_{1-x}$Ge$_x$ and pseudobinaries such as Hg$_{1-x}$Cd$_x$Te and GaAs$_{1-x}$Sb$_x$, where x is the fractional concentration. These alloys have been widely used and studied; however, detailed information about their elastic constants is scarce both experimentally and theoretically. One reason for the lack of rigorous calculation is that the elastic constants of these systems are more complex; existing theories are not as accurate, particularly for disordered alloys. Another reason may be attributed to the fact that most properties of these alloys, including their elasticity, were thought to be reasonably well approximated by the concen-

1. ELASTIC CONSTANTS AND THEIR ALLOYS

TABLE XI
BOM(3) CALCULATIONS[a]

	f	n	m	u_0	C_{44}	ζ	ω
C	1.336	2.901	3.872	20.814	47.691	0.135	1,531
Si	1.262	3.251	5.346	6.484	7.472	0.580	562
Ge	1.302	3.886	6.264	5.942	6.171	0.658	333
AlP	1.193	3.43	5.77	5.722	4.936	0.541	452
AlAs	1.369	3.559	5.507	6.498	4.812	0.490	387
AlSb	1.249	3.253	5.669	4.402	3.248	0.619	343
GaP	1.276	3.550	5.575	6.422	5.738	0.542	373
GaAs	1.290	3.445	5.511	5.855	4.724	0.549	281
GaSb	1.318	3.415	5.704	4.923	3.670	0.621	239
InP	1.210	3.812	6.153	4.929	3.398	0.604	304
InAs	1.199	3.456	5.724	4.442	2.755	0.579	216
InSb	1.245	3.411	5.943	3.871	2.367	0.657	188
ZnS	0.968	3.062	5.848	3.618	3.120	0.630	343
ZnSe	1.040	3.097	5.671	3.727	2.611	0.580	247
ZnTe	1.178	3.019	5.573	3.371	2.301	0.601	212
CdTe	1.065	3.198	6.368	2.621	1.403	0.690	165
HgTe	1.059	2.937	6.262	2.523	1.563	0.730	154

[a] The results for the parameters f, n, m, and u_0 obtained from the fitting of the bond energy, bond length, bulk modulus, and shear coefficient $C_{11} - C_{12}$ of Table V using the BOM(3) described in the text. Also listed are the calculated C_{44}, internal displacement parameter ζ, and the TO optical phonon mode ω at Γ. All the elastic constants are in units of 10^{11} dynes/cm^2, u_0 is in eV, and ω is wave number in 1/cm.

tration-weighted averages of their constituents. Because of the rudimentary state of the theory, we shall deal only with the simplest elastic constant, the bulk modulus. Our focus is on the difference between the alloy bulk modulus B and the concentration-weighted averaged value $\bar{B}$, i.e., $\Delta B = B - \bar{B}$. There are several fundamental questions that can be addressed. Is ΔB positive or negative? Do the sign and the magnitude of ΔB depend on the state of order? How can we calculate the bulk modulus in a disordered system? Analysis of these questions constitutes the content of this section.

a. Ordered Alloys

Long-range ordering has been found to exist in many epitaxially grown III-V semiconductor alloys (Kuan *et al.*, 1985; Jen *et al.*, 1986; Kuan *et al.*, 1987; Ihm *et al.*, 1987; Klem *et al.*, 1987; Huang *et al.*, 1988; Gomyo *et al.*, 1987 and 1988; Norman *et al.*, 1987; Shahid *et al.*, 1987). Wei and Zunger (1989) have given a rather complete list of these ordered alloys, most of which

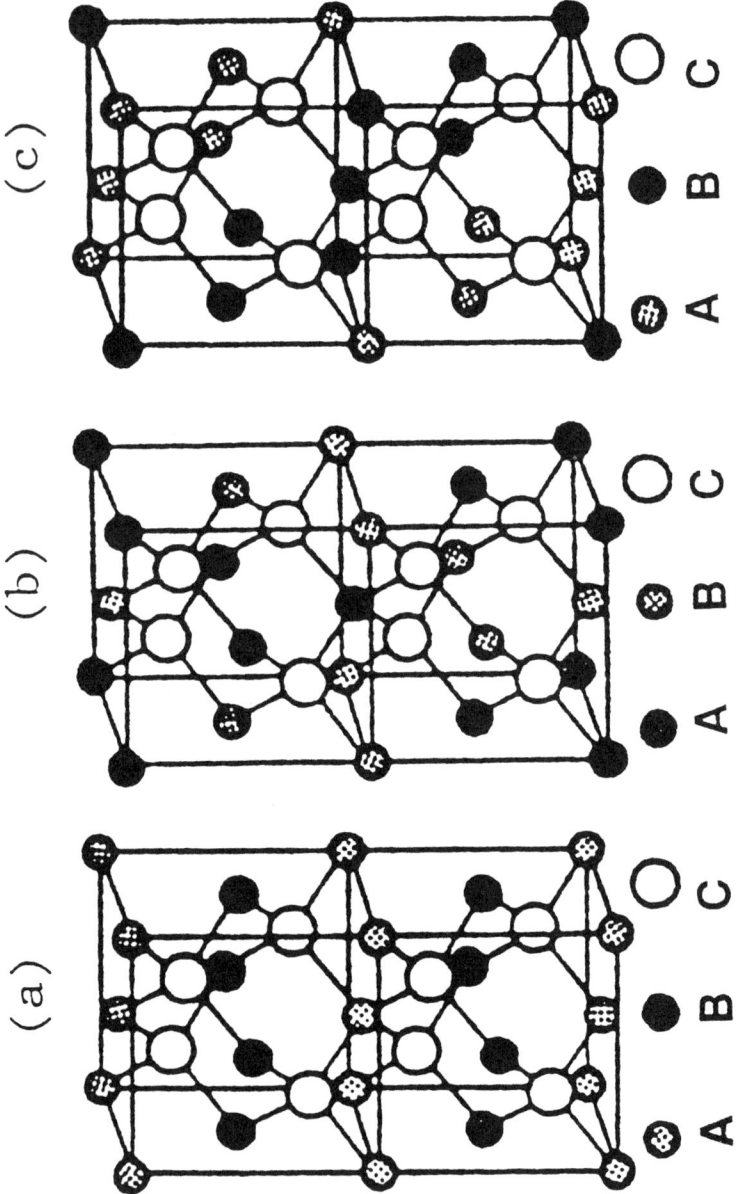

FIG. 4. Three ABC_2 structures studied: (a) CuAuI structure ordered in [001] direction. (b) Chalcopyrite structure ordered in [201] direction. (c) CuPt structure ordered in [111] direction. (Extracted from Wei and Zunger, 1989.)

are of the form ABC_2 existing in three different types of structures, CuAuI, chalcopyrite, and CuPt; all have their atomic planes stacked as ACBCACBC, but along three different directions, (100), (201), and (111), respectively, as shown in Fig. 4. There have been several first-principles calculations made to study the structural properties of these compounds; these focused mainly on the cohesive energies and bond lengths. These results have been compared with TB calculations by Yeh et al. (1990). As indicated in the preceding section, one virtue to fitting the TB model for the bulk semiconductors is to use it for interpolating alloy properties. Table XII lists the results for the bulk moduli of a number of III-V and II-VI alloys derived from full TB band structure calculations using the parameters given in Table VII. Also listed are the average values and percentage deviations from the mean $\Delta B/\bar{B}$. Note that all ΔB values are negative and that most of the magnitudes are small, except for Ga_2AsSb and Ga_2PSb; the latter has the largest difference in the constituent compounds among the alloys listed. Although the magnitudes of ΔB get larger for systems with larger differences in the bond lengths, the dependence does not seem to be a simple function of the bond length difference. The uniformly negative ΔB values also appeared in the first-principles local-density functional calculations for ordered GaAsSb alloys by Ferreira et al. (1989), as shown in Table XIII.

TABLE XII

ORDERED ALLOYS: TB ELASTIC CONSTANT CALCULATIONS[a]

Alloy	B_{Ch}	B_{Ca}	ZB_{Cp}	$\bar{B}$	$\Delta B/\bar{B} \times 100$		
					Ch	Ca	Cp
AlGaAs	7.695	7.693	7.689	7.009	−0.18	−0.20	−0.25
AlGaP	8.858	8.858	8.854	8.872	−0.15	−0.15	−0.20
GaInSb	5.226	5.202	5.156	5.312	−1.61	−2.07	−2.92
AlInAs	6.705	6.691	6.661	6.761	−0.83	−1.03	−1.48
InGaAS	6.610	6.579	6.508	6.742	−1.96	−2.42	−3.47
InAlP	7.876	7.860	7.774	7.924	−0.61	−0.08	−1.88
GaInP	8.007	8.035	8.878	8.195	−1.44	−1.95	−3.87
GaAsP	8.328	8.291	8.294	8.417	−1.05	−1.50	−1.46
GaAsSb	6.314	6.198	6.157	6.741	−6.34	−8.05	−8.66
GaPSb	6.584	6.297	6.188	7.468	−11.84	−15.68	−17.14
HgCdTe	4.470	4.472	4.471	4.485	−0.33	−0.28	−0.31
HgZnTe	4.890	4.887	4.632	4.925	−0.71	−0.76	−5.93
CdZnTe	4.611	4.604	4.338	4.650	−0.85	−1.00	−6.72

[a] Bulk moduli (in 10^{11} dynes/cm^2) of ordered alloys calculated using the full TB band-structure method described in Section 10. The three structures are chachopyrite (Ch), CuAuI (Ca), and CuPt(Cp) types. $\bar{B}$ is the average value of the constituent compounds, and $\Delta B = B - \bar{B}$.

The reason for the negative values of ΔB, in a very qualitative argument, is that the bulk moduli of semiconductors scale inversely as high powers of the lattice constant, and, at the same time, the alloy lattice constant is approximated well by the mean value, by Vegard's law (1921). This implies that the value of B at the mean lattice constant should lie below the straight-line average. Since the TB results for the bulk moduli constants should not be qualitatively different from the valence force-field model (VFF) predictions, most of the key physics for the bowing of B should be contained in a VFF analysis. The major effects in the VFF can in turn be realized from the following simple analysis.

Consider the local structure of a CuAuI or chalcopyrite crystal ABC_2. Focus on a local tetrahedral cluster A_2B_2C with the two A atoms and two B atoms on the vertices of the tetrahedron and the C atom near the center. Let the coordinates of the two A atoms be $(-1, 1, -1)d/\sqrt{3}$ and $(-1, 1, -1)d/\sqrt{3}$ and the two B atoms be at $(1, 1, 1)d/\sqrt{3}$ and $(1, -1, -1)d/\sqrt{3}$. Let the force constants be k_A and k_B, and equilibrium bond lengths be d_A and d_B for the AC and BC bonds, respectively. To attain equilibrium, the central C atom is displaced by $(\varepsilon, 0, 0)d/\sqrt{3}$. We further define mean values $\bar{d} = (d_A + d_B)/2$ and $\bar{k} = (k_A + k_B)/2$, relative differences $\delta_0 = (d_A - d_B)/\bar{d}$ and $\Delta_0 = (k_A - k_B)/\bar{k}$, and $d = \bar{d}(1 + \delta)$. Then the AC bond is stretched by an amount $\bar{d}(\delta + \varepsilon/3 - \delta_0/2)$ from its equilibrium value, and similarly, the BC bond is compressed by $\bar{d}(\delta - \varepsilon/3 + \delta_0/2)$. The strain energy for any arbitrary δ and ε is given by $\Delta E = \bar{d}^2[k_A(\delta + \varepsilon/3 - \delta_0/2)^2 + k_B(\delta - \varepsilon/3 + \delta_0/2)^2]$. When ΔE is minimized with respect to δ and ε, one finds $\delta = 0$ and $\varepsilon = 3\delta_0/2$, and the minimum ΔE is zero. If the crystal expands uniformly with δ having a fixed small value, then ε becomes $\varepsilon = 3\delta_0/2 - 3\Delta_0\delta$ and $\Delta E = 2\bar{k}(1 - \Delta_0^2/4)\bar{d}^2\delta^2$. Thus, the effective spring constant is

$$k_{eff} = \bar{k}(1 - \Delta_0^2/4), \quad (123)$$

which is smaller than the average value $\bar{k}$. This weakening of the restoring force constant in the alloy is due to the internal displacement, represented by ε in the preceding model, which provides an extra degree of freedom for relaxation in response to the external stress. The bulk modulus B is proportional to k_{eff}/d, so the alloy bulk modulus minus the mean B is then given by

$$\Delta B = \bar{B}(\delta_0\Delta_0 - \delta_0^2 - \Delta_0^2/4), \quad (124)$$

where we recall the definitions $\delta_0 = (d_1 - d_2)/\bar{d}$ and $\Delta_0 = (k_1 - k_2)/\bar{k}$. Since and δ_0 and Δ_0 tend to have different signs, the bond length difference gives an extra negative contribution to ΔB (the first two terms). If both the bond-stretching force constant α and the bond-angle restoring force β in the VFF are included, the equilibrium value of ΔE is no longer zero, but the deviation ΔB can also be shown to be similar to Eq. (123) and is given by

$$\Delta B = \bar{B}\left[\delta_0\left(\frac{3\Delta\alpha + \Delta\beta}{3\alpha + \beta}\right) - \delta_0^2 - 3\frac{(\Delta\alpha)^2}{(\alpha + 2\beta)(3\alpha + \beta)}\right]\bigg/4, \quad (125)$$

where $\Delta\alpha = (\alpha_1 - \alpha_2)$ and $\alpha = (\alpha_1 + \alpha_2)/2$ and similarly for $\Delta\beta$ and β. Equation (125) reduces to (124), if β is set equal to zero.

The above descriptions illustrate two mechanisms for the negative ΔB values, the $1/a^q$ scaling of B, with q ranging from 3.5 to 9, and more degrees of freedom for internal relaxation. Quantitative results should be described by TB, because in addition to the strain energy, there is also some chemical effect built into the TB theory. Although these ordered compounds have been found from epitaxial growth, the bulk moduli are probably difficult to measure, because these alloys are not single bulk crystals and because the ordering is only partial. It is interesting to note that the B values for a SiC/AlN alternating layer superlattice along (100) and for the constituent compounds have been calculated by Lambrecht and Segall (1990) using the LMTO; their percentage deviation from the mean $\Delta B/\bar{B}$ was found to be about -2%, which falls in the range of the ternary alloys in Tables XII and XIII.

Not all the mechanisms considered above apply to the ordered compounds of the elemental semiconductors, because internal relaxation under pressure may not be allowed, e.g., if the structure is assumed to be zincblende. Unfortunately, a simple analysis of the elastic constants of the 4-4 compounds cannot yet be made, because the tight-binding and VFF parameters have not yet been extended to deal with the atomic pairs not existing in the

TABLE XIII

ORDERED ALLOYS: BULK MODULI[a]

Structure	GaAs	GaSb	Ga$_2$AsSb			Ga$_4$As$_3$Sb		Ga$_4$AsSb$_3$	
	zb	zb	CA	CH	CP	LU	FA	LU	FA
B	7.46	5.18	6.10	5.92	5.96	6.52	6.58	5.40	5.31
$\Delta B/\bar{B} \times 100$			-3.5	-6.3	-5.7	-5.4	-4.5	-6.1	-7.7

[a]Calculated bulk moduli for GaAs, GaSb, and GaAsSb ordered alloys by Ferreira et al. (1989), and the corresponding percentage deviation from the concentration-weighted average. The structures are zincblende (zb), CuAl (CA), chachopyrite (CH), CuPt (CP), luzonite (LU), and famatinite (FA).

constituent crystals. However, several first-principles calculations have been made on the ordered SiC and SiGe (Martins and Zunger, 1986; Qteish and Resta, 1988; van Schilfgaarde, 1990). The main results are listed in Table XIII; the theoretical results were calculated for the zincblende structure. The plane-wave pseudopotential (PP-PW) calculation of Martins and Zunger (1986) for SiC gave a -21% value for $\Delta B/\bar{B}$, which is in reasonable agreement with the experimental value of -17%. This is consistent with the qualitative argument based on the $1/d^q$ ($q \geq 3.5$) scaling of B. The theory also yielded a negative formation energy, which also agrees with experiment. For SiGe, the theoretical calculations cited in Table XIV gave positive formation energies, which are consistent with the fact that no ordered bulk compounds of SiGe have been grown. However, some weak ordering has been found in the epitaxial SiGe films (Ourmazd and Bean, 1985). The calculated values of $\Delta B/\bar{B}$ for the zincblende SiGe are either slightly above or just below zero. These differences, however, fall within the uncertainties of the present first-principles theory. The best conclusion that can be drawn from these results is that the B value for SiGe should be very close to the mean value.

b. Disordered Alloys

Disordered binary alloys $A_{1-x}B_x$ of diamond semiconductors and pseudobinary alloys $A_{1-x}B_xC$ of zincblende semiconductors AC and BC are considered in this section; they are not amorphous materials, as they still

TABLE XIV

BULK MODULI OF ORDERED BINARY ALLOYS AB OF THE DIAMOND SEMICONDUCTORS A AND B FROM THEORIES AND EXPERIMENT

SiC	PP-PW[a]	Experiment		
$B(C)$	50.3	44.23		
$B(Si)$	9.53	9.92		
$B(SiC)$	23.4	22.4		
$\Delta B/\bar{B}(\%)$	-21	-17		
SiGe	PP-PW[a]	PP-PW[b]	ASA[c]	FP-LMTO[c]
$B(Si)$	9.53	9.8	8.80	9.58
$B(Ge)$	7.75	7.7	6.25	7.05
$B(SiGe)$	8.73	8.7	7.38	8.31
$\Delta B/\bar{B}(\%)$	1	0	2	0

[a]Martins and Zunger (1986).
[b]Qteish and Resta (1988).
[c]van Schilfgaarde (1990).

possess their constituent diamond and zincblende lattices, respectively, as characterized by their crystal diffraction patterns. We shall consider the pseudobinaries first. The alloying atoms A and B in these alloys belong to a fcc sublattice, and the C atoms to the other sublattice. However, the positions of the A and B atoms are not necessarily locked precisely on the lattice sites. The extended x-ray absorption fine-structure spectroscopy (EXAFS) data (Mikkelsen and Boyce, 1982 and 1983; Boyce and Mikkelsen, 1985; Balzarotti et al., 1985) have consistently shown a bimodal distribution of the bond lengths in these alloys, although the average lattice constant follows the Vegard (1921) law $a = (1 - x)a_{AC} + xa_{BC}$. Figure 5 shows an example of the results for the bond lengths in $Ga_{1-x}In_xAs$ deduced by Mikkelsen and Boyce

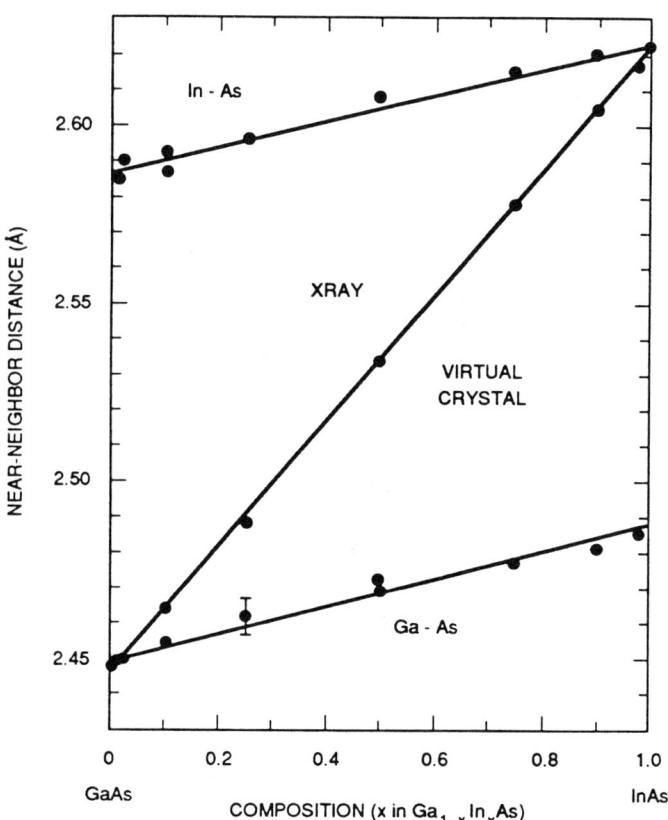

FIG. 5. Near-neighbor bond lengths (GaAs and InAs in the $Ga_{1-x}In_xAs$ alloy) as a function of composition x, measured by EXAFS (Mikkelsen and Boyce, 1982).

(1982) from their EXAFS data. In a first approximation, the crystal structure of an alloy can be viewed as having the A and B atoms on their fcc sublattice with an average lattice constant, while the C atoms are distorted away from their lattice sites, in a way similar to the local structures of the three ordered superlattices considered earlier. The difference is that there is no long-range superlattice ordering of the A and B atoms in the disordered state. This simple crystal picture is only a first approximation; the EXAFS experiments just cited also indicate that the sublattice of the A and B atoms is less than a perfect fcc. There are also theoretical calculations (e.g., Sher et al., 1987; Wei et al., 1990) that suggest a certain degree of short-range ordering in these alloys, namely that the arrangement of the A and B atoms is not random. Figure 6 shows an example of the calculated deviations (Sher et al., 1987) of the probabilities from random distribution, $\Delta p_n = p_n - p_n^0$ for $Ga_{1-x}In_xAs$ as a function of alloy concentration x, where p_n is the probability of having n Ga atoms and 4-n In atoms on the vertices of a local tetrahedral cluster in the alloy, and $p_n^0 = {}_4C_n(1-x)^n x^{4-n}$, where ${}_4C_n$ is a binomial coefficient, is the corresponding value for the random distribution.

The structural energy needed for calculating the elastic constants of a disordered alloy is an ensemble average of the total energy over the distribution of the alloying atoms under strains. It has been demonstrated (Ferreira et al., 1989) that the total energy of a semiconductor can be decomposed into the sum of multisite correlation energies, from the single-site, the pair, and up to a cluster containing a handful of sites. In other words, the multisite correlation energies converge to zero quickly at a manageable number of sites. This implies that the structural energy of an alloy is an average of these multisite correlation energies. Connolly and Williams (1983), working on metal alloys, proposed that these multisite correlation energies be deduced from the ordered systems that are composed of the same atoms. This scheme allows a direct application of the first-principles theory in the calculation of the energy parameters. These energetics can then be used in the alloy statistics such as in the cluster variational method of Kikuchi (1951) or in the Monte Carlo calculations to deduce the distribution functions or the average properties. This theory has been carried out extensively for semiconductor alloys by Ferreira et al. (1989), and respectable results have been obtained for the phase diagrams and alloy equilibrium properties. For this theory to fit the elastic constants requires detailed dependences of the multisite energies under different strains that have yet to be worked out. Also, the validity of using the energy parameters deduced from ordered alloys in the disordered systems needs to be examined further.

There is a different cluster approach, which directly relates the alloy Hamiltonian to the distribution function (Gautier et al., 1975; Chen et al., 1987; Berera et al., 1988; Dreysse et al., 1989). In this approach one focuses

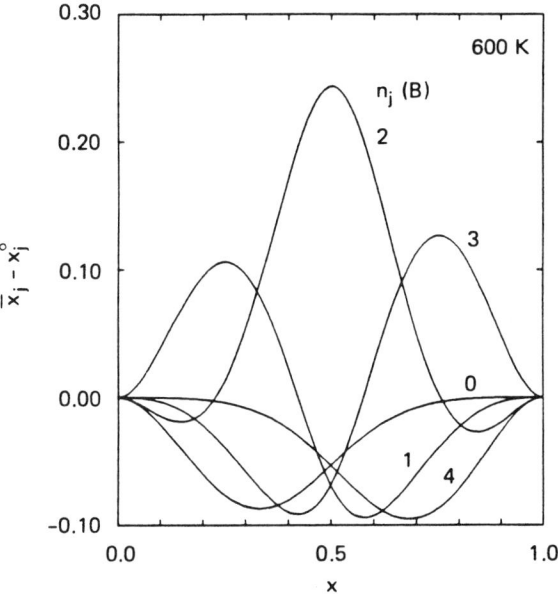

FIG. 6. Cluster populations relative to those in a random alloy $x_j - \bar{x}_j$ for clusters with $n_j = 0, 1, 2, 3, 4$ B atoms for a $Ga_{1-x}In_xAs$ alloy equilibrated at 600 K.

a particular cluster in an alloy ensemble. The average energy per cluster can be written (Chen et al., 1988) as

$$\langle \varepsilon \rangle = \sum_\eta \sum_m \left(\varepsilon_n + \frac{1}{2} h_{nm} \right) p_n P_{nm}. \quad (126)$$

on where ε_n is the energy of a cluster detached from a given alloy configuration, and h_{nm} is the interaction energy, which is the change of energy of the combined system when the cluster is put back into the alloy. In Eq. (126), p_n is the probability that a cluster is of the type n, specified by the number of A and B atoms and their arrangements, and P_{nm} is a conditional probability that the surrounding environment is in state m when the cluster is in state n. The factor 1/2 in Eq. (126) is to eliminate the double counting in the total average alloy energy $\langle E \rangle = M \langle \varepsilon \rangle$, where M is the ratio between the size of the alloy and the cluster. If one writes $\langle \varepsilon \rangle = \Sigma \, p_n \varepsilon(n)$, then an effective cluster energy can be defined as

$$\varepsilon(n) = \sum_m \left(\varepsilon_n + \frac{1}{2} h_{nm} \right) P_{nm}. \quad (127)$$

This procedure is particularly useful when the interaction energy is short-ranged. One can start with a small cluster and a given probability distribution, then calculate the effective cluster energies from Eq. (126). These energies are then used in a statistical theory to deduce a cluster distribution, which in turn is used to calculate a new set of $\varepsilon(n)$ and distribution functions, and the process is iterated until it converges. It should be pointed out that to compute the total energy of an alloy quantum mechanically, one needs to solve the Schroedinger equation for a Hamiltonian that does not have the lattice translational symmetry so indispensable in traditional band-structure theory. If the fluctuation of the alloy potential from the virtual crystal approximation (VCA), where the alloy potential is approximated as the concentration-weighted average, is small, then the next leading correction to VCA can be obtained from perturbation theory. This should work for most semiconductor alloys except for systems with large potential fluctuations such as $Hg_{1-x}Cd_xTe$ (Chen and Sher, 1982; Spicer *et al.*, 1982; Hass *et al.*, 1983). A more general but more difficult approach is to extend the present molecular coherent potential approximation (MCPA) (Hass *et al.*, 1983) to clusters and to achieve a triple self-consistency (Chen *et al.*, 1987): consistency between cluster distribution and Hamiltonian, between the Hamiltonian and electron density, and between the self-energy operator Σ in the cluster CPA theory and the potential fluctuations. To date this theory has been carried out only for metal alloys, and then only within the single-site KKR-CPA with a random distribution (Schwartz and Bansil, 1975; Gyorffy and Stocks, 1978). Major work is needed to determine if this approach can achieve the same degree of rigor for disordered alloys as self-consistent density functional theory, which has been successfully used in dealing with crystalline semiconductors.

The above idea has been applied to an elastic medium model to deduce a mean field theory for the internal strain and bulk modulus in semiconductor alloys (Chen *et al.*, 1988). This theory starts by assuming that the alloy has an effective lattice constant and effective modulus. When part of the effective alloy medium is replaced by a specified cluster, there will be strain energy introduced. It was shown that this strain energy can be taken as the effective energy $\varepsilon(n)$ for that cluster, the probability distribution p_n within a statistical theory can then be deduced. The internal strain energy is calculated as $E = M\langle\varepsilon\rangle = M \Sigma p_n\varepsilon(n)$. When the alloy is under an external pressure δP, the effective cluster energy will change by an amount $\delta\varepsilon(n)$, which implies a change of the total energy by an amount $\delta E = M\langle\delta\varepsilon(n)\rangle$. Then the bulk modulus of the alloy can be obtained from $\Delta E = \frac{1}{2}(\delta P)^2 V/B$, where V is the alloy volume. The mean-field nature of this approach is evident from the fact that the calculation requires knowledge of the alloy lattice constant and elastic constants that are only assumed and are required to be calculated self-consistently. To illustrate this self-consistency procedure, let us consider the following simple spring model for a random pseudobinary alloy $A_{1-x}B_xC$. The cluster corresponds to the four bonds surrounding an "impurity" atom A or B, and the environment of the cluster corresponds to the 12 bonds that connect inwardly to the cluster and outwardly to a rigid lattice of the effective alloy. It is worth mentioning that there have been detailed analyses of the valence-field force models for the strain energies of semiconductor alloys in regard to the range allowed for lattice relaxation (Martins and Zunger, 1984; Chen and Sher, 1985). It was found that by neglecting the bond-angle forces, one can use a shorter range of lattice relaxation to obtain the correct mixing enthalpies and bond lengths. The simple model considered here works amazingly well for these properties.

Let the spring constants for the pure AC and BC compounds be k_A and k_B, respectively, and the effective alloy spring constant be k, with similar notations for the bond lengths d_A, d_B, and d. When an A atom is embedded in the medium, all 16 bonds under consideration will relax, and the strain energy is given by

$$\varepsilon(A) = \frac{1}{2} k_1 (d - d_A)^2, \tag{128}$$

where

$$k_1 = 4k_A k / (3k_A + k). \tag{129}$$

A similar energy $\varepsilon(B)$ is obtained, when a B atom is embedded. The effective bond length d is obtained from a minimization of the average cluster energy $E = (1 - x)\varepsilon(A) + x\varepsilon(B)$ with respect to d, which yields

$$d = [(1 - x)k_1 d_A + x k_2 d_B] / [(1 - x)k_1 + x k_2]. \tag{130}$$

When the alloy is compressed, the alloy bond length is reduced to $d(1 - e)$, where e is a macroscopic strain corresponding to the external pressure. The pressure-induced strain energy for the 16 bonds in the medium is $\delta E = 8k(de)^2$, and $\delta E = 2k(de)^2$ for each cluster. Embedding an A atom in this compressed medium, one finds the total strain energy for the 16 bonds to be

$$E_A = \frac{1}{2} k_1 (d - d_A - 4de)^2. \tag{131}$$

To obtain the extra cluster energy $\delta\varepsilon(A)$ induced by the pressure, we subtract $\varepsilon(A)$ of Eq. (127) and the background energy for the surrounding 12 bonds from E_A to give

$$\delta\varepsilon(A) = -4k_1(d - d_A)de + 8k_1(de)^2 - 6k(de)^2. \tag{132}$$

Similarly, the following expression for $\delta\varepsilon(B)$ is obtained when the embedded atom is B:

$$\delta\varepsilon(B) = -4k_2(d - d_B)de + 8k_2(de)^2 - 6k(de)^2. \tag{133}$$

Thus, the change of the average cluster energy that is due to the pressure is given by $\delta E = \langle\delta\varepsilon(n)\rangle = (1 - x)\delta\varepsilon(A) + x\delta\varepsilon(B)$, which, when equated to $2k(de)^2$, leads to the following self-consistent equation for the effective spring constant k : $k = (1 - x)k_1 + xk_2$. The k can now be solved analytically when both the expression for k_1 in Eq. (129) and the similar expression for k_2 are used. The result is

$$k = \langle k\rangle[1 - 3x(1 - x)(\delta k/\langle k\rangle)^2], \tag{134}$$

where $\langle k\rangle = (1 - x)k_A + xk_B$ is the mean spring constant and $\delta k = k_A - k_B$ the difference. It is interesting to compare this result for the 50/50 alloy, i.e., $k = \bar{k}(1 - \frac{3}{4}\Delta_o^2)$ with the value $\bar{k}(1 - \Delta_o^2/4)$ in Eq. (123) for the ordered alloys in the CuAuI and chalcopyrite structures. The alloy spring constant is slightly below the straight-line average, and the bowing is larger for a disordered alloy than for the corresponding ordered compound. Using the effective spring constant of Eq. (134), we find that the effective bond lengths for most alloys also bow slightly below their mean value,

$$d = \langle d\rangle + 4x(1 - x)(d_A - d_B)(k_A - k_B)k/[(3k_A + k)(3k_B + k)] \tag{135}$$

because the spring constant tends to increase as the bond length decreases.

To compare the calculations above with experimental data for pseudobinary alloys, we were able to find results for GaAlAs (Landolt-Bornstein, 1988), CdZnTe, CdMnTe, and HgCdTe (Quadri et al., 1986). For GaAlAs, the following linear x dependences were measured (Landolt-Bornstein, 1988): $C_{11} = 11.85 + 0.14x$, $C_{12} = 5.38 + 0.32x$, and $C_{44} = 5.94 - 0.05x$. This lack of detectable bowing is expected, because of the nearly equal bond lengths of the two constituent compounds and small differences in the elastic constants. The bulk moduli in the three II-VI alloy systems mentioned were obtained from high-pressure x-ray diffraction data. For HgCdTe, the results are similar to those for the GaAlAs in that both the bond lengths and the bulk moduli of

TABLE XV

MEASURED ELASTIC CONSTANTS IN 10^{11} DYNES/CM^2 OF SiGe ALLOYS BY BUBLIK ET AL (1974)

Alloy	C_{11}	C_{12}	C_{44}
$Si_{0.28}Ge_{0.72}$	16.1 ± 0.8	8.35 ± 0.8	8.55 ± 0.4
$Si_{0.54}Ge_{0.46}$	17.0 ± 0.8		
$Si_{0.46}Ge_{0.36}$	17.1 ± 0.8		

HgTe and CdTe are so close that the differences in B between the alloys and the pure crystals were beyond the experimental resolution. However, a 5% Zn in CdZnTe alloy was found to give a 15% increase in the B value from the pure CdTe value and a 10% Mn in CdMnTe gave a 21% decrease (Quadri et al., 1986). These significantly large changes in the B values caused by smaller concentrations cannot be explained from the above considerations.

The qualitative model considered above does not apply to the binary alloys $A_{1-x}B_x$, because in these alloys both the A and B atoms can be found in both sublattices, and the local bond length arrangement is more complicated than the pseudobinary alloys. However, one can expect that there are still more degrees of relaxation in the disordered binaries than in the ordered compounds. Therefore, one would conclude that the bulk modulus of the disordered 50-50 SiGe alloy would have a smaller value than those tabulated in Table XIII for the ordered compounds. At least one would not expect the alloy B values to be significantly larger than the mean values $\bar{B}$. However, the only experimental data available (Bublik et al., 1974), Table XV, show that all three elastic constants for these alloys at three different concentrations exceed the values for Si, and that $\Delta B/\bar{B}$ is as large as 20%, despite the fact that the bond length difference between Si and Ge is only about 3% and the measured alloy lattice constants are only bowed slightly below the average. This, and the unexplained results for the II-VI alloys, point to the need for a more systematic study of the elastic properties of semiconductor alloys, both experimentally and theoretically.

IV. Dislocations and Hardness[8]

Hardness has proven to be a useful probe of the mechanical properties of the brittle semiconductors. Here we will use the term hardness to refer specifically to Vickers' hardness, unless otherwise noted. In the Vickers'

---

[8] Much of this section is adapted from "Final Report" (AFOSR-F49620-85-0023) by M. A. Berding (1988), SRI International, Menlo Park, California.

hardness measurement, a square pyramidal indenter is used, and the hardness number is given by the applied load divided by the area of the indentation (i.e., units of pressure). Hardness has been found to be an intrinsic property of the material, because it is relatively independent of the applied load. One advantage of hardness measurements for semiconductors is that, in contrast to bending tests, only small samples are necessary for conventional Vickers' hardness measurements, or for nanoindenter measurements (Fang *et al.*, 1990), and relatively thin epitaxial films can be probed. Additionally, unlike conventional tests used to measure yield stress, hardness measurements can be made at room temperature, which is far below the usual plastic regime for most semiconductors. As such, the hardness measurement provides a convenient and usable probe.

The question remains, though, as to the interpretation of the hardness measurement in semiconductors: Just what property or properties of a semiconductor are we measuring when we measure hardness? In metals, an empirical relationship is found between the hardness H and yield stress Y, such that $H = 3Y$.

In metals, this relationship can be justified on the basis of continuum theory, as discussed in McClintock and Argon (1966). In semiconductors such a simple relationship between H and Y is not necessarily appropriate for several reasons. During deformation in metals, many slip planes can be active because the Peierls barriers for dislocation motion in most directions are low. In contrast, because the bonds in semiconductors are strongly covalent, the Peierls barriers are high and dislocation in the $(111)1/2\langle\bar{1}10\rangle$ slip system dominate.

To date, there is no complete quantitative theory of hardness in the semiconductors in which the temperature dependence, photoplastic effect, and the alloy hardening effect are included. Sher *et al.* (1985) proposed a model of hardness for the semiconductor compound that gives good quantitative agreement with experiment, but this model does not provide an explanation for several of the observed dependences of hardness. This model of hardness in semiconductors differs from more conventional interpretations and suggests that hardness is dominated by dislocation–dislocation interactions, as opposed to dislocation activation and motion terms. We discuss the results of an improved quantitative model of hardness below.

12. SLIP SYSTEMS

In the dislocation interaction hardness model, Vickers' hardness is found to be dominated by the interaction energy of an idealized array of dislocations that has been generated by the indenter. The idealized array can be considered as a first approximation to the more realistic dislocation tangles found experimentally, leading in higher order to an expansion in dislocation configurations. In this idealized array, no account is taken of the true slip systems active in the semiconductors. Experiments (Hirsch *et al.*, 1985) have demonstrated that, in Vickers' hardness, slip occurs primarily on the $\{111\}\frac{1}{2}\langle\bar{1}10\rangle$ glide set, where the threefold symmetry of slip and rosette lines occurs at the intersection of the $\{111\}$ planes with the (111) surface.

For indentation on the (111) plane, dislocations can glide on the (111) plane parallel to the surface or on one of the three other $\{111\}$ planes with a total of four active slip planes. Although the detailed analysis differs from that given previously (Sher *et al.*, 1985), the contribution to the hardness from the interaction energy is comparable to that previously calculated. This contribution to H is directly proportional to the shear coefficient.

13. PEIERLS ENERGY

The Peierls energy is difficult to calculate precisely because of dislocation charge effects and reconstruction at the dislocation core. In the context of the hardness measurement we calculate the Peierls energy in order to evaluate the importance of this contribution to the Vickers' hardness number. Although it is generally agreed that dislocations in semiconductors move through the generation and propagation of double kinks, in the hardness measurement, the region about the indenter is grossly plastically deformed. Because the dislocation velocity is low at room temperature (see below), the large dislocation pile-up model proposed by Sher *et al.* (1985) may be appropriate. If the dislocation separation is small, dislocation motion through kink processes will be suppressed and the dislocations will propagate as a complete unit.

To get from Configuration A to Configuration B in Fig. 7, we must break a row of bonds. Since the long-range strain fields should be comparable in the two configurations as well as in intermediate configurations, the Peierls force can be calculated from local energy considerations only. The energy to break a bond at the dislocation core is approximately given by

$$U_b = 2\sqrt{V_2^2 + V_3^2} + 2\varepsilon_{met} - V_0 + \frac{n}{4}(\varepsilon_h^a - \varepsilon_h^c), \tag{136}$$

where V_2 is the covalent energy, V_3 the ionic energy, ε_{met} the metallization energy, V_0 the bond overlap energy, n is 1 for III-V and 2 for II-VI compounds, and ε_h^a and ε_h^c the hybrid energy for the anion and cation, respectively. The first two terms in Eq. (136) account for the loss of the

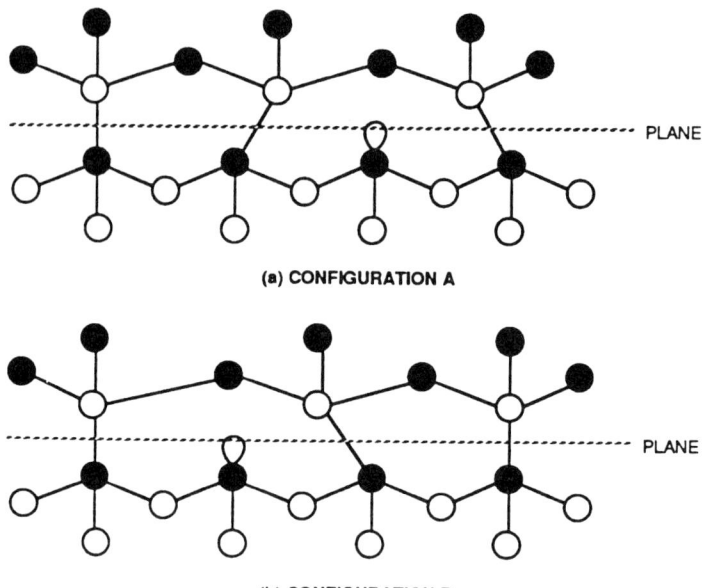

Fig. 7. Atom configurations during the slip of a dislocation. (a) Configuration A; (b) Configuration B.

bonding energy of the two electrons in the breaking bond, the third term accounts for regaining the repulsive interaction energy of the bond, and the fourth term accounts for the energy gain to transfer electrons back from the cation to the anion. We note that the electron orbitals of the atoms at the dislocation core are left in the sp^3 hybrids after the bond breaking. The expression in Eq. (136) represents a theoretical maximum of the Peierls energy, since no reconstruction at the core has been included.

To calculate the Peierls energy per unit length, we consider a primary dislocation in the $\langle \bar{1}10 \rangle$ direction in a zincblende compound. The number of bonds per unit length in $\langle \bar{1}10 \rangle$ is given by $1/b$, where b is Burger's vector. Thus, the Peierls energy per unit length is given by

$$E_p = \frac{U_b}{b} = \sqrt{\frac{3}{8}} \frac{U_b}{d}, \qquad (137)$$

where d is the bond length.

We can now calculate the Peierls force, or the force per unit length necessary to move a dislocation over the potential barrier, as illustrated in

Fig. 8. The Peierls energy is related to the Peierls force through

$$E_p = F_p L, \qquad (138)$$

where

$$L = \frac{b}{2} \qquad (139)$$

is the distance between Configuration A and Configuration B. Solving for F_p in terms of U_b and d, we arrive at:

$$F_p = \frac{3}{4} \frac{U_b}{d^2}, \qquad (140)$$

or

$$\tau_p = \sqrt{\frac{3}{2}} \frac{3}{8} \frac{U_b}{d^3}. \qquad (141)$$

Values for U_b and τ_p are summarized in Table XVI.

Now we incorporate the Peierls energy into the hardness model for low temperature where the full barrier must be surmounted. The Sher model is based on energy considerations. The Vickers' hardness number is given by the applied force divided by the area of indentation. Multiplying the numerator and denominator by h, the depth of indentation, we have

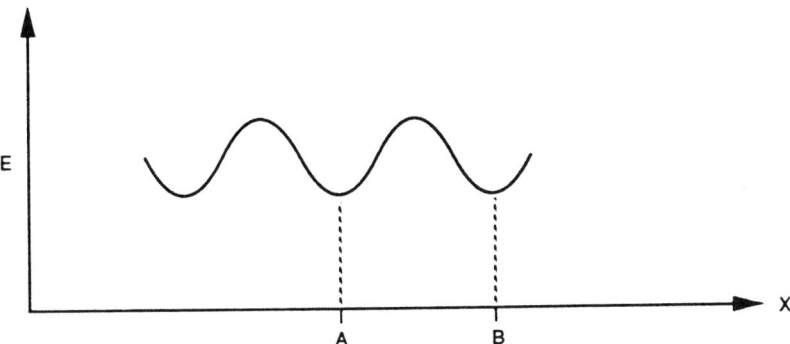

FIG. 8. Schematic of the dislocation potential as a function of its position.

TABLE XVI

CALCULATED PEIERLS STRESS AND HARDNESS FOR VARIOUS ZINCBLENDE SEMICONDUCTORS, WITH EXPERIMENTAL HARDNESS VALUES FOR COMPARISON[a]

	U_b	τ_p	H_p	H_{int}	$H_p + H_{int}$	H_{exp}
C	11.35	23,300	1940	9244	11,184	10,000
Si	5.95	3430	286	1098	1384	1370
Ge	6.67	3440	286	893	1179	1000
Sn	5.66	1930	161	—	—	—
AlP	6.04	3410	284	—	—	—
GaP	6.15	3500	292	903	1195	940
InP	5.78	2640	220	548	768	520
AlAs	5.90	3000	250	—	—	505
GaAs	6.03	3000	256	750	1006	580
InAs	5.64	2340	195	469	664	430
AlSb	5.08	2840	237	524	761	400
GaSb	5.36	2180	182	553	735	450
InSb	4.92	1670	139	365	504	230
ZnS	5.14	3000	250	515	765	—
CdS	4.77	2750	229	288	517	—
HgS	5.68	2590	216	—	—	—
ZnSe	4.92	2500	209	462	671	137
CdSe	4.59	1890	157	254	411	—
HgSe	4.523	1840	154	232	386	—
ZnTe	4.42	1720	143	374	517	82
CdTe	4.01	1350	113	222	335	60
HgTe	3.99	1360	113	230	343	25

[a]The U values are in eV, and the others in kg/mm².

$H = E/(W^2 h)$, where $E = Fh$ is the energy of indentation, and h is the depth of the indentation. Including the interaction energy only we have

$$H = H_{int} = \frac{G \cot \theta}{6\pi(1-v)}\left[-\ln\left(\frac{\cot \theta}{2}\right) + \frac{4}{3} + \sin^2 \frac{\theta}{2}\right], \quad (142)$$

where θ is one half the indenter angle. To include the Peierls energy, we consider the total energy necessary to move the dislocations from their initial to final positions in the idealized model. The hardness is then given by:

$$H = H_{int} + H_p, \quad (143)$$

where

$$H_p = \frac{E_p}{W^2 h} \quad (144)$$

is the Peierls contribution to the hardness, and E_p is the total Peierls energy expended. The total length of dislocation to be moved is calculated to be

$$L_T = \frac{1}{6}\frac{W^3}{b^2}\cos^2\theta. \quad (145)$$

The total Peierls energy is given by

$$E_p = \frac{U_b}{b}L_T = \frac{1}{6}\frac{U_b W^3}{b^3}\cos^2\theta. \quad (146)$$

Thus, we have

$$H_p = \frac{1}{3}\frac{U_b}{b^3}\cos\theta\sin\theta. \quad (147)$$

For $\theta = 45°$,

$$H_p = \frac{1}{6}\frac{U_b}{b^3}. \quad (148)$$

Values of H_p are summarized in Table XVI. Several features of H_p should be noted. First, we have used a zero-temperature value of the Peierls energy. Because hardness measurements are typically done at room temperature, one should take the thermal energy into account; this will reduce the values of H_p from those listed in Table XVI. Also shown in Table XVI are H_{int}, $H_{int} + H_p$, the best theoretical estimate for H, and H_{exp}. Note that, like H_{int}, H_p is independent of the applied load, in agreement with experiment. Also note that H_p improves the agreement between theory and experiment for the hard, nonpolar materials. For the softer, more ionic materials, H is overestimated by the theory. The overestimation of H may be because of neglect of dislocation velocity effects and their temperature dependence, as discussed below.

14. Temperature Dependence

Here we summarize the experimental results and discuss a tentative theory of the temperature dependence of the hardness.

Several recent studies on the temperature dependence of hardness serve to illustrate the behavior. Results for GaAs and Ge are shown in Fig. 9. The (111) and (100) faces of GaAs have been examined by Hirsch et al. (1985) and Guruswamy et al. (1986), respectively. Results for the Knoop hardness on the (100) face of n-type Ge are also shown (Roberts et al., 1986). The (100) face of GaAs and the Ge show a definite temperature dependence with a relatively temperature-independent region for T < 450 K and an exponential temperature dependence for $T > 550$ K:

$$H = H_0 e^{U/kT}, \qquad (149)$$

with

$$U \cong 0.24 \text{ eV} \qquad (150)$$

for (100) GaAs. The results for GaAs (111) appear to follow a similar behavior.

The temperature dependence of hardness suggests that two different mechanisms may determine hardness in the two temperature regimes. At low temperature, the hardness is nearly independent of temperature and may be limited primarily by dislocation interactions. Dislocation mobility is low at low temperatures, and the tendency for dislocation pile-up is high. At elevated temperature, the dislocation mobility is increased, so that dislocations move more readily under an applied stress. Therefore, at higher temperatures, dislocation pile-up is reduced and the hardness is limited by lattice friction, which shows a strong temperature dependence.

V. Concluding Remarks

The experimental methods available to measure elastic constants vary greatly in their accuracy and in the size of samples required. Generally, those that measure the velocity of sound are quite accurate, some yielding elastic constants to one part in 10^6. These methods are also capable of measuring higher-order elastic constants, a subject not treated in this paper. However, the samples required for these measurements must be large, of the order of several centimeters, and must be perfect bulk single crystals. Many semiconductors, alloys in particular, are only grown as thin films on disparate

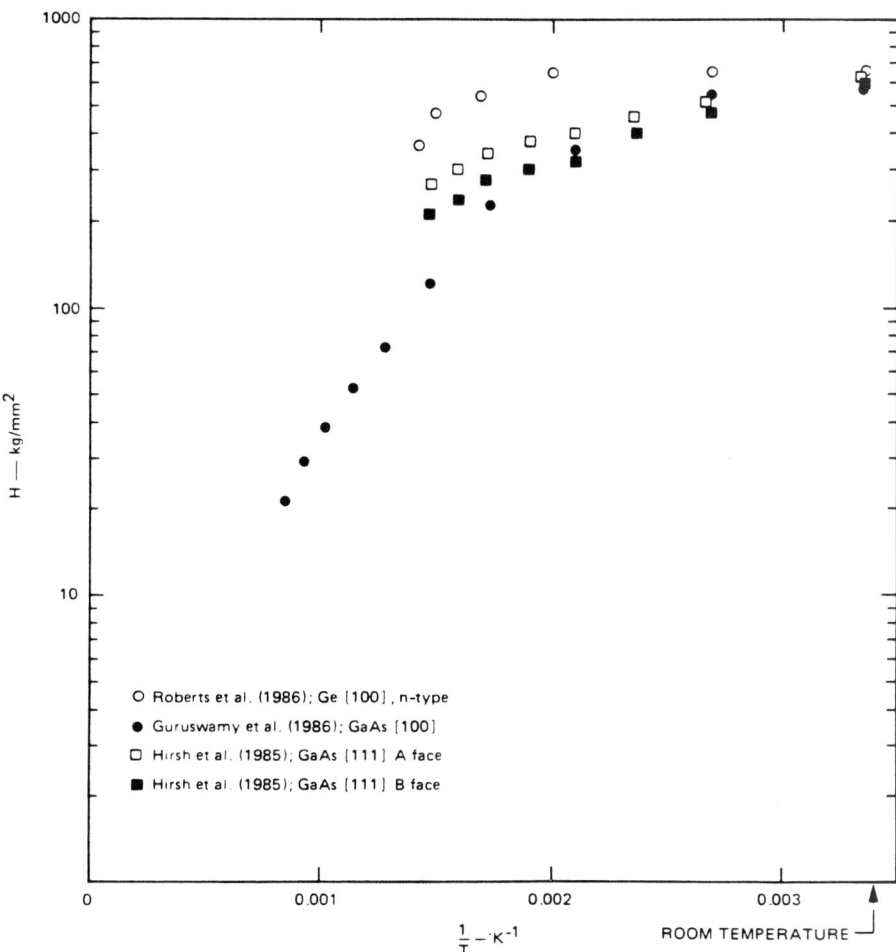

FIG. 9. Measured hardness of Ge and several GaAs samples as a function of temperature.

substrates. For these examples, the velocity-of-sound methods fail, and the less accurate Raman and Brillouin scattering techniques become the methods of choice. Their accuracy is about 1 to 4%, which is adequate for many practical applications.

Most group IV, III-V compound, and II-VI compound semiconductors have been studied, and their elastic constants tabulated. A few remain to be examined, and several should be reexamined because different experimenters do not agree on the results. The situation in the pseudobinary alloys is quite different. Few alloy systems have been adequately studied; those studied have mostly fallen into the class of materials in which the bond lengths of the constituents nearly match. More interesting results are expected from alloys with a bond length mismatch. Such studies would yield a wealth of information on mechanisms responsible for correlations in these alloys, and perhaps even on those responsible for producing the ordered alloys that have been grown recently.

We have emphasized the utility of various parameterized models for treating nonideal situations. However, the most powerful new theoretical developments are in the area of first-principles theories. The advent of self-consistent local density theories (Hohnberg and Kohn, 1964) more than 20 years ago, and the advances in methods to solve the Schroedinger equation, are making real applications, as evident from the excellent structural and elastic properties produced by these theories (see Table II and Anderson et al., 1985). Recent progress in the full-potential LMTO method (Methfessel and van Schilfgaarde et al., 1990), in the non-self-consistent approach (Harris, 1985; Foules and Haydock, 1989; van Schilfgaarde et al., 1991), and the Carr-Parrinello (1985) quantum molecular dynamics extension has increased solution speeds into the realm in which it is practical to attack many mechanical property problems. Moreover, LDA has now been extended to include many-body corrections (Hybertsen and Louie, 1987), so properties that are sensitive to conduction bands can now be computed accurately. As computational speeds continue to increase, these methods will evolve into practical engineering tools.

Acknowledgments

This work was supported in part by AFOSR Contract F49620-88-K-0009 and ONR Contract N00014-88-C0096. We are especially indebted to M. A. Berding for permitting us to include in Section IV unpublished ideas from a report she wrote. We also wish to acknowledge contributions from J. Heyman.

References

Anderson, O. K., Jepson, O., and Glötzel, D. (1985). *In* "Highlights of Condensed Matter Theory," (F. Bassiani *et al.*, eds.). North Holland, Amsterdam.
Ashcroft, N. W., and Mermin, N. D. (1976), p. 445. "Solid State Physics." Saunders College, Philadelphia.
Bachelet, G. B., Hamman, D. R., and Schluter, M. (1982). *Phys. Rev.* **B26**, 4199.
Balzarotti, A., Czyzk, M. T., Kisel, A., Motta, N., Podgorny, M., and Zimnal-Starnawska, M. (1985). *Phys. Rev.* **B31**, 7526.

Barnes, J. M., and Hiedemann, E. A. (1957). *J. Acoust. Soc. Am.* **28**, 1218–1221; **29**, 865.
Baroni, S., Giannozzi, P., and Testa, A. (1987). *Phys. Rev. Lett.* **58**, 1861.
Benedek, G. B., and Fritsch, K. (1966). *Phys. Rev.* **149**, 647–662.
Benson, G. C., and Kiyohara, O. (1974). *J. Acoust. Soc. Am.* **55**, 1184–185.
Berera, A., Dreysse, H., Wille, L. T., and de Fontaine, D. (1988). *Phys. Rev.* **F18**, 149.
Bernard, J. E., Ferreira, L. G., Wei, S.-W., and Zunger, A. (1988). *Phys. Rev.* **B38**, 6338.
Beyer, R. T., and Letcher, S. V. (1969). "Physical Ultrasonics," Vol 32 in *Pure and Applied Physics*. (H. S. W. Massey and K. A. Brueckner, consulting editors), pp. 47–50. Academic Press, New York.
Bhatia, A. B., and Noble, W. J. (1953). *Proc. Roy. Soc. London, Ser. A* **220**, 356–368, 369–385.
Blackman, M. (1959). *Phil. Mag.* **3**, 831.
Bloomfield, P. E., Ferren, R. A., Radice, P. F., Stefanou, H., and Sprout, O. S. (1978). *Nav. Res. Rev.* (May), 1–15.
Bolef, D. I., and Miller, J. G. (1971). In "Physical Acoustics," (W. P. Mason and R. N. Thurston, eds.), Vol. VIII, pp. 96–201, Academic Press, New York.
Bolef and Miller (1974). Table 2.
Bolef, D. I., deKlerk, J., and Gosser, R. B. (1962). *Rev. Sci. Instrum.* **33**, 631.
Born, M., and Oppenhimer, J. R. (1927). *Ann. Phys.* **4**, **84**, 457.
Born, M., and Wolf, E. (1970). "Principles of Optics," 4th ed., pp. 593–610. Pergamon Press, Oxford. A concise treatment of this subject appears in Chapter 12.
Boyce, J. B., and Mikkelsen, J. C. Jr. (1985). *Phys. Rev.* **B31**, 6903.
Breazeale, M. A., and Ford, J. (1965). *J. Appl. Phys.* **36**, 3486. For the pure longitudinal mode directions and for isotropic materials, the nonlinear wave equation has been solved here.
Breazeale, M. A., Cantrell, J. H. Jr., and Heyman, J. S. (1981). "Methods of Experimental Physics, Vol. 19, Ultrasonics," 67–135. Academic Press, New York.
Brewer, R. (1965). *Appl. Phys Lett.* **8**, 165.
Brewer, R. G., and Rieckhoff, K. E. (1964). *Phys. Rev. Lett.* **13**, 334–336.
Brugger, K. (1964). *Phys. Rev.* **133**, A1611.
Bublik, V. T., Gorelik, S. S., Zaitsev, A. A., and Polyakov, A. Y. (1974). *Phys. Stat. Soli. B* **66**, 427.
Callaway, J., and March, N. H. (1984). *Solid State Phys.* **38**, 135.
Cantrell, J. H., Jr. (1982). *J. Testing and Eval.* **10**, 223.
Cantrell, J. H., and Breazeale, M. A. (1977). *J. Acoust. Soc. Am.* **61**, 403.
Carr, M., and Parrinello, M. (1985). *Phys. Rev. Lett.* **55**, 2471.
Ceperly, D. M., and Alder, B. J. (1980). *Phys. Rev. Lett.* **45**, 566.
Chadi, D. J. (1978). *Phys. Rev. Lett.* **41**, 1062.
Chadi, D. J. (1979). *Phys. Rev.* **B19**, 2074.
Chadi, D. J. (1984). *Phys. Rev.* **B29**, 785.
Chadi, D. J., and Cohen, M. L. (1973). *Phys. Rev.* **B8**, 5747.
Chen, A.-B., and Sher, A. (1982). *J. Vac. Sci. & Technol.* **21**(1), 138.
Chen, A.-B., and Sher, A. (1985). *Phys. Rev.* **B32**, 3695.
Chen, A.-B., van Schilfgaarde, M., Krishnamurthy, S., Berding, M. A., and Sher, A. (1987). In "Ternary and Multinary Compounds" (S. K. Deb and A. Zunger, eds.). Materials Research Society, Pittsburgh, Pennsylvania.
Chen, A.-B., Berding, M. A., and Sher, A. (1988). *Phys. Rev.* **B37**, 6285.
Chern, E. J., Cantrell, J. H., and Heyman, J. S. (1981). *J. Appl. Phys.* **52**, 3200–3204.
Chiao, R. Y., Townes, C. H., and Stoicheff, B. P. (1964). *Phys. Rev. Lett.* **12**, 592–595.
Cohen, M. L. (1985). *Phys. Rev.* **B32**, 7988.
Connolly, J. W.D., and Williams, A. R. (1983). *Phys. Rev.* **B27**, 5169.
Conradi, M. S., Miller, J. G., and Heyman, J. S. (1974). *Rev. Sci. Instrum.* **45**, 358–360.
Cousins, C. S.G., Gerward, L., Staun, L. J., Selsmark, B., and Sheldon, B. J. (1987). *J. Phys.* **C20**, 29.

Dreysse, H., Berera, A., Wille, L. T., and de Fontaine, D. (1989). *Phys. Rev.* **B39**, 2442.
Elmore, W. C., and Heald, M. A., (1969). "The Physics of Waves," pp. 98–104. McGraw-Hill, New York.
Fang, S., Farthing, L. J., Tang, M.-F.-S., and Stevenson, D. A. (1990). *J. Vac. Sci. Technol.* **A8**(2), 1120.
Ferreira, L. G., Wei, S.-H., and Zunger, A. (1989). *Phys Rev.* **B40**, 3197.
Fleury, P. A. (1970). In "Physical Acoustics" (W. P. Mason and R. N. Thurston, eds.), Vol. 6, Chap. 1. Academic Press, New York.
Foulkes, M., and Haydock, R. (1989). *Phys. Rev.* **B39**, 12520.
Froyen, S., and Harrison, W. A. (1979). *Phys. Rev.* **B20**, 2420.
Gautier, G., Ducatelle, F., and Giner, J. (1975). *Phil. Mag.* **31**, 1373.
Goldstein, H. (1965). "Classical Mechanics," p. 347. Addison-Wesley, Reading, Massachusetts.
Gomyo, A., Suzuki, T., Kobayashi, K., Kawata, S., Hino, I., and Yuasa, T. (1987). *Appl. Phys. Lett.* **50**, 673.
Gomyo, A., Suzuki, T., and Iijima, S. (1988). *Phys. Rev. Lett.* **60**, 2645.
Green, R. E., Jr. (1973). "Treatise on Materials Science and Technology, Vol. 3, Ultrasonic Investigation of Mechanical Properties," pp. 7–8, 11–25. Academic Press, New York.
Guruswamy, S., Hirth, J. P., Faber, K. T. (1986). *J. Appl. Phys.* **60**, 4136.
Gyorffy, B., and Stocks, G. M. (1978). *Inst. Phys. Conf. Ser.* **39**, 394.
Harris, J. (1985). *Phys. Rev.* **B31**, 1770–1790.
Harrison, W. A. (1980). "Electronic Structure and the Properties of Solids." Freeman and Co., San Francisco.
Harrison, W. A. (1983a). *Phys. Rev.* **B27**, 3592.
Harrison, W. A. (1983b). "The Bonding Properties of Semiconductors," SRI International, Menlo Park, California.
Hass, K. C., Ehrenreich, H., and Velicky, B. (1983). *Phys. Rev.* **B27**, 1088.
Hedin, L., and Lunquist, B. I. (1971). *J. Phys.* **C4**, 2064.
Hiedemann, E. A., and Hoesch, K. H. (1934). *Z. Phys.* **90**, 322–326.
Hiedemann, E. A., and Hoesch, K. H. (1937). *Z. Phys.* **107**, 463–473.
Heyman, J. S. (1976). *Proc. IEEE Ultrason. Symp., Annapolis, Maryland*, p. 113.
Hirsch, P. B., Pirouz, P., Roberts, S. G., and Warren, P. D. (1985). *Philo. Mag.* **B52**, 759.
Hohenberg, P., and Kohn, W. (1964). *Phys. Rev.* **136**, B864.
Huang, D., Chyi, J., Klem, J., and Morkoc, H. (1988). *J. Appl. Phys.* **63**, 5859.
Hybertsen, M. S., and Louie, S. G. (1987). *Phys. Rev. Lett.* **58**, 1551.
Ihm, Y. E., Otuska, N., Klem, K., and Morkoc, H. (1987). *Appl. Phys. Lett.* **51**, 2013.
Jen, H. R., Cherng, M. J., and Stringfellow, G. B. (1986). *Appl. Phys. Lett.* **48**, 1603.
Johnson, G. C., and Mase, G. D. (1984). *J. Acoust. Soc. Am.* **75**, 1741.
Keating, P. N. (1966). *Phys. Rev.* **145**, 637.
Khimunin, A. S. (1972). *Acustica* **27**, 173–181.
Kikuchi, R. (1951). *Phys. Rev.* **81**, 988.
Kittel, C. (1986). "Introduction to Solid-State Physics," Table 5. John Wiley & Sons, Inc., New York.
Kleiman L. (1962). *Phys. Rev.* **128**, 2614.
Klein, W. R., Cook, B. D., and Mayer, W. G. (1965). *Acustica* **15**, 67–74. This work is a theoretical analysis and defines Q.
Klem, J., Huang, D., Morkoc, H., Ihm, Y. E., and Otuska, N. (1987). *Appl. Phys. Lett.* **50**, 1364.
Kohn, W., and Sham, L. J. (1965). *Phys. Rev.* **140**, A 1133.
Krahauer, H., Posternak, M., and Freeman, A. J. (1979). *Phys. Rev. Lett.* **43**, 1885.
Krischer, C. (1968). *Appl. Phys. Lett.* **13**, 310.
Kuan, T. S., Kuech, T. F., Wang, W. I., and Wilkie, E. L. (1985). *Phys. Rev. Lett.* **54**, 201.

Kuan, T. S., Wang, W. I., and Wilkie, E. L. (1987). *Appl. Phys. Lett.* **51**, 51.
Lambrecht, W. R.L., and Segall, B. (1990). Private communication.
Landolt-Bornstein (1982). "Numerical Data and Functional Relationships in Science and Technology" (K.-H. Hellwidge, ed.), Vol. 17. Springer-Verlag, Berlin.
Landolt-Bornstein (1988). "Numerical Data and Functional Relationships in Science and Technology" (K.-H. Hellwidge, ed.), Vol. 22. Springer-Verlag, Berlin.
Landau, L. D., and Lifshitz, E. M. (1959). "Theory of Elasticity." Pergamon Press, New York.
Landau, L. D., and Lifshitz, E. M. (1986). "Theory of Elasticity." 3rd ed. Pergamon Press, New York. Equation 16 of text is in slightly different form than given on p. 2. Using $x_i = a_i + u_i$, and substituting into Eq. (16), one obtains the strain tensor found in this reference. This reference stresses physics concepts to derive the important equations.
Love, A. E.H. (1944). "A Treatise on the Mathematical Theory of Elasticity." Dover, New York.
Mann, J. B. (1967). "Atomic Structure Calculations, 1: Hartree-Fock Energy Results for Elements from Hydrogen to Lawrencium." Clearinghouse for Technical Information, Springfield, Virginia.
Martin, R. M. (1970). *Phys. Rev.* **B1**, 4005.
Martins, J. L., and Zunger, A. (1984). *Phys. Rev.* **B30**, 6217.
Martins, J. L., and Zunger, A. (1986). *Phys. Rev. Lett.* **56**, 1400.
Mayer, W. G., and Hiedemann, E. A. (1958). *J. Acoust. Soc. Am.* **30**, 756–760.
Mayer, W. G., and Hiedemann, E. A. (1959). *Acta Crystallogr.* **12**, 1.
McClintock, F. A., and Argon, A. S. (1966). "Mechanical Behavior of Materials." Addison-Wesley, Reading, Massachusetts.
McSkimin, H. J. (1950). *J. Acoust. Soc. Am.* **22** 413–418.
McSkimin, H. J. (1957). *IRE Trans. Ultrason. Eng.* **5**, 25.
McSkimin, H. J. (1961). *J. Acoust. Soc. Am.* **33**, 12.
McSkimin, H. J., and Andreatch, P. (1964). *J. Appl. Phys.* **35**, 3312–3319.
Methfessel, M., and van Schilfgaarde, M. (1990). Private communication.
Methfessel, M., Rodriguez, C. O., and Andersen, O. K. (1989). *Phys. Rev.* **B40**, 2009.
Michard, F., and Perrin, B. (1978). *J. Acoust. Soc. Am.* **64**, 1447–1456.
Mikkelsen, J. C., Jr., and Boyce, J. B. (1982). *Phys. Rev. Lett.* **49**, 1412.
Mikkelsen, J. C., Jr., and Boyce, J. B. (1983). *Phys. Rev.* **B28**, 7130.
Miller, J. G. (1973). *J. Acoust. Soc. Am.* **53**, 710–713.
Mitra, S. S., and Massa, N. E. (1982). *In* "Handbook on Semiconductors" (T. S. Moss, ed.), Vol. 1, Chap. 3. North-Holland, Amsterdam.
Murnaghan, F. D. (1951). "Finite Deformation of an Elastic Solid." Wiley, New York. This reference uses matrices throughout and is mathematically more formal.
Musgrave, J. P., and Pople, J. A. (1962). *Proc. Roy. Soc. (London)* **A268**, 474.
Nielsen, O. H., and Martin, R. M. (1983). *Phys. Rev. Lett.* **50**, 697.
Nielsen, O. H., and Martin, R. M. (1985a). *Phys. Rev.* **B32**, 3792.
Nielsen, O. H., and Martin, R. M. (1985b). *Phys. Rev.* **B32**, 3780.
Nomoto, O. (1942). *Proc. Phys. Math. Soc. Japan* **24**, 380–400, 613–639. This work covers an experimental investigation into the overlap region.
Norman, A. G., Mallard, R. E., Murgatroyd, I. J., Brooker, G. R., More, A. H., and Scott, M. D. (1987). *In* "Microscopy of Semiconductor Materials," *Inst. Phys. Conf. Ser.* **87**, 77.
O'Donnell, M., Busse, L. J., and Miller, J. G. (1981). In "Methods of Experimental Physics, Vol. 19, Ultrasonics," pp. 29–65. Academic Press, New York.
Ourmazd, A., and Bean, J. C. (1985). *Phys. Rev. Lett.* **55**, 765.
Papadakis, E. P. (1963). *J. Acoust. Soc. Am.* **35**, 490–494.
Papadakis, E. P. (1964a). *J. Appl. Phys.* **35**, 1474.
Papadakis, E. P. (1964b). *J. Acoust. Soc. Am.* **36**, 414–422.

Papadakis, E. P. (1966). *J. Acoust. Soc. Am.* **40**, 863–876.
Papadakis, E. P. (1967). *J. Acoust. Soc. Am.* **42**, 1045.
Papadakis,. E. P. (1972). *J. Acoust. Soc. Am.* **52**, 843.
Papadakis, E. P. (1976). *In* "Physical Acoustics" (W. P. Mason and R. N. Thurston, eds.), Vol. XII, pp. 277–374. Academic Press, New York.
Patrick, R. S., Chen, A.-B., and Sher, A. (1987). *Phys. Rev.* **B36**, 6585.
Patrick, R. S., Chen, A.-B., Sher, A., and Berding, M. A. (1988). *J. Vac. Sci. and Technol.* **6**, 2643.
Perdew, J. P., and Zunger, A. (1981). *Phys. Rev.* **B23**, 5048.
Phillips, J. C. (1973). "Bonds and Bands in Semiconductors." Academic Press, New York.
Pollard, H. F. (1977). "Sound Waves in Solids," pp. 308–309. Pion Limited, London.
Prosser, W. H., and Green, R. E., Jr. (1985). *In* "Proceedings of the 1985 SEM Spring Conference on Experimental Mechanics," pp. 340–346. Society for Experimental Mechanics.
Qteish, A., and Resta, R. (1988). *Phys. Rev.* **B37**, 1308.
Quadri, S. B., Skelton, E. F., and Webb, A. W. (1986). *J. Vac. Sci. and Technol.* **A4**(4), 1971, 1974.
Raman, C. V., and Nath, N. S.N. (1935a). *Proc. Indian Acad. Sci. Sect. A, Part I* **2**, 406–412.
Raman, C. V., and Nath, N. S.N. (1935b). *Proc. Indian Acad. Sci. Sect. A, Part II* **2**, 413–420.
Raman, C. V., and Nath, N. S.N. (1936a). *Proc. Indian Acad. Sci. Sect. A, Part III* **3**, 75–84.
Raman, C. V., and Nath, N. S.N. (1936b). *Proc. Indian Acad. Sci. Sect. A, Part IV* **3**, 119–125.
Raman, C. V., and Nath, N. S.N. (1936c). *Proc. Indian Acad. Sci. Sect. A, Part V* **3**, 459–465.
Roberts, S. G., Warren, P. D., and Hirsch, P. B. (1986). *J. Mater. Res.* **1**, 162.
Rogers, P. H., and Van Buren, A. L. (1974). *J. Acoust. Soc. Am.* **55**, 724–728.
Schwartz, L., and Bansil, A. (1975). *Phys. Rev.* **B10**, 3261.
Scott, W. R., and Bloomfield, P. E. (1981). *Ferroelectrics* **32**, 79–83. Lead attachment to KYNAR films is covered in this paper.
Seki, H., Granato, A., and Truell, R. (1956). *J. Acoust. Soc. Am.* **28**, 230–238.
Shahid, M. A., Mahajan, S., Laughlin, D. E., and Cox, H. M. (1987). *Phys. Rev. Lett.* **58**, 2567.
Sher, A., Chen, A.-B., and Spicer, W. E. (1985). *Appl. Phys. Lett.* **46**, 54.
Sher, A., van Schilfgaarde, M., Chen, A.-B., and Chen, W. (1987). *Phys. Rev.* **B36**, 4279.
Sher, A., Berding, M. A., van Schilfgaarde, M., Chen, A.-B., and Patrick, R. S. (1988). *J. Crystal Growth* **86**, 15.
Simondet, F., Michard, F., and Toquet, R. (1976). *Opt. Commun.* **16**, 411–416.
Slater, J. C., and Koster, G. F. (1954). *Phys. Rev.* **94**, 1498.
Spicer, W. E., Silberman, J. A., Morgan, P., Lindau, I., Wilson, J. A., Chen, A.-B., and Sher, A. (1982). *Phys. Rev. Lett.* **47**, 948.
Truell, R., Elbaum, C., and Chick, B. B. (1969). "Ultrasonic Methods in Solid State Physics." Academic Press, New York.
Tu, L. Y., Brennan, J. N., and Sauer, J. A. (1955). *J. Acoust. Soc. Am.* **27**, 550.
van Schilfgaarde, M. (1990), unpublished.
van Schilfgaarde, M., and Sher, A. (1987a). *Phys. Rev.* **B36**, 4375.
van Schilfgaarde, M., and Sher, A. (1987b). *Appl. Phys. Lett.* **51**, 175.
van Schifgaarde, M., Methfessel, M., and Paxton, A. T. 1991. In preparation.
Van Vechten, J. A. (1969). Phys. Rev. **187**, 1007.
Vasile, C. R., and Thompson, R. B. (1977). *In* "IEEE Procedings, 1977 Ultrasonics Symposium" (J. deKlerk and B. R. McAvoy, eds.), p. 84.
Vegard, L. (1921). *Z. Phys.* **5**, 17.
Voight, W. (1928). "Lehrbuch der Kristallphysik." Teubner, Leipzig. This contraction of the indices is also covered in Landau and Lifschitz (1986) and Green (1975), page 6. The subscripts are contracted as $11 \to 1$, $22 \to 2$, $33 \to 3$, $23 \to 4$, $31 \to 5$, $12 \to 6$.
Wei, S.-H., and Krahauer, H. (1985). *Phys. Rev. Lett.* **55**, 1200.
Wei, S.-H., Ferreira, L. G., and Zunger, A. (1990). *Phys. Rev.* **B41**, 8240.

Wei, S.-W., and Zunger, A. (1989). *Phys. Rev.* **B39**, 3279.
Wigner, E. P. (1934). *Phys. Rev.* **46**, 1002.
Williams, A. O., Jr., (1970). *J. Acoust. Soc. Am.* **48**, 285 (1970).
Williams, J., and Lamb, J. (1958). *J. Acoust. Soc. Am.* **30**, 308–313.
Wimmer, E., Krahauer, H., Weinert, M., and Freeman, A. J. (1981). *Phys. Rev.* **B24**, 864.
Wood, D. M., Wei, S.-W., and Zunger, A. (1988). *Phys. Rev.* **B37**, 1342.
Yeh, C.-Y., Chen, A.-B., and Sher, A. (1990). *Phys. Rev.* **B43**, 9138.
Zallen, R. (1982). In "Handbook on Semiconductors," (T. S. Moss, ed.), Vol. 1, Chap. 1. North-Holland, Amsterdam.

CHAPTER 2

Fracture of Silicon and Other Semiconductors

David R. Clarke

MATERIALS DEPARTMENT
UNIVERSITY OF CALIFORNIA
SANTA BARBARA, CALIFORNIA

I. INTRODUCTION	80
II. CRACK EXTENSION: THE MECHANICS OF FRACTURE	83
1. *Theoretical Cleavage Strength*	83
2. *Uniform Loading*	85
3. *Steady-State Conditions*	88
4. *R-Curve Behavior*	90
III. FRACTURE MECHANISMS	90
5. *Models for the Crack Tip*	91
6. *Direct Observations of Crack Tips*	94
7. *Mechanisms of Crack Advance*	98
8. *Cleavage versus Plasticity*	100
9. *Dislocation Models for Fracture Resistance*	103
IV. FRACTURE OF SEMICONDUCTORS	106
10. *Crystallographic Aspects*	106
11. *Fracture Surfaces*	108
12. *Fracture Surface Energy*	110
13. *Dependence on Energy Gap*	112
14. *Strain Rate Effects*	113
15. *Effect of Illumination*	113
V. THE BRITTLE-TO-DUCTILE TRANSITION	116
16. *Strain Rate Effect*	118
17. *Doping Effects on the Transition Temperature*	120
18. *X-Ray Topographic Studies*	123
19. *In-Situ TEM Studies*	125
20. *Modelling the Brittle-to-Ductile Transition*	129
21. *Summary*	132
VI. ENVIRONMENTAL EFFECTS	133
VII. CLOSING REMARKS	135
ACKNOWLEDGMENTS	136
REFERENCES	136
APPENDIX: FRACTURE RESISTANCE MEASUREMENT BY INDENTATION	140

I. Introduction

There has long been a practical interest in identifying the conditions under which semiconductor materials fracture. As with other materials, the thrust of much of this work has been largely empirical and restricted to discovering the maximum conditions under which fracture can be avoided. However, in recent years, as the physics of fracture has developed as a scientific field in its own right (Thomson, 1986; Lawn and Wilshire, 1975; see also the special issue on physics and chemistry of fracture, *J. Phys. Chem. Solids*, **48** (11), pp. 965–1157), there has been a growing recognition that some of the fundamental aspects of the fracture process itself can be investigated using semiconductors. The principal reason lies in their very different fracture behavior at low and high temperatures. At low temperatures, including room temperature, the majority of semiconductors behave in a seemingly ideal brittle manner, whilst at high temperatures they behave as ideal plastic solids. Their behavior thus mimics, in many respects, at low temperatures that a ceramics and at high temperatures that of metals. Furthermore, the low density of dislocations typical of many semiconductors makes it possible to investigate the mechanisms by which individual dislocations mediate fracture mechanisms.

Whilst the physics of fracture are peripheral to the majority interested in devices, an understanding of the mechanisms of fracture is nevertheless important in designing device structures and semiconductor processing. Thus, at low temperatures, below the brittle-to-ductile transition (BDT), stresses introduced by fabrication, processing, and handling can only be relieved by creep deformation, by atomic migration in the metallurgical features, or by fracture of the semiconductor itself. At high temperatures such fracture probably will not occur, but the stresses may be well be relieved by the generation of dislocations in the semiconductors and subsequent plastic flow at sites, such as at steps or trenches, where the stress gradients are particularly high. The consequences of this behavior are of importance in electronic devices; in opto-electronic devices, where dislocations can act as recombinant centers; and also in the increasing use of silicon, at least, as an active component in mechanical devices (McLoughlin and Willoughby, 1987; Petersen, 1982).

The general, macroscopic fracture behavior of semiconductors as a function of temperature was delineated in early work on whiskers, circa 1956. As described in this review, considerably more detailed work has followed in the intervening years providing a more comprehensive picture of the effects of crystallography, doping and temperature, but the essential phenomenon is much the same. Pearson *et al.* (1957) studied the deformation and fracture behavior of fine (1–30 μm diameter) silicon, germanium, and zinc oxide whiskers at different temperatures. They demonstrated that at room temperatures the whiskers fractured after only elastic deformation, whereas at higher temperatures they would yield plastically without fracturing. For silicon the transition occurred above about 600°C. Between 600 and 650°C the whiskers either flowed or fractured, or fractured after a small amount of flow. Fewer measurements were made on germanium and zinc oxide whiskers, but Pearson *et al.* reported that the germanium whiskers yielded above 350°C, whereas the zinc oxide whiskers sublimed before a temperature was reached for plastic flow. In addition to identifying a transition from brittle to ductile behavior with increasing temperature, Pearson *et al.* also made the remarkable observation that the transition was dependent on prior deformation. They found that silicon whiskers that had been plastically deformed at 800°C and subsequently reloaded at lower temperatures yielded rather than fracturing at temperatures of 450°C and above, namely well below the BDT temperature established for previously undeformed whiskers.

The fracture of semiconductors can be characterized in a number of different ways. The fracture planes have been ascertained by either easy cleavage of bulk material or wafers or by indenting the surface of a material with a diamond pyramid and observing the trace of the cracks created. An example is shown in Fig. 1. Subsequent observation by x-ray topography or transmission electron microscopy reveals whether dislocation plasticity accompanies the fracture. Alternatively, but less convincing, etching the material with a solution that preferentially attacks sites of dislocations can be used to detect the presence of dislocations generated during deformation and fracture. Evaluating quantitatively the resistance to fracture, often termed its fracture toughness, K_{IC}, has proved to be more difficult. In large part this is because the fracture resistance of semiconductors is so low compared with most other solids — i.e., they are much more brittle — that the measurements are difficult to carry out. Nevertheless, careful measurements using samples with deliberately introduced, straight cracks tested under standard mechanics test configurations (ASTM, 1965; Evans, 1974) have been performed for a number of semiconductors as described later. In addition, tests utilizing the cracks introduced by indentation (Fig. 1) have been developed in the last decade to give fracture toughness values for samples that are smaller than those prescribed by the mechanical testing community.

In this contribution we seek primarily to review what is presently known about the fracture properties of semiconductors. The majority of work has, perhaps not surprisingly, been devoted to the fracture of silicon, and that dominance is reflected here. However, we have also tried to utilize some of the simple ideas presented to predict the fracture behavior of those semiconductors that have not yet been investigated. It is for this reason that some of the pertinent aspects of the theoretical models for understanding fracture are also

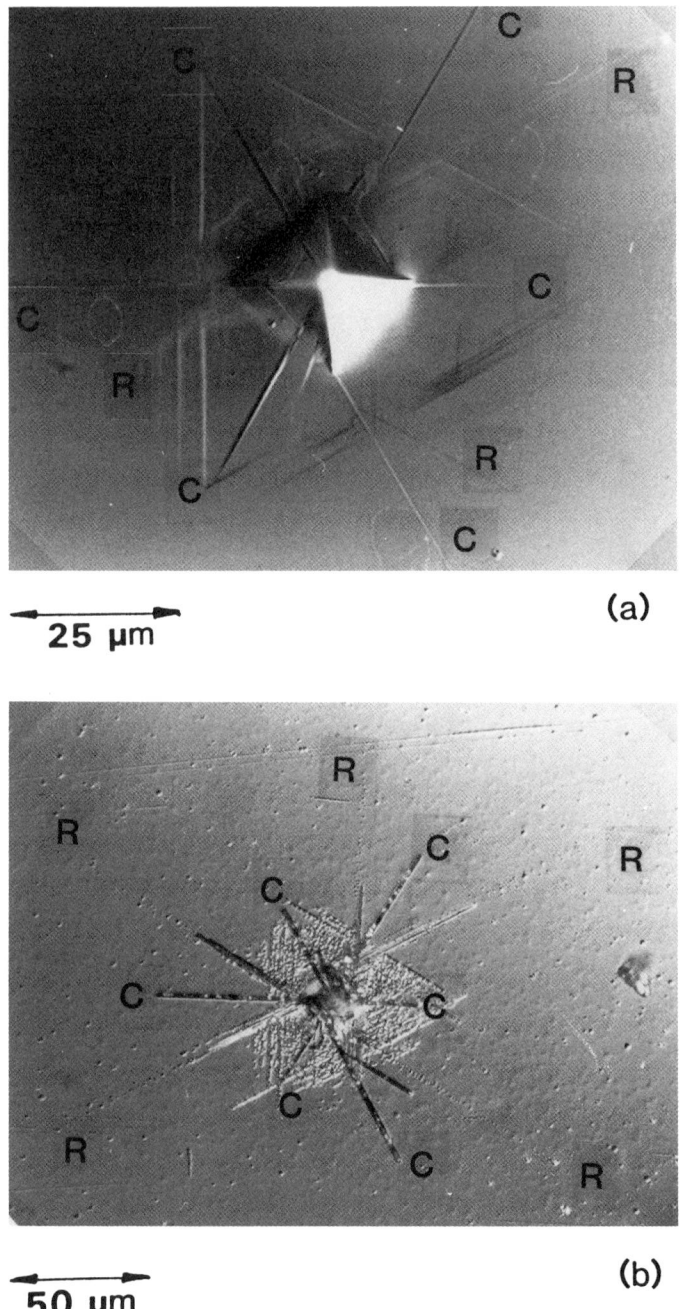

FIG. 1. Traces of radial cracks, C, introduced by indentation into a (111) As surface of GaAs (a) before, and (b) after etching to reveal the associated dislocation structure introduced by indentation (etch pits). The symbols R mark the extent of dislocation glide away from the indentation impression. Micrograph courtesy of P. Warren.

reviewed in the following sections. A final point, the focus of this review is on fracture rather than strength. Despite the everyday familiarity of the notion of strength, it is not an intrinsic materials property. The measured strength of a piece of a material is actually dependent on the presence of flaws or cracks in the material and their geometrical configuration with respect to the local stresses. It is also dependent on an intrinsic materials property, the fracture resistance. This will be seen in a following section.

II. Crack Extension: The Mechanics of Fracture

Although it is not the objective of this contribution to discuss the detailed mechanics of fracture, we do summarize the essential aspects here so as to place the work within a single context. Of the concepts involved, the key ones appropriate to the fracture of semiconductors are: conditions for crack extension in both a uniform and nonuniform loading, crack driving forces, steady-state crack growth, and R-curve behavior. However, we start by addressing the theoretical cleavage strength.

1. THEORETICAL CLEAVAGE STRENGTH

Although the theoretical cleavage strength of a semiconductor, or indeed any solid, is never reached, a rather simple model for calculating it provides an insight into the materials parameters affecting fracture resistance. The model is due to Orowan (1949) and assumes that the stress required to pull apart rigidly two blocks of crystal along their cleavage plane is given by a simple sinusoidal relation:

$$\sigma = \sigma_{th} \sin \frac{\pi(z - a)}{2z_0}, \qquad (1)$$

where z_0 is the atomic displacement at which the strength corresponds to the theoretical strength, σ_{th}; z is the displacement; and a is the lattice spacing. Assuming the material to be isotropic and the displacement $(z - a)$ to be small, the differential of the stress-displacement curve is, from Hooke's law, the elastic modulus, E:

$$E = \frac{a\pi\sigma_{th}}{2z_0} \cos \frac{\pi(z - a)}{2z_0}. \qquad (2)$$

On rearranging, the theoretical strength is given by

$$\sigma_{th} = \frac{2Ez_0}{\pi a}. \tag{3}$$

In order to evaluate the constant z_0, Orowan argued that the minimum work done to stress the material to failure would be that required to produce two new fracture surfaces, each with surface energy γ. From the definition of the surface energy this leads to the expression

$$2\gamma = \int_a^{a+2z_0} \sigma \, dz, \tag{4}$$

leading to

$$z_0 = \frac{\pi \gamma}{2\sigma_{th}}. \tag{5}$$

Eliminating the constant z_0 gives an expression for the theoretical cleavage strength:

$$\sigma_{th} = \sqrt{\frac{E\gamma}{a}}. \tag{6}$$

Although more sophisticated arguments can be made, the essential materials requirements for high strength are deduced in the above analysis. High strength is associated with materials having small lattice spacings, high elastic modulus, and high surface energy.

It must be emphasized again that the actual strength of a brittle solid, such as a semiconductor at room temperature, is determined by the presence of flaws or other stress concentrators. Thus, while whiskers of silicon have been stressed to strengths of at least 5 GPa (Pearson et al., 1957), fine, lithographically produced mechanical levers of silison have been reported to exhibit strengths of up to 10 GPa (Johansson et al., 1988), and silica gel (Siton) polished wafers can exhibit biaxial strengths of 2.8 ± 1.2 GPa (Hu, 1982), these extraordinary strengths are rarely achieved except on the most ideal surfaces before any subsequent processing and certainly cannot be relied on. The attainable strength is usually limited by the stress at which flaws or incipient cracks will extend, as described in the following section.

2. UNIFORM LOADING

Our starting point, as indeed it was for Griffith in 1920, is the application of the first law of thermodynamics to the virtual extension of a pre-existing crack. The system considered by Griffith was closed, consisting of a crack of length c in an elastically homogeneous body subject to a spatially uniform biaxial tensile stress field, σ, by an externally applied load. In establishing the conditions for crack extension Griffith explicitly considered, for the first time, that the total energy, U, of the system, including the crack, consisted of two terms, that produced by the applied loading, U_A, and the work to create fresh surface, U_s. The applied loading energy represents the potential energy of the system to do work, to extend the crack, and is given by $U_A = (U_E - W)$, where U_E is the elastic strain energy stored in the system and W is the work done by the system.

For the geometry of a through crack in a thin sheet of material of unit thickness, the individual terms can be simply stated: The term U_s is equal to the area of the crack multiplied by the surface energy per unit area, γ. For a linearly elastic material, the work done by the loading system, W, is twice the elastic strain energy, but of opposite sign. The elastic strain energy is more difficult to calculate, but by arguing that a crack could be represented by an elliptical hole, Griffith was able to utilize an earlier result due to Inglis, who had calculated the distribution of stresses and strains around an elliptically-shaped hole in a loaded sheet. With these terms, Griffith was able to express the total energy of the cracked plate in terms of the crack length and the materials properties:

$$U = -\frac{\pi c^2 \sigma^2}{E} + 4c\gamma. \tag{7}$$

As in other examples of stability theory, the system is at equilibrium when its overall energy has a stationary value with respect to any virtual change in the crack length, c:

$$\frac{dU}{dc} = \frac{dU_A}{dc} + \frac{dU_s}{dc} = 0. \tag{8}$$

Applying this condition led Griffith to the statement of equilibrium that bears his name:

$$\sigma = \left[\frac{2E\gamma}{\pi c}\right]^{1/2}. \tag{9}$$

This differs from the theoretical cleavage strength (Eq. (6)) by inclusion of the crack length rather than the interplanar spacing. As mentioned earlier, Griffith used the stress solution for an elliptical hole to model that of a crack. The notion of a crack as an ellipse has conditioned several generations of materials scientists to think that a crack has a well-defined tip radius. The energy arguments of Griffith, as well as subsequent thermodynamic models, however, have only one physical dimension in them — the length of the crack.

The dependence of strength on the flaw size was confirmed experimentally by Griffith using glass, and it has subsequently become deeply ingrained in the protocol of the materials science community as a way of determining the critical flaw size introduced by processing or machining. Indeed, it has been demonstrated to hold for silicon wafers containing different flaw sizes (Yasutake *et al.*, 1986; Wong and Holbrook, 1987). However, the Griffith relationship is only valid if there is a dominate flaw and there are no residual stresses present.

In general, testing a series of samples yields a statistical variation in fracture strength. In terms of the Griffith relationship this variation can be understood on the basis that any material contains a population density of flaws of varying size, and that an observed variation in strength is attributable to a variation from one sample to the next in the size of the largest flaw. Thus, from a practical point of view any attempts to increase the reliability (as expressed for instance using a Weilbull modulus) of a material have been directed towards improving the reproducibility of processing or minimizing handling damage in order to produce a tighter distribution of flaw sizes. As an example, the use of Siton polishing to polish away flaws and dislocation damage induced by grinding and cutting leads to a higher strength (Hu, 1982).

To continue with the energetics of crack growth, the thermodynamic arguments introduced by Griffith can be generalized to any system irrespective of the crack and loading geometry. Defining the quantities:

$$G = -dU_A/dc, \qquad (10)$$

where G is the strain energy release rate and the resistance force, R, to fracture,

$$R = +dU_s/dc, \qquad (11)$$

enables the Griffith equilibrium condition to be stated in terms of the (thermodynamic) forces acting on the crack:

$$G = G_c = R. \qquad (12)$$

The concept of the force balance acting on a crack front was introduced by Irwin (1958), who showed that the strain energy release rate corresponded to a thermodynamic force acting per unit length of crack front,[1] and that previously derived equations for mode I and mode II loaded cracks could be explicitly re-expressed in terms of this crack driving force.

Another widely used function to express the forces acting on a crack, and one that is used extensively in the semiconductor fracture literature, is the stress intensity factor, K, again introduced by Irwin. For elastic materials under plane stress conditions, the stress intensity factor is simply related to the strain energy release rate by

$$G = K^2/E$$

For elastic materials under plane stress conditions, the strain energy release rate and the resistance force may be expressed in terms of stress intensities, K, and material toughness, T:

$$K = \sqrt{GE} \quad \text{and} \quad T = \sqrt{RE}. \tag{13}$$

In terms of surface energy the latter equations can be rewritten as

$$K = \sqrt{2\gamma E} \quad \text{and} \quad T = \sqrt{2\gamma E}.$$

One of the principal advantages of the stress intensity formalism is that (for a given mode of fracture) the stress intensities are additive quantities, whereas the strain energy release rates, G, are not. This is a mathematical convenience when considering the effects of superimposed stress fields, such as those due to the presence of residual stresses.

Equation (12) represents the *necessary* conditions for crack extension and failure. As in any other thermodynamic system, instability occurs when the second derivative of the total energy is less than zero. Thus, the *sufficient* condition for equilibrium crack extension is that the second derivative of the energy be less than zero. In terms of the forces on the crack, the sufficient condition derives when the following inequality is satisfied:

$$dG/dc > dR/dc. \tag{14}$$

In the case of a spatially homogeneous material, the crack resistance is expected to be independent of the crack area. This reduces the condition for

---

[1]Other generalized energy functions have been introduced to incorporate effects such as inelastic and creep behavior on fracture, for instance, the J integral and the C^* integral.

crack extension to the simpler one that the strain energy release rate decreases as the crack extends. However, if the material is not homogeneous, a variation in crack resistance force with length, termed *R*-curve behavior, may result.

In terms of the stress intensity factor, the instability condition may be expressed as $dK/dc > 0$. Thus, the necessary and sufficient conditions for unstable crack extension are:

$$K_{tip} > T_0 \quad \text{and} \quad dK_{tip}/dc > dT_0/dc \qquad (15)$$

in the crack-tip frame of reference, where T_0 is the intrinsic fracture resistance of the material, or equivalently in the applied loading system frame of reference, by

$$K_{app} > T \quad \text{and} \quad dK_{app}/dc > dT/dc. \qquad (16)$$

T in this equation is the (measurable) fracture resistance at crack length c.

Recently, a number of investigators (Rice, 1978; Maugis, 1985; Chudnovsky and Moet, 1985) have argued that the irreversibility in crack extension imposes an additional constraint on the conditions for the propagation of a crack. A most thorough analysis, one within the framework of irreversible thermodynamics, has been introduced by Rice (1978). He shows that for a non-negative entropy production associated with the extension of a crack, the following inequality holds:

$$(G - 2\gamma)v \geq 0, \qquad (17)$$

where v is the crack speed. The 2γ term in Eq. (17) properly refers to the reversible work of separating the fracturing surfaces per unit area of surface. This is the same as the true "surface energy" under conditions of chemical equilibrium with an adsorbing species in a bulk phase at a constant chemical potential (Section VI). In general, this may be identified with the Dupre energy of adhesion for the separation of two dissimilar materials. The condition for crack growth of Eq. (17), although apparently not widely appreciated in the literature, is of particular relevance to the discussion of environmentally affected crack growth (see Section VI) since it relates both the crack velocity and the energy to create new surface in the environment in which crack growth is occurring.

3. STEADY-STATE CONDITIONS

Invoking the condition of steady-state propagation of a crack allows, as will be seen below, a separation of the effects that occur at the crack tip from those that occur at the rest of the crack. This is of importance in establishing the role of a variety of phenomenon, ranging from intermolecular and surface forces to crack bridging effects, on fracture. Such crack bridging can arise, for instance in device structures, from constraints on crack opening from unbroken metallized lines spanning the crack. The assumption of steady state is tantamount to operationally assuming that there exists a crack-tip "core" region that moves along with the crack and where the details of the atomic arrangements and forces are invariant with crack position, and hence can be neglected when determining conditions of steady state.

Separation of these terms can be accomplished by writing the surface work term, U_s, as

$$U_s = 2\gamma c + \Gamma, \tag{18}$$

where Γ is the energy associated with the core. Substituting into Eq. (8) for the equilibrium condition for crack growth gives

$$-\frac{dU}{dc} = -\frac{dU_A}{dc} - \frac{d(2\gamma c + \Gamma)}{dc}. \tag{19}$$

The crack length dependence of the term Γ is unknown. However, if the assumption is made that in steady state the term Γ is independent of crack length, the second term on the right will be the constant 2γ and the equation reduces to the Griffith equilibrium condition:

$$G = G_A - 2\gamma. \tag{20}$$

Thus, the detailed knowledge of any short-range (compared to the crack length) crack-tip interactions can be neglected if steady state pertains. The concept of steady state has been invoked by Cook (1986) in order to show that the strain energy release rate at the threshold stress intensity for slow crack growth (Section VI) is a direct measure of the equilibrium surface energy. The assumption of steady state, and with it the possibility of neglecting the details of the actual processes involved, covers not only the detailed physics of lattice trapping,[2] relaxation phenomena, but also the contributions lumped together under the term "disjoining pressures"[3] used to describe the short-range interactions between crack surfaces in close proximity when fracture occurs in a liquid environment. It is also pertinent to the effects on fracture of super-band gap excitations (Section IV.14).

There are a number of precedents for this type of propagating "core"

---

[2]The crack analogy of the trapping of a dislocation in a Peierls barrier.

[3]This includes, amongst many others, screened van der Waals interactions, solvation forces, electrostatic interactions, and structural forces (Clarke *et al.*, 1986; Clarke, 1987).

assumption in different fields. In considering lattice trapping of a crack, Thomson (1980) partitions the energy in such a manner that the atoms at the crack tip are in a core and that their energy can be neglected in the change in overall energy of a long crack. Perhaps the most widely used analogy is in the field of dislocation mechanics. The energy of the dislocation core is neglected compared to changes in the elastic strain energy when calculating the energies of dislocation motion, alternative dislocation configurations, and dislocation interactions. Another analogy is in the field of wetting. As de Gennes (1985) has noted, Thomas Young realized that the equilibrium contact angle for a partially wetting drop can be derived from the far field surface energies, without consideration of any surface force effects at the core of the contact line of the drop. In a situation paralleling that of fracture, de Gennes shows that the spreading of a liquid drop over a solid surface in vacuum ("dry spreading") can be well described by neglecting any energy dissipation in the core region local to the range of any surface forces, at the contact line of the spreading drop, and that the driving force for spreading depends only on the excess energy above that necessary to maintain drop equilibrium.

4. R-Curve Behavior

The concept of a crack resistance that increases with the length of the crack — R-curve behavior — is well established in the fracture mechanics literature (Broek, 1982). It has been recognized for many years that metal structures under plane stress conditions can exhibit such a crack-length dependence on fracture toughness. Although it is highly unlikely that single crystal semiconductors will exhibit any R-curve behavior in their brittle regime, such effects have not been investigated. However, R-curve behavior is a predicted feature of even the simplest theoretical models based on the formation of a process zone, such as that created by the emission of dislocations from a crack tip. Thus, above the brittle-to-ductile transition temperature, the fracture resistance can be expected to vary with crack advance. The actual dependence on crack extension has yet to be measured in any of the studies of the brittle-to-ductile transition (see Section V). There is, nevertheless, circumstantial evidence that after the first dislocations are emitted, the fracture resistance increases dramatically, so much so that wholesale plastic deformation occurs.

III. Fracture Mechanisms

Under the influence of a residual stress, a thermally induced stress, or a stress from handling, any flaw, crack, or step on the surface of a semiconduc-

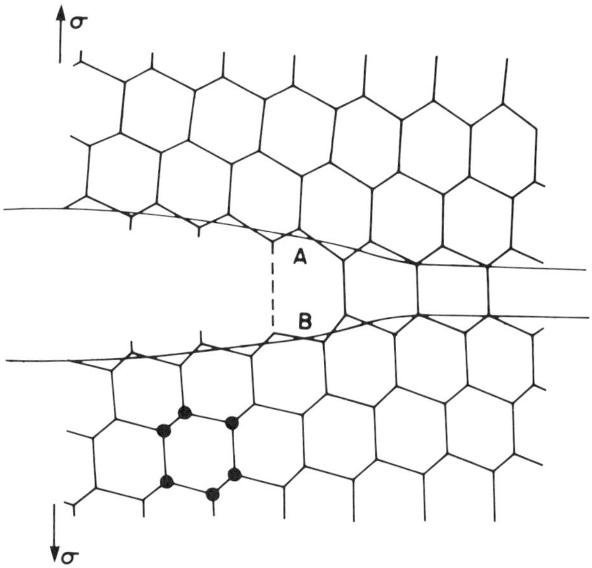

FIG. 2. Computed atomic configuration around an equilibrium (111) crack tip in silicon projected onto the (0$\bar{1}$1) plane. The two superimposed lines indicate the crack profile computed on the basis of lienar elasticity. The bond AB is strained to 20% and is taken to represent the crack tip. Redrawn following Lawn and Sinclair (1972).

tor can act as a stress-concentrating flaw. At the end of the crack, residual stresses or any externally applied stress to the semiconductor are amplified geometrically, and the tensile stresses may reach the cohesive stress of the material. (Within the approximation of linear elastic fracture mechanics, the stress varies as $1/\sqrt{r}$, where r is distance from the crack tip). Alternatively, the stress may reach the ideal shear strength of the material and dislocations may be nucleated to relieve the stress. At room temperature and in the absence of any dislocation plasticity, fracture in brittle solids, such as semiconductors, is thought to occur by a cleavage process; the crack advances by the breaking of successive atomic bonds along the cleavage plane. This is shown in Fig. 2 for the computed structure of a crack tip in silicon.

5. Models for the Crack Tip

As the actual positions of atoms at a crack tip have, at least until recently, been beyond the capabilities of existing microscopies, the prevailing emphasis of many of the models have been on the conditions for crack extension rather than on the details of the atomic arrangement. This lack of specificity has not been too limiting to date since it has been shown, within the linear elastic approximation, by Willis (1967), and later for the more general nonlinear elastic model by Rice (1968), that contrasting descriptions of blunt and sharp cracks described below both lead to the same conditions for crack extension. It remains an open question whether the situation is similar to that found in simulations of grain boundaries, where the energy of a grain boundary appears to be relatively insensitive to the exact details of the interatomic potentials used, whereas the detailed atomic positions are very dependent (Wolf, 1984).

Prior to detailed computer simulations, two contrasting modes for the atomic structure of the crack tip in brittle solids have evolved. The first, now attributed to Griffith although it strictly has its origin in the work of Inglis, is referred to as the Griffith–Inglis model. It is a continuum description and assumes that the material behaves in a linear elastic manner on all length scales down to the very crack tip. It is the extension of the assumption that a crack can be considered to be an elliptical hole and leads to an abrupt 90-degree bifurcation of the plane of atoms at the end of the crack (and hence gives it a blunt shape). The crack shape behind the tip then depends only on the applied stress intensity and the elastic modulus of the material. Apart from the problem that the Griffith–Inglis model leads to a singularity in the stress field at the tip, it also assumes that the atomic cohesive forces are equivalent to a "hard-wall" interaction. There is thus no length scale associated with the cohesive forces, which is inconsistent with the existence of the surface energy used in the Griffith equation.

The contrasting model, one in which the crack surfaces separate continuously, stems from the work of Elliott (1947), who noted that the tip of the crack cannot have an elliptical geometry. Rather, it consists of three distinct regions. Well ahead of the crack the atomic bonds are subject to small strains, and so the displacements are linear elastic. Well behind the crack the bonds are stretched far apart; the atoms can exert little force across the crack, and the traction-free surface assumption of linear elastic fracture mechanics remains essentially valid. In between, there is a third region where the cohesive forces are well beyond their linear elastic range but nevertheless remain significant to provide a degree of bonding across the crack faces. This idea of a cohesive zone was extended mathematically by Barenblatt (1962) in describing a linear continuum mechanics theory in which arbitrary cohesive forces acted between the crack faces in the cohesive zone.[4] In the Barenblatt model, the width of the cohesive regions depends upon the details of the interatomic force laws, but is assumed to be independent of the applied stress

---

[4]The Barenblatt model was later reformulated in terms of continuous distributions of dislocations, thereby providing a rigorous continuum description of a crack consistent with both dislocation and continuum mechanics (Bilby and Eshelby, 1968).

on the crack. A recent paper (Chan *et al.*, 1987) shows that a number of the inadequacies of the Griffith–Inglis and the Barenblatt models can be resolved by a consistent treatment of the intermolecular and surface forces in the vicinity of the crack tip. The predicted crack profile coincides with that of the Griffith–Inglis crack away from the crack tip and has a form similar to that of the Barenblatt cohesive zone at the tip.

Discrete calculations performed on the computer with a variety of interatomic force laws have all substantiated the existence of a cohesive zone and a general crack shape consistent with that predicted by Barenblatt. This is not altogether surprising: In the majority of the simulations the calculations are rendered manageable by assuming that the atoms in the immediate vicinity of the crack tip, which are allowed to move individually in response to the local interatomic potential, are embedded in a linear elastic continuum — a refined but nevertheless conceptually similar calculation to that performed by Barenblatt. The extent of the cohesive zone, where it has been calculated, has been found to be quite restricted in size for reasonable force laws; for instance in the work of Gehlen and Kanninen (Gehlen and Kanninen, 1970; Gehlen *et al.*, 1972), the zone size is only two or three atomic spacings. The advantage of these computations is that they can provide detailed predictions of the atomic positions around the crack tip. Hitherto, the calculations have been primarily performed to calculate the conditions for fracture, but the atomic positions as a function of applied stress intensity up to K_{IC} will undoubtedly be needed for comparison with experiment.

A rather detailed treatment of an equilibrium crack in silicon, diamond, and germanium has been presented by Lawn and Sinclair (1972) using noncentral potentials to describe the forces between silicon atoms. They considered two regions, one in the immediate vicinity of the crack tip wherein the interatomic forces were treated explicitly, and the other in which the surrounding atoms were considered to behave as if they were part of an elastic continuum. The boundary condition between the two regions was chosen such that the displacement of the inner region matched that predicted for the continuum material calculated for a Griffith crack using anisotropic elasticity theory. The calculation then consisted of repositioning the atoms in the inner region until the minimum energy configuration was attained for each of a number of stresses applied to the outer region. An example of the resulting atomic positions is shown in Fig. 2. There is no clear point at which the atomic bonds can be considered to have failed, but using a rupture strain of 20 percent, the bonds to the right of A–B may be said to be unbroken, whereas those to the left are broken. Superimposed on Fig. 2 is the crack profile calculated using continuum elasticity for the same elastic constants and applied stress intensity. The comparison indicates that except for right at the crack tip, the continuum and atomistic calculations agree. In this, and a later study (Sinclair, 1974) using more flexible boundary conditions and varying forms for the interatomic potentials, there was no evidence for any shear instability, suggesting that cracks propagate in silicon in a fully brittle manner. Sinclair's calculation also gave an estimate for the conditions under which the crack would be lattice-trapped about the threshold of $0.408 < K_{IC} < 0.455$. This value is similar to that obtained from the elastic constants and the $\{111\}$ surface energy 1.46 Jm^{-2} calculated using the same potentials. However, both the fracture resistance and the calculated surface energies are lower than measured, being approximately half those actually measured.

6. Direct Observations of Crack Tips

In contrast to the voluminous literature devoted to linear elastic fracture mechanics, the modes of fracture in different materials and the mechanisms of imparting fracture resistance to materials, there have been few attempts to observe directly the atomic structure of crack tips in solids. For many decades, until the widespread use of transmission electron microscopy, this was understandable since the crack tip, and the enclave in which linear elasticity is thought to break down, was beyond the capabilities of most imaging techniques. There was thus little opportunity to put to the test which of the contrasting models best described the crack-tip geometry. This situation is now beginning to change.

The experiments of Hockey (Hockey and Lawn, 1975; Lawn et al., 1980), marked the first direct observations of crack tips in brittle solids. He used diffraction contrast imaging in the transmission electron microscope to study indentation produced cracks in Si, SiC, and Al_2O_3 and provided the first confirmation that there are no dislocations at crack tips in these materials when fractured at room temperature. An example of his observations is the crack in silicon reproduced in Fig. 3. Since the time of Hockey's original experiments, newer generations of electron microscopes with higher resolution have been developed, and imaging technique has improved. Together, they have made is possible to observe directly the lattice structure of most materials, including the majority of semiconductors. As a result, the techniques of lattice fringe imaging, which have proved so useful in other areas of materials physics, are finally being brought to bear on the observation of the crack tip.

Figures 4 and 5 are examples of the type of direct, high-resolution observation of the crystal lattice in the immediate vicinity of crack tips that is now attainable using the transmission electron microscope. Figure 4 is such a lattice fringe image of a crack tip in a Sialon ceramic — a Al, Mg, O substituted silicon nitride (Clarke and Faber; 1987). The crack is viewed so

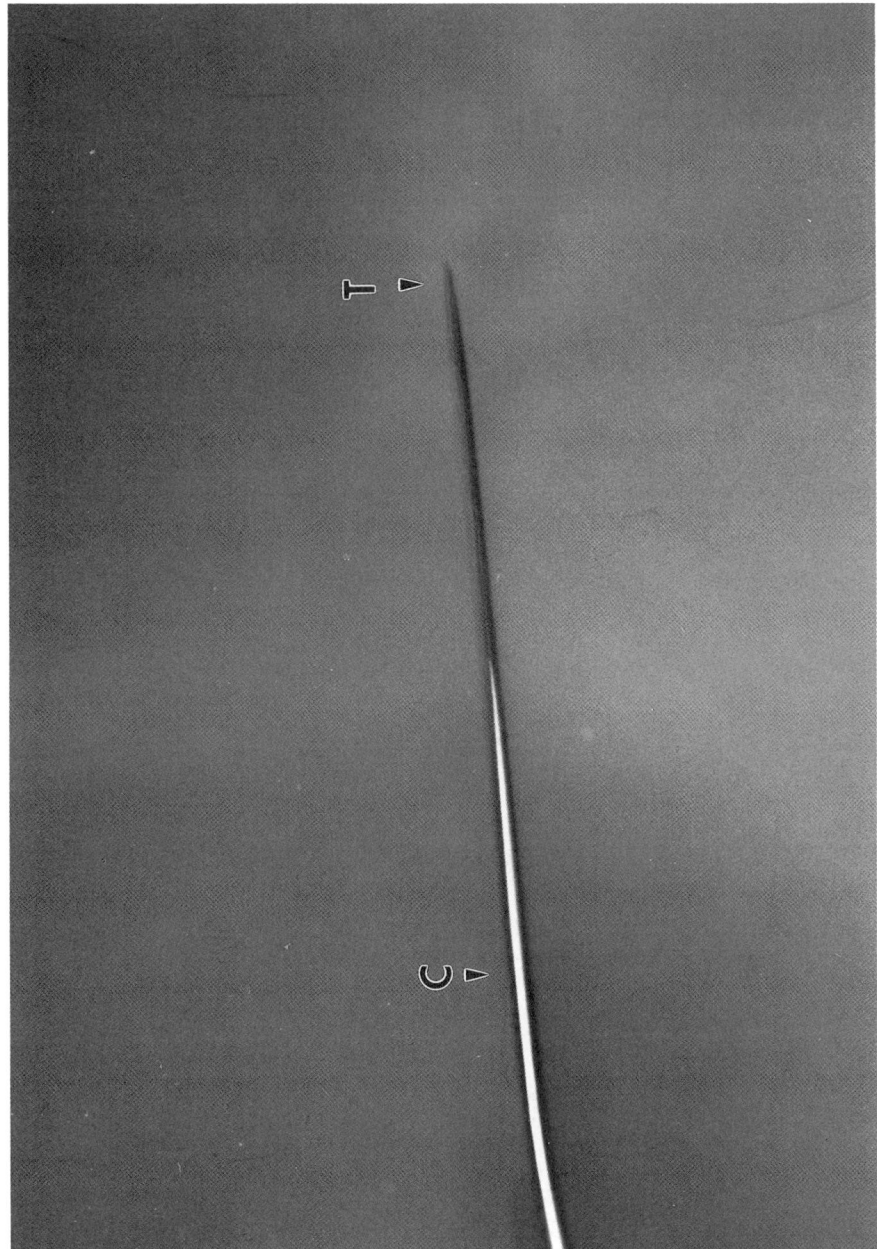

Fig. 3. Low-magnification transmission electron micrograph of a crack tip, T, in silicon illustrating that no dislocations are present in its vicinity. Reproduced courtesy of Hockey.

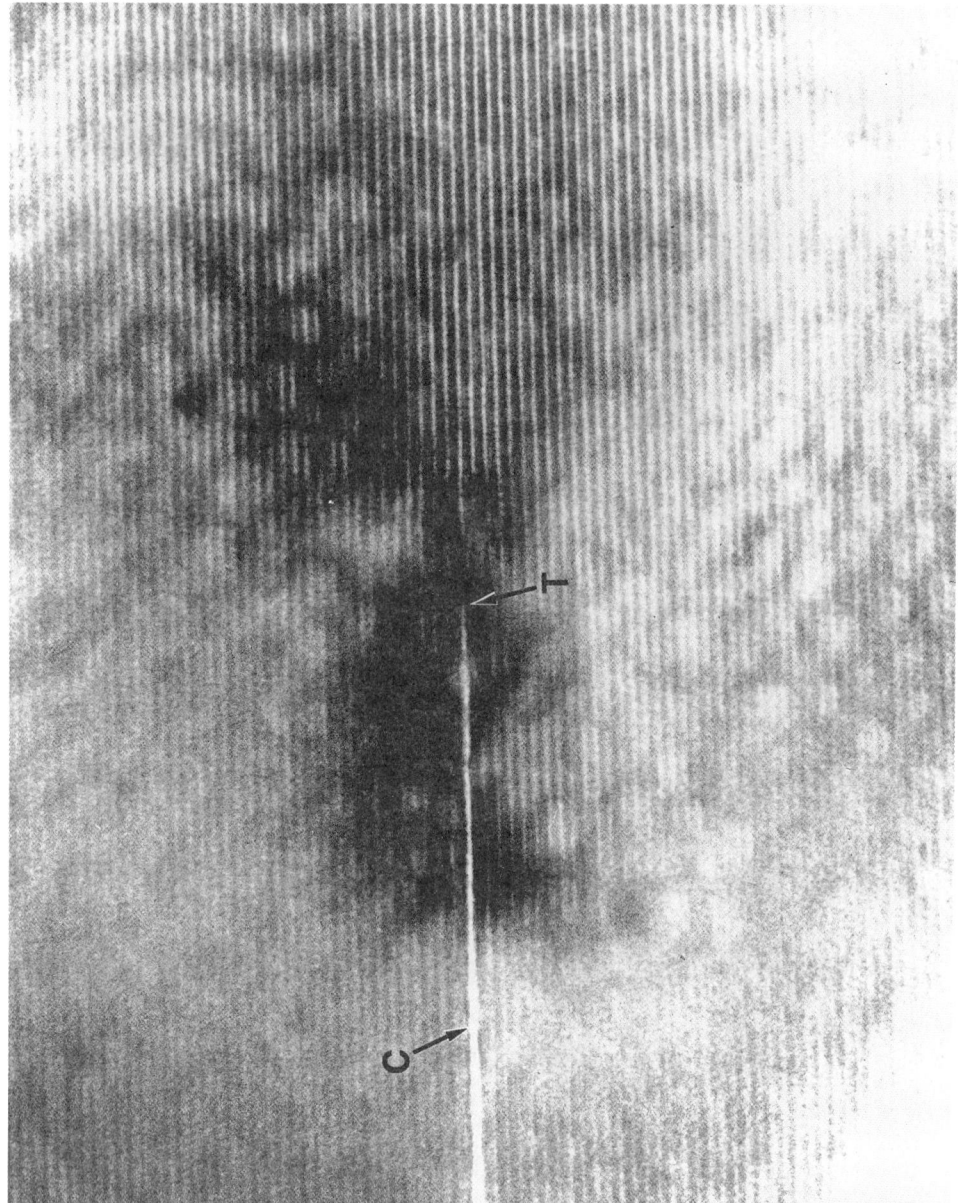

Fig. 4. Lattice fringe image of crack tip in Sialon ceramic. The crack, C, is seen edge-on with the tip at the location T. The fringes correspond to the spacing of the lattice planes.

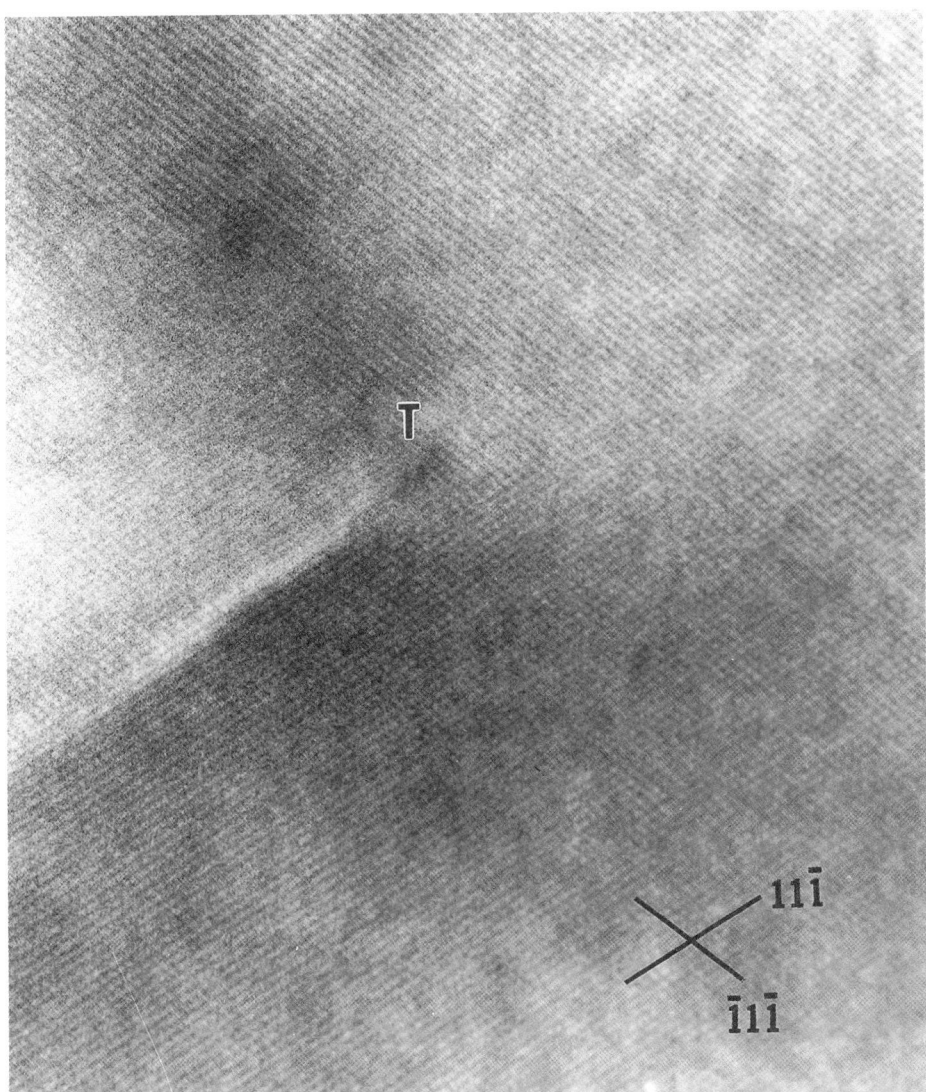

Fig. 5. Lattice fringe image of the tip of a {111} cleavage crack in silicon.

that the crack plane, C, is seen edge-on. The tip, T, is clearly identifiable. The lines correspond to the individual lattice planes of the crystal lattice. The diffraction contrast (the bright and dark contrast) around the tip provides direct evidence that the crack is under load; if it had not been, there would be no such local variation in contrast. Figure 5 is a still higher-resolution lattice fringe image, in this case of a {111} cleavage crack in silicon. The crossed lattice fringes correspond to the two {111} lattice planes having a spacing of 0.3 nm. This photomicrograph is drawn from a recent investigation of cracks produced by indentation, and the brittle-ductile transition in silicon by the author. Quantitative analysis of the atom positions in such images is underway. They provide the first opportunity to test the contrasting continuum elasticity models and the discrete computations, such as those of Lawn and Sinclair (1972), performed for the same (111) cleavage crack in silicon.

7. Mechanisms of Crack Advance

The precise mechanism by which the bonds break is far from understood, although it is recognized that the crack front cannot move *en masse*, but rather its motion is localized by the movement of kinks along the crack front. It has been suggested that the kinks migrate by quantum tunnelling (Gilman and Tong, 1971), and although this may well be the case, no quantum effects have yet been distinguished. To date kinetic theory has been adequate in modelling crack propagation, but it begs the question as the underlying mechanism by which the bonds break and the crack extends. Some light may be thrown on it by illumination effects discussed later in Section IV.15.

The rate of crack advance may be expressed in terms of the kinetics of an activated bond rupture process by considering the individual bond breaking and bond healing processes within the framework of kinetic theory. The energy as a function of crack position is shown schematically in Fig. 6.

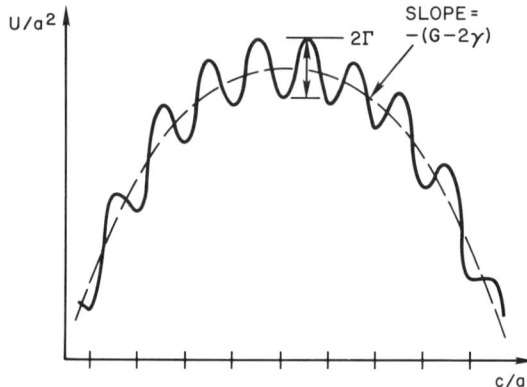

FIG. 6. Schematic of the potential energy, U, for an advancing crack of length c, normalized to the lattice spacing a. The height of the local barrier to advancement by one lattice spacing is 2Γ.

Following Cook's recent re-analysis (Cook, 1989), of the original formulation of the problem by Lawn (1975) and Thomson (1980), the crack velocity may be written as

$$v = \omega_0 \cdot a \cdot [\exp(-\beta U_+^*) - \exp(-\beta U_-^*)], \tag{21}$$

where ω_0 is an attempt frequency for bond rupture, a is the lattice spacing parallel to the crack advance, $U_\pm^*$ are the activation energies for forward bond breaking and backward bond healing, respectively, and $\beta = 1/k_B T$. The crack velocity depends on the mechanical energy release rate, since the activation energies are functions of this release rate. Under the conditions for which the difference between the mechanical energy release rate, G, and the surface energy, 2γ, are small with respect to the energy barriers for bond breaking (or healing), the activation energies may be written as

$$U_\pm^* = \frac{2\Gamma}{\pi N}\left(1 \mp \frac{\pi}{2}\left[\frac{G-2\gamma}{2\Gamma}\right]\right), \tag{22}$$

where $N \sim 1/a^2$ represents the number of bonds per unit of crack area, and 2Γ is the energy barrier per unit area.

Combining Eqs. (21) and (22) provides an explicit expression for crack velocity in terms of the mechanical energy release rate:

$$v(G) = \omega_0 \cdot a \cdot \exp\left(-\frac{2\beta\Gamma}{\pi N}\right) \sinh \beta\left(\frac{G-2\gamma}{N}\right). \tag{23}$$

This equation exhibits a number of interesting characteristics that are borne out by experimental studies of fracture in brittle materials, primarily ceramics. Although detailed studies have not also been carried out on semiconductors, it is expected that they, too, will exhibit similar behavior. Firstly, the crack velocity is a rapidly increasing function of G. Secondly, there is a threshold, corresponding to $G = 2\gamma$, below which the crack will not advance and above which it will accelerate. The threshold, and hence the crack velocity at any applied mechanical energy release rate, will depend on the chemical environment in which the fracture takes place by its effect on the surface energy. At very high velocities, far from equilibrium, the crack driving force approaches that for intrinsic fracture corresponding to bond rupture in vacuum. Thirdly, under ideal circumstances at least, the crack may be lattice-trapped. Under very low driving forces, comparable to the height of the barriers, the crack experiences the discrete nature of the lattice and may become trapped, exhibiting only thermally induced fluctuations, in position along its length. Such a situation is analogous to the trapping of dislocations by the Peierls barrier, an effect observed in covalent solids, such as silicon, at relatively low temperatures.

8. Cleavage versus Plasticity

The question of whether a crystal fractures in a brittle manner, by cleavage, or fails by a predominantly ductile mode by the emission of dislocations from the crack tip (Fig. 7) depends in large part on the competition between bond breaking by tension and bond breaking by shear at the crack tip. The earliest argument posed was that the ratio of the maximum tensile stress to maximum shear stress in the vicinity of the crack tip determined whether a solid would be brittle or inherently ductile. If the tensile stress was the lower of the two, fracture would occur by cleavage, whereas if the theoretical shear stress was attained first, dislocations would be generated and plastic flow would ensue. The generation of dislocations is not only thought to lead to the onset of plasticity, but also causes a "blunting" of the crack tip, thereby reducing the stress at the crack tip. The stress-controlled criterion of failure was subsequently refined by Kelly *et al.* (1967) to include the effects of constraint appropriate to that predicted to occur at the crack tip. Computations of the tensile and shear stresses, using the best then available, indicated that more brittle solids had lower ratios to these stresses than ductile materials, such as silver and gold. However, for none of the materials was the calculated tensile stress lower than the shear stress. More recent calculations (Kelly and Macmillan, 1986) of these stresses have lowered the ratios, but even those for diamond and silicon would predict that the shear stress is lower than the tensile stress, and hence they would fail in a ductile manner.

The issue as to whether a solid was brittle or ductile was re-examined by Rice and Thomson (1974). They investigated, using conventional elasticity theory, the stability of a stressed crack relative to the nucleation, and growth, of a dislocation loop having a Burgers vector that would blunt the crack. Crystals for which the emission is spontaneous would be considered ductile, whereas if the energy barrier is large for such a process they would be brittle. Their calculations predicted that elements such as C (diamond), Si, Ge, and Zn and compounds such as MgO and Al_2O_3, will have a barrier against dislocation emission, and hence will be brittle. Using the same approach, Rice and Thomson calculated that there is no barrier to dislocation emission in the conventionally ductile, face-centered cubic metals, such as Pb, Cu, Au, Ag, and Al. From their analysis, brittle and ductile materials can be distinguished according to the value of the parameter $\mu \mathbf{b}/\gamma$, where μ is the shear modulus of the material and $\mathbf{b}$ is the Burgers vector of the emitted dislocation. A number of refinements to the basic Rice-Thomson analysis have been made over the years, principally by Rice and his co-workers (Rice,

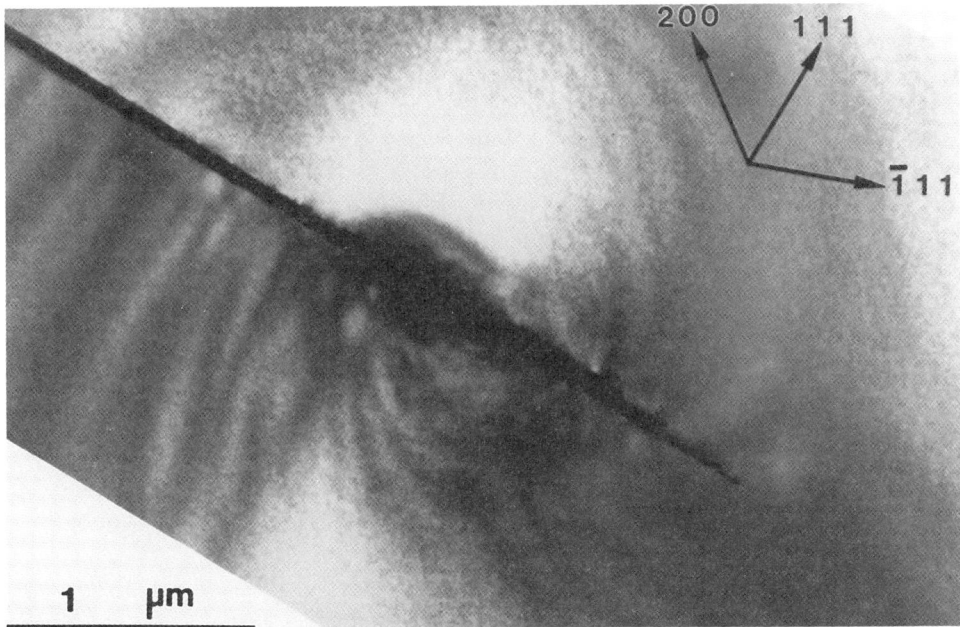
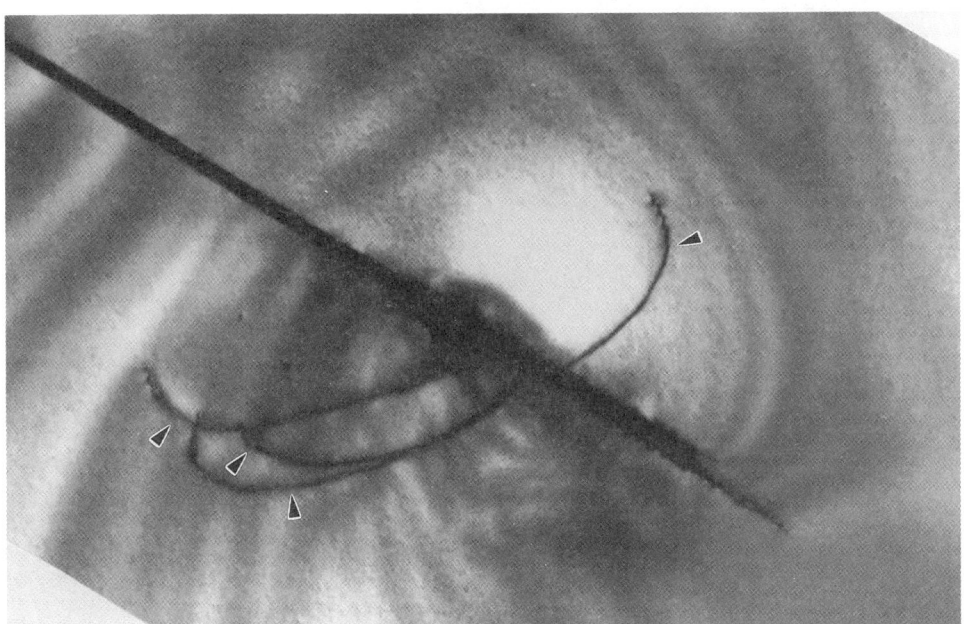

FIG. 7. Dislocations (arrowed) emitted from a {111} cleavage crack tip in silicon when stressed below (top) and above (bottom) the brittle-to-ductile transition temperature. Transmission electron micrograph.

1985; Anderson and Rice, 1986; Rice and Anderson, 1987), and now all fcc metals except aluminum and iridium are predicted to be ductile, and all bcc metals except niobium are predicted to be brittle. As will be noted in a later section, Chiao and Clarke (1989) utilizing their measurements made of *in-situ* dislocation emission in silicon, conclude from a Rice-Thomson model that the energy barrier to dislocation emission is comparable to $k_B T$ at the brittle-to-ductile transition temperature. Furthermore, they also observed after the emission of dislocations having a Burgers vector component perpendicular to the cleavage plane the corresponding blunting steps envisaged in Fig. 8.

Since the development of the physically based, analytical approaches of Kelly *et al.* (1967) and Rice and Thomson (1974), considerable effort has been devoted to modelling crack growth using computer simulations. The earliest devoted to the fracture of silicon (and germanium and diamond) by Lawn and Sinclair (1972), and described above, concluded that crack growth would be perfectly brittle (at zero Kelvin) with no dislocation emission. More recent computations using a variety of interatomic potentials and boundary condi-

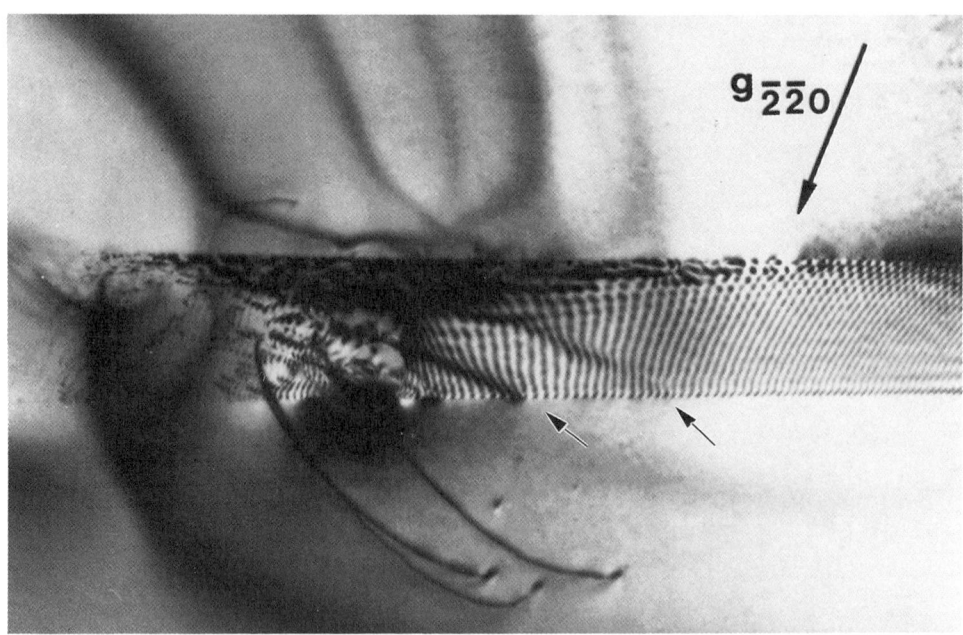

FIG. 8. Steps (arrowed) created by the emission of the dislocations from the crack tip are revealed on the surface of the crack by shifts in the moiré fringes. The emitted dislocation loops are also visible in this transmission electron micrograph.

tions have confirmed Lawn and Sinclair's finding. However, none of these computations has yet been performed at temperatures appropriate to the transition to ductile behavior.

9. Dislocation Models for Fracture Resistance

As mentioned earlier, the prevailing evidence is that single crystals of the majority of semiconductors are truly brittle at room temperature with no dislocations being generated from the crack tip during crack propagation. The work of fracture then corresponds to that required to separate the material along the cleavage plane. However, if the semiconductor contains grown-in dislocations, or contains precipitates that can nucleate dislocations in the stress field of an approaching crack, additional energy may be expended by moving the crack with respect to the dislocations, and the material will exhibit a higher fracture resistance. This occurs in rather dramatic fashion at the brittle-to-ductile transition (BDT), when the crack emits dislocations and the semiconductor becomes resistant to fracture and instead deforms plastically. The additional fracture resistance, over and above the material's intrinsic fracture resistance, arises from the interaction between the elastic fields of the dislocations and that of the crack.

In the absence of any dislocations, the fracture resistance of the material is T_0. Any increase in fracture resistance above this intrinsic toughness, whether in the macroscopically brittle or ductile fracture regime, is given in dislocation models by the shielding (and/or antishielding) afforded by the distribution of dislocations ahead and around the sharp crack tip. As the crack is loaded by an externally applied load, the applied stress intensity, K_{app}, rises until at some value the crack-tip stress intensity equals that at which a dislocation is emitted from the crack tip, K_{e1}. In the absence of any other dislocations in the material affecting the stresses at the crack tip, this occurs when

$$K_{app} = K_{tip} = K_{e1}. \tag{24}$$

The Kelly *et al.* (1967) criterion may be generalized in such stress intensity terms to express whether the bonds at the crack tip fail by cleavage or slip. If, as the applied loading on the crack is increased, the stress intensity, $K_{app} = K_{tip}$ attains the intrinsic fracture resistance T_0 of the material first, bond rupture will occur and the crack will advance by cleavage. If however, the condition for the emission of a dislocation loop, namely that the energy barrier is less than $k_B T$ and given by K_{e1}, is reached first, the bonds will break by shear.

The interaction between the elastic field of a row of screw dislocations coplanar with a simple, mode-III crack (Fig. 9) has been calculated in closed form, within the linear-elastic approximation, as

$$\tau_i = \tau_{app} + \tau_{image} + \tau_{inter}, \tag{25}$$

$$\tau_i = \frac{K_{app}}{\sqrt{2\pi x_i}} - \frac{\mu b}{2\pi}\left[\frac{1}{2x_i} + \sum_{j=i}\left(\frac{1}{x_i + \sqrt{x_i x_j}} + \frac{1}{x_j - x_i}\right)\right], \tag{26}$$

(a)

(b)

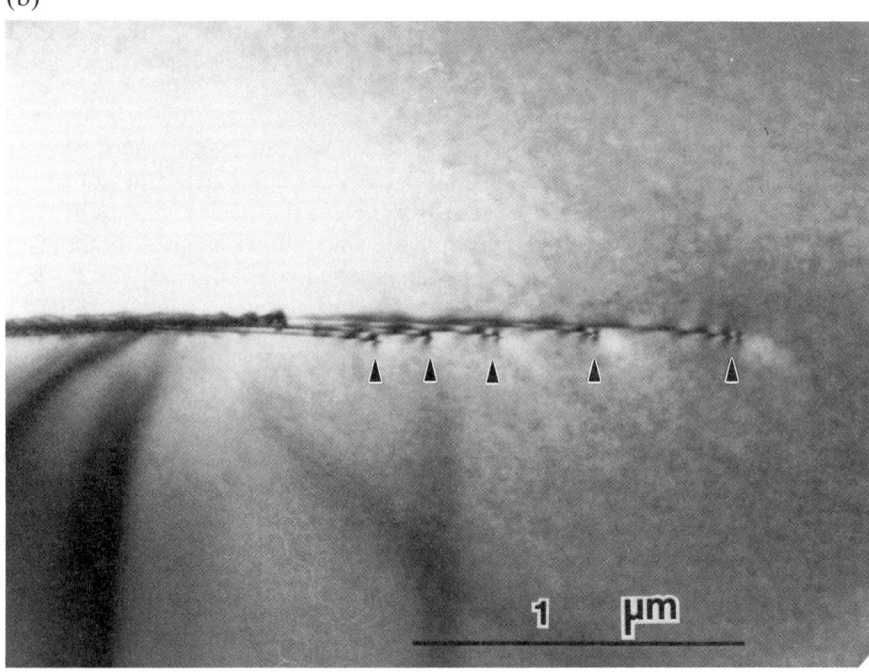

FIG. 9. Dislocation arrangement ahead of a mode-III crack. (a) Schematic of the equilibrium spacing predicted by elasticity theory, and (b) observed ahead of a crack tip in silicon after emission from a {111} cleavage crack above the brittle-to-ductile transition.

where τ_i is the net stress acting on the ith dislocation located a distance x_i from the crack. The three terms of this equation in turn represent the stress arising from the stress applied to the sample, the image stress, and the interaction between the other dislocations in the pile-up.

The overlap of the elastic fields causes a force to be exerted on the dislocation, and by Newton's second law, a reaction force to act on the crack tip. Depending on the sign of the dislocation stress field, it can either increase or decrease the driving force on the crack tip, thereby "shielding" or "antishielding" the crack from the applied loading. An emitted dislocation contributes to the fracture toughness measured by exerting a back stress intensity onto the crack tip, K_{shd}, thereby shielding the crack tip from the applied loading. For instance, in the case of a single screw dislocation lying coplanar with a crack-tip, such as one of the dislocations in Fig. 9, an analytical expression for the shielding contribution is obtainable and has a value of

$$K_{shd} = -\mu b/\sqrt{(2\pi x)}.$$

Additional dislocations in the vicinity of the crack will affect the overall stress fields felt by the crack tip and by the other dislocations, and can therefore alter the net force experienced by the crack tip by altering the shielding contribution. The second possible effect of dislocation emission is to alter the shape of the crack tip ("blunting") by the shear associated with the Burgers vector component perpendicular to the crack plane. This contribution is represented by the stress intensity term K_{blt}.

The effect of these two terms is to alter the effective crack driving force, and hence the net stress intensity, K_{tip}, acting in the crack-tip frame from that applied by an external loading system, K_{app}. Both the shielding and blunting terms tend to reduce the value of K_{tip}, so the crack-tip stress intensity may be expressed in equilibrium as

$$K_{tip} = K_{app} + K_{shd} + K_{blt} = T_0. \tag{27}$$

In modes II and III, where, by definition, K_{blt} is zero, complete shielding occurs when $K_{shd} = -K_{app}$, leading to a value, measured in the crack-tip frame, of the stress intensity $K_{tip} = T_0$.

Since both the crack shielding and blunting contributions act to lower the crack-tip stress intensity, they have the effect of increasing the resistance to crack growth. Thus, alternatively, their contribution may be expressed in terms of the net crack resistance, T, as follows:

$$T = T_0 - K_{shd} - K_{blt}. \tag{28}$$

Thus, the emission of dislocations from the crack tip increases the applied loading required to propagate the crack, and hence has the effect of increasing the measured fracture resistance of the material. However, it should be reiterated that unstable crack propagation will always occur when the inequalities of Eq. (14) (and equivalently, Eq. (15)) are satisfied, irrespective of the details of the actual processes occurring ahead of the crack tip.

The equations and relationships presented up until this point are completely general and apply equally whether the crack and dislocations are moving relative to one another or not, or indeed whether or not the dislocations are in their equilibrium (zero net force) positions. Thus, the equations can be used in developing models that include not only the crack velocity, but also the dynamics of the dislocations as they interact with each other and with the moving crack. Such factors are the essential ingredients of the BDT.

For the case of the brittle-to-ductile transition, which the experimental studies (Section V) clearly indicate is mediated by dislocations, the problem is reduced to developing a consistent dislocation model that gives the fracture resistance as a function of crack extension. Such a comprehensive model does not yet exist, but the crack-length dependence of the fracture resistance would be the so-called R-curve for toughening due to dislocation emission. As recognized since the work of St. John (1975), the problem is compounded by the fracture resistance being strain-rate dependent.

IV. Fracture of Semiconductors

10. Crystallographic Aspects

Of all the fracture properties of semiconductors, the best known are the cleavage planes and directions. In large part this is because cleavage is a rapid and practical method for separating circuits on wafers and can be used as an alternative to sawing. It has also become a routine method of producing parallel surfaces that can act as mirrors in solid-state lasers, such as those made from GaAs–GaAlAs. The principal cleavage planes and directions are listed in Table I. Also included in Table I are the principal slip planes for dislocation glide. The primary cleavage planes are predictable from the crystal structure of the solid since cleavage generally occurs along the planes having the highest atomic density. Thus, all diamond cubic materials cleave on the close-packed $\{111\}$ planes, and the zincblende materials on the $\{110\}$ planes. Although the $\{111\}$ planes in the zincblende structure have the highest atomic density, cleavage onto these planes would lead to the formation of polar surfaces. There would thus be an additional energy expenditure in cleaving the surfaces associated with the electrostatic energy

TABLE I

Material	Principal Planes	Cleavage Direction	Principal Planes	Slip Direction
Si, Ge	{111} {110}	⟨011⟩ ⟨110⟩	{111}	⟨011⟩
Diamond	{111}	⟨011⟩	{111}	⟨011⟩
Cubic III–V	{110}	⟨110⟩	{111}	⟨110⟩

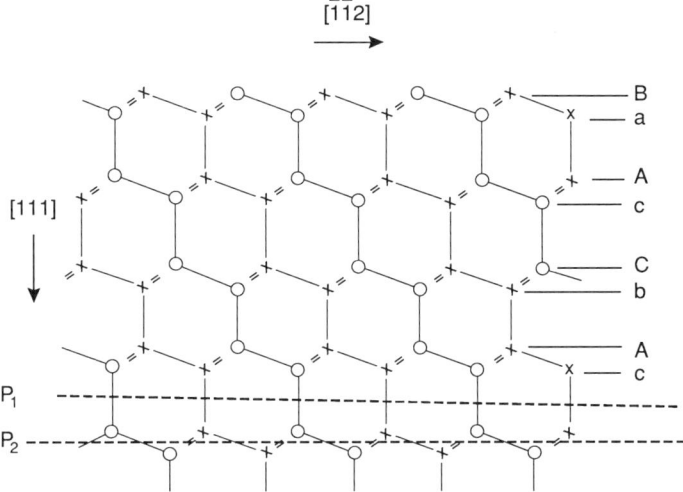

FIG. 10. Schematic of the two alternative crack paths P_1 and P_2 in the diamond cubic crystal structure. It is not yet known which corresponds to the cleavage path.

required to separate the charges. By cleaving on the less densely packed {110} planes, which are, on average, neutral, this Coulombic contribution is avoided. In a number of cases, including silicon and diamond, cleavage can be induced to occur on planes other than the primary cleavage plane. Thus, in silicon and diamond, for instance, cleavage can also take place on the {110} planes. This tendency to cleave along the secondary cleavage planes is particularly pronounced in thin pieces of material, such as wafers. Indeed, a number of people in the semiconductor industry, presumably experienced only with wafers, claim that silicon cleaves preferentially on the {110} planes.

Although the cleavage planes of all the semiconductor materials are known, it has yet to be established on which of the crystallographically equivalent planes fracture occurs. Thus, in reference to Fig. 10, it is not known whether during fracture the crack passes along the path P_1 or P_2. If a dislocation defect model is assumed to describe the advance of a cleavage crack, then quite different defects will be required to cause fracture to occur along P_1 and P_2, and simultaneously leave the fracture surfaces having the atomic reconstructions observed by LEED and STM (Huang et al., 1990; Haneman and Lagally, 1989).

11. Fracture Surfaces

The properties of the surfaces created by fracture at low temperatures have been studied in considerable detail because cleavage has been the principal method of producing surfaces for UHV studies. The interested reader is referred to recent reviews (Chadi, 1983; Haneman, 1987), of this subject for greater detail, as well as for details of the atomic reconstructions that take place.

The fracture surface of a cleaved brittle solid would ideally be perfectly flat with no steps or roughness. However, even in the absence of any crack tip plasticity, the surface can exhibit topographic features. The commonest cause is a result of the propagating crack interacting with stress waves reflected from the surfaces of the material. These can cause the local stress field at the tip of the crack to be perturbed, thereby altering the trajectory of the crack by kinking it. Another cause of an uneven fracture surface is if the load varies as the crack advances. Since the crack velocity is such a strong function of mechanical release rate (see Eq. (23)), variations in applied load can accelerate and decelerate the crack. Indeed, a number of investigators have deliberately applied an oscillating load to a material during crack propagation in order to be able subsequently to measure the crack velocity from the markings left on the fracture surface (Kerkhof, 1970; Green et al., 1976). The interaction of the crack with pre-existing, grown-in dislocations can also perturb the crack path, leaving behind steps on the surface. The step height corresponds to the Burgers vector, or some multiple, of the dislocation line intersected by the crack. An example is shown in the micrograph of Fig. 11 recorded under reflection high-energy electron diffraction conditions (Bleloch et al., 1989). (This particular microscopy mode results in a foreshortening of the image but enables an accurate measure to be made of the vertical height). In this photograph, the zigzag features are cleavage steps on the (110) surface of a cleaved GaAs crystal with the steps all running within a few degrees of the [001] direction. From this information it is possible to conclude that the sides of the steps probably correspond to one of the other {110} cleavage planes. The image contrast of the horizontal line passing through point D indicates that a screw dislocation intersects the surface at that point.

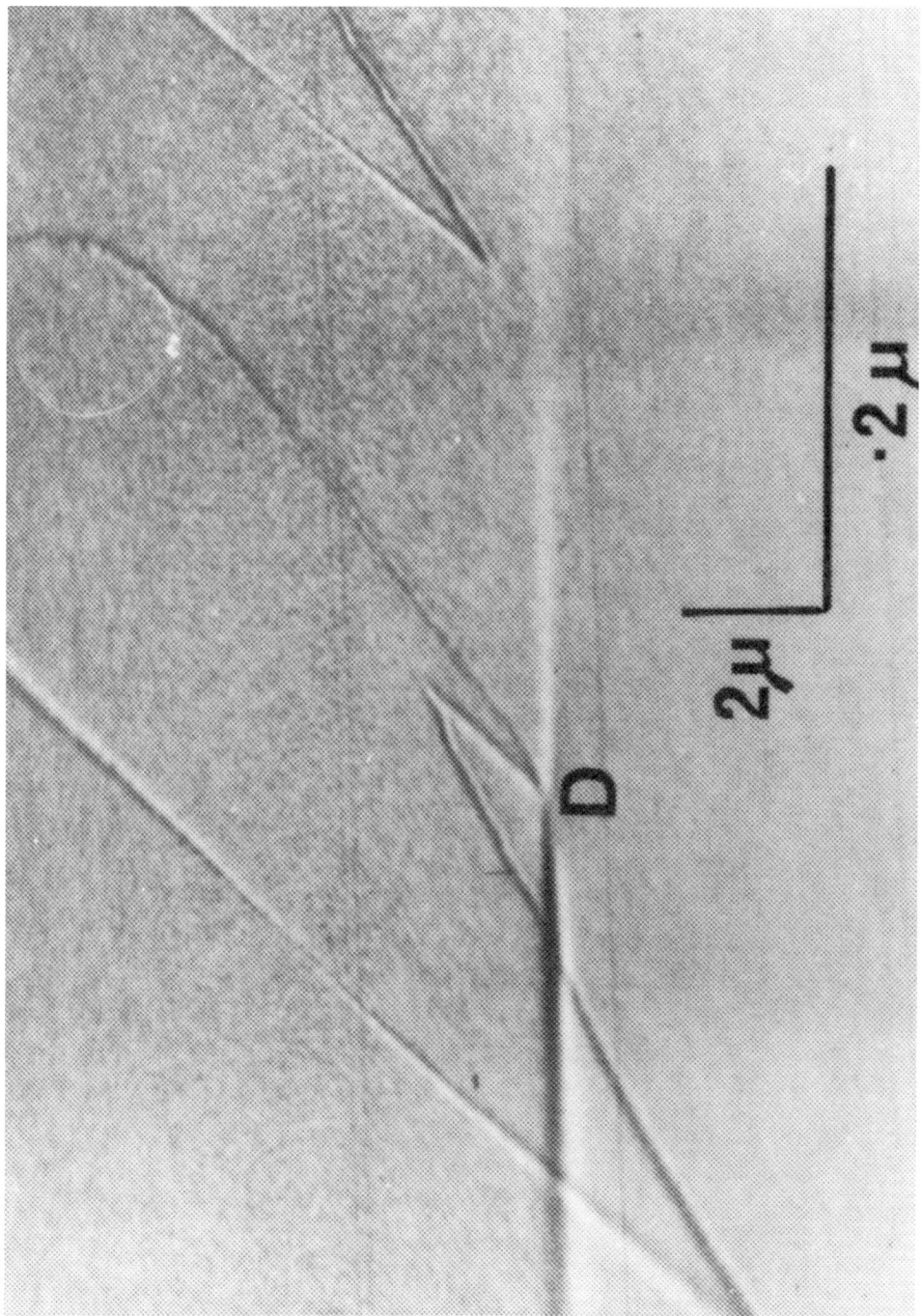

FIG. 11. Zigzag cleavage steps interacting with an emergent screw dislocation at D on the (110) surface of a cleaved GaAs crystal. The image was recorded in the RHEED imaging mode at 100 keV using a 880 resonance diffraction condition. Photograph courtesy of A. Howie.

Surface steps can also be created by the emission of dislocations having a Burgers vector component perpendicular to the fracture plane. This is not expected to be a major source of surface steps, except at temperatures close to the BDT where some crack-tip plasticity may occur. Both Brede and Haasen (1988) and Michot (1988) have reported that etching reveals the presence of dislocations on the fracture surface just prior to the BDT. Also, the moiré patterns recorded from between the crack faces of silicon in the *in situ* TEM studies referred to earlier, (Chiao and Clarke, 1989), indicate that surface steps were created when dislocations were emitted from (111) cracks.

12. Fracture Surface Energy

A number of sophisticated computations, including recent *ab initio*, first-principles electronic structure calculations, have been made of the surface energy of solids. Whilst these are important, considerable insight into the surface energy of the covalent semiconductors can be obtained from the rather simple bond calculation that follows. The calculation, which appears to have first been performed by Harkins (1942) in estimating the surface energy of diamond, relies on the fact that in covalently bonded crystals, nearest-neighbor interactions dominate and the energy associated with long-range interactions can be ignored. The surface energy at 0 K is then simply one-half of the energy required to rupture the number of bonds across a cross-section of one square meter. For fracture on the (111) planes in

TABLE II

Approximate Calculated Surface Energies of {100}, {110} and {111} Surfaces[a]

Material	γ (Jm^{-2})		
	{100}	{110}	{111}
InSb	1.1	0.75	0.6
GaAs	2.2	1.5	1.3
InAs	1.4	1.0	0.84
GaSb	1.6	1.1	0.91
InP	1.9	1.3	1.1
AlSb	1.9	1.3	1.1
AlAs	2.6	1.8	1.5
GaP	2.9	2.0	1.7
AlP	3.4	2.4	2.0
Si	2.13	1.51	1.46
Ge	1.84	1.30	1.07
Diamond	—	—	5.35

[a]Surface energies for III–V crystals from calculations of Cahn and Hanneman (*Surface Science* 1964, **1**, p. 387). {111} fracture surface energies for Si, Ge, and diamond from Lawn and Sinclair (1972). Other values from Jaccodine (*J. Elecrochem Soc.* **110**, 524).

TABLE III

Measured Fracture Surface Energies

Material	Cleavage Planes	Fracture Energy, Jm^{-2}	Technique[i]	Reference
Si	{111}	1.23	DCB	a,b
	{110}	1.81	DCB	c
Ge	{111}	1.06	DCB	b
Diamond	{111}	~6.0	I	d
GaAs	{110}	0.76	DCB	e
		0.73	IBF	f
InP	{110}	0.86	IBF	f
InAs	{110}	0.43	I	f
GaP	{110}	1.5	IBF	g
	{111}	2.0	IBF	g
	{100}	2.6	IBF	g
SiC	{11$\bar{2}$0}	23	DCB	h
	{1$\bar{1}$00}	23	DCB	h

[a]Gilman, J.J. (1960). *J. Appl. Phys.* **31**, 2208.
[b]Jaccodine, R.J. (1963), *J. Electrochem. Soc.* **110**, 524.
[c]Bhaduri, S.B. and Wang, F.Y. (1986), *J. Mater. Sci.* **21**, 2489.
[d]Fields, J. in "The Physical Properties of Diamond," Academic Press.
[e]Michot, G., George, A., Chabli-Brenac, A., and Molva, E. (1988), *Scripta Met.* **22**, 1043.
[f]Yasutake, K. *et al.* (1988), *Jap. J. Appl. Phys.* **27**, 2238.
[g]Hayoshi, K. *et al.* (1982), *Mater. Lett.* **1**, 116.
[h]Henshall, J.L., and Brookes, C.A. (1985), *J. Mater. Sci. Lett.* **4**, 783.
[i]DCB: double cantilever beam; I: indentations; IBF: indentation bend fracture.

diamond cubic and zincblende compounds, three bonds have to be broken per unit cell. Thus, the number of bonds per unit area that must be ruptured is $\sqrt{3}/4a_0^2$. The energy per bond can be obtained either from the sublimation energy of the compound or from tables of the appropriate bond energies (see, for instance; Cotton and Wilkinson, (1988)). The surface energies calculated in this manner are tabulated in Table II for a number of elemental and compound semiconductors. The surface energy calculated in this manner can, of course, only be an approximation. It ignores, for instance, the energy associated with surface distortions or reconstructions, both of which are known to occur in semiconductors.

With approximate values for the surface energies, it is possible to estimate the fracture resistance using Eq. (13). This assumes that the materials cleave in an ideally brittle manner and the work of fracture is expended solely in creating the two fracture surfaces. The required modulus in Eq. (13) is that in the fracture plane in the cleavage direction. For comparison with the values in Table II, the experimentally determined, in the few instances that they have been measured, fracture surface energies are listed in Table III. A number of

13. Dependence on Energy Gap

One of the principal results of the thermodynamic approaches to the fracture is, as indicated by Eq. (13), that the fracture resistance of a brittle material is proportional to the square root of its surface energy. It has also been suggested that the energy band gap, the elastic constants, and the surface energy are all linearly related in semiconductor solids. If this is indeed so, a linear relationship might thus be expected between the low-temperature fracture resistance and the band gap. The data, collected from a variety of sources, suggest that such a relationship does exist, at least for the III-V compounds (Fig. 12). A more systematic analysis and considerably more experimental data are required before the relationship can be tested thoroughly. Also, very large doping levels will probably be required in order to create changes that can be distinguished from the experimental errors in determining the fracture resistance.

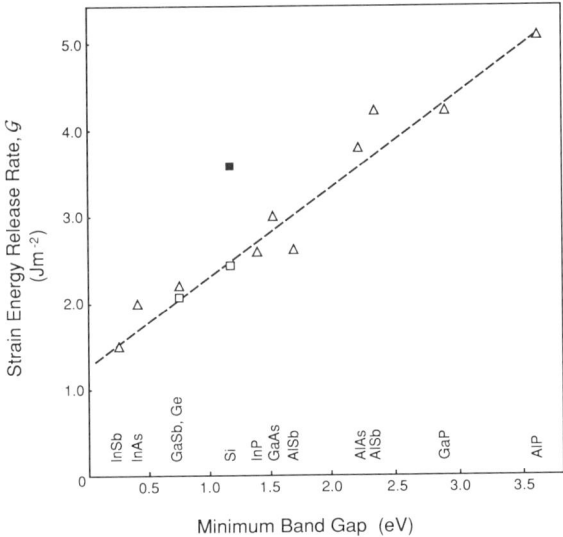

FIG. 12. Calculated strain energy release rate for elemental and compound semiconductors as a function of band gap minimum energy. Filled symbol refers to {110} fracture and open symbols to {111} fracture. Band gap data obtained from Landolt-Bernstein, Vol. 22, Subvolume a, "Semiconductors," published by Springer-Verlag, 1987.

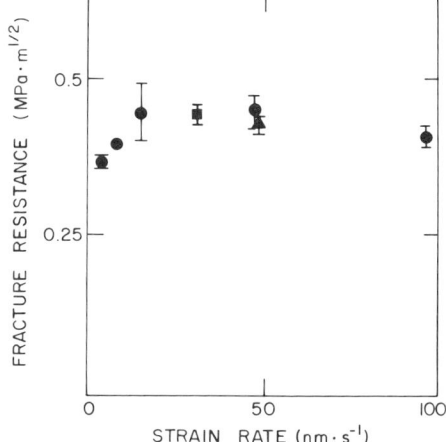

FIG. 13. Fracture resistance of single crystal GaAs as a function of strain rate. The tests were performed in helium (Δ), forming gas (■), and argon (●). Reprinted with permission from *Scripta Met.* **22**, 1043–1048, "Fracture Toughness of Pure and In-Doped GaAs," © 1988, Pergamon Press plc.

14. STRAIN RATE EFFECTS

In the brittle fracture regime, no significant strain rate dependence of the fracture resistance of either elemental or compound semiconductors has yet been reported. Although, in the absence of environmental effects (Section VI) there is no reason to suppose that there will be a strain rate dependence of the fracture resistance, few systematic experiments have been undertaken. The results of one of the few is reproduced in Fig. 13, taken from the work of Michot *et al.* (1988), on the fracture of GaAs single crystals. There is a slight indication of a strain rate effect, but the differences have yet to be unequivocally attributed to a material response rather than to experimental variation from test to test.

15. EFFECT OF ILLUMINATION

There are a number of intriguing reports of illumination affecting the length of cracks produced by indentation in compound semiconductors. Although there is no reason to doubt the observations themselves, it is likely that the effects are not manifestations of the fracture resistance of the materials being altered by irradiation.[5] Rather, as will be described, they are

---

[5] Heating on irradiation can cause existing cracks to extend and materials to fracture. Such thermal shock effects are distinct from the effects discussed here.

more likely to be a result of the illumination affecting the relative balance of energy dissipation between plastic flow and crack extension.

A growing body of literature indicates that dislocation mobility in both elemental and compound semiconductors can be affected by light illumination or electron-beam irradiation (Maeda et al., 1977, 1983; Ahlquist and Carlsson, 1973; Maeda and Takeuchi, 1983; Fujita et al., 1988; Küsters and Alexander, 1982). Pronounced effects have been reported for II-VI compounds, particularly CdTe, and have led to the term "photoplasticity" being coined for the phenomena. In these semiconductors, illumination with bandgap excitation decreases dislocation mobility and their hardness appears to increase. In marked contrast, illumination of the III-V semiconductors causes an increased dislocation mobility, and as a result the flow stress decreases. A similar, albeit weaker, softening has also been reported in elemental semiconductors (Küsters and Alexander, 1982). This softening, sometimes referred to as radiation-enhanced dislocation glide (REDG), has been shown to be an optical effect and not simply the result of heating caused by the radiation.

The most complete work has been carried out on GaAs single crystals by Maeda and his colleagues (Maeda and Takeuchi, 1983; Fujita et al., 1988). In a number of papers they have investigated the effect of laser illumination on dislocation mobilities by deforming single crystals of n-GaAs at different temperatures whilst exposing the crystals to either light or 30 keV electrons. Above a relatively high temperature (650 K for β dislocations and 550 K for α dislocations), the dislocation mobilities are unaffected by illumination. However, below these temperatures the mobilities of both types of dislocation are significantly enhanced by illumination. The apparent activation energies derived from the dislocation velocities are much lower below the transition temperatures, being approximately 0.29 eV for α-dislocations and 0.6 eV or β-dislocations (compared with 1.0 eV and 1.7 eV, respectively, above the transition). The detailed mechanism for the observed mobilities being enhanced is not clear. However, there is agreement that the effect is independent of the type of the illumination (presumably above some, as yet to be determined, threshold intensity) and is electronic in origin, since any super-band-gap excitation appears to cause the effect.

The fracture of GaAs under illumination conditions has, up to now, been studied using the indentation technique. As described in the appendix, the fracture resistance of an ideally brittle material can be ascertained from the indentation load, the hardness, the Young's modulus and the observed crack length. Thus, in the experiments on Si doped, n-type GaAs singe crystals, Fujita and colleagues (1988) measured the hardness and the crack lengths when indenting on the (111) Ga face and the ($\bar{1}\bar{1}\bar{1}$) As face in the dark and under illumination with an argon ion laser (514 nm). As has been reported in earlier studies (Hirsch et al., 1985; Yasutake et al., 1988), the (111) Ga faces were measured to have a greater hardness than the $(\bar{1}\bar{1}\bar{1})$ As faces at all temperatures. No change was observed under illumination with an incident power density of 11 kW/m^2, a finding similar to that made on GaP crystals some years earlier. The crack lengths were, however, quite different when the (111) Ga face was indented and illuminated, being significantly shorter. From the smaller crack lengths, and using the standard crack length–fracture resistance relationship for indentation fracture toughness, the fracture resistance was calculated to have doubled under illumination. No change in crack length was observed when indenting on the $(\bar{1}\bar{1}\bar{1})$ As face, and hence the fracture toughness is calculated to be unchanged.

An effect of illumination on the fracture resistance of CdTe has also been reported by Ahlquist and Carlsson (1973). They used the same indentation technique as Maeda, but only performed their measurements at room temperature and at considerably lower power densities (a mW He–Ne laser focussed down onto the indent area). When indenting on the (110) plane under room lighting, few if any cracks formed at loads of 20 and 30 N. However, after indenting under the laser illumination, long cracks formed emanating from the hardness impression. As with both the GaP and GaAs experiments, the hardness (measured from the size of the impression) remained unaffected. The micrographs presented by the authors unfortunately did not extend to the ends of the cracks, so no measurement can be made of the crack length and the fracture toughness calculated. Interestingly, the authors did carry out etch-pit studies before and after indentation to reveal both the grown-in dislocations and the density of dislocations produced during indentation. From these measurements, they were able to establish that under illumination conditions, the density of dislocations produced was lower than that under darkness. They attributed the difference in cracking to be the light decreasing the dislocation mobility. Whilst the light may well be affecting the dislocation mobility, it does not in itself explain why under conditions of darkness few, if any, cracks form.

The explanation is more likely to be that advanced by Fujita *et al.* (1988) in discussing their observations on GaAs. They argued that the observed crack length was the result of shielding of incipient cracks, formed within the plastic deformation zone of the indentation, by dislocations created by the indentation. (An example of such dislocations produced by indentation in GaAs is shown in Fig. 1).

Following Hirsch *et al.* (1985), cracks are believed to form on indentation in GaAs, by dislocation pile-ups at Lomer–Cottrell locks at the intersection of slip planes. The propagation of these crack nuclei is then driven by the indentation stress field and shielded by the surrounding dislocations. The suggestion by Fujita *et al.* is that illumination affects the mobility of the dislocations, thereby altering the shielding they afford to the crack, and hence changing the crack driving force and hence the crack length.[6] In support of their explanation, Fujita *et al.* note that on indenting on the (111) Ga face, the cracks would be shielded by the α dislocations, and these are the dislocations whose mobility is most marked by illumination. Thus, crack extension would be suppressed. Indenting on the ($\bar{1}\bar{1}\bar{1}$) As face, the β dislocations shield the cracks. However, their mobility even under illumination is low, so there would be little apparent change in the shielding and hence crack length.

One is led to conclude that whilst there is an effect of illumination on dislocation mobilities, no such effect has yet been unequivocally demonstrated on the intrinsic fracture resistance of either elemental or compound semiconductors. Experiments with single, controlled cracks under well-defined crack driving forces, such as in a double cantilever beam geometry, in dislocation-free crystals will be required in order to explore illumination effects fully.

V. The Brittle-to-Ductile Transition

In common with many other materials, metallic as well as ceramics, semiconductors deform in a brittle manner at low temperatures and in a ductile manner at high temperatures. For instance, in tension, pure iron is brittle below about $-160°C$ and ductile above. Alloying, to form different steels, raises this transition temperature. Although all the semiconductors studied to date are brittle at room temperature, the temperature at which they become ductile has not been established. The exception is silicon, which has been the object of a number of rather thorough studies in the last 15 years investigating the effects of straining rate and doping on the brittle-to-ductile transitions (BDT).[7] The first, and in many respects the pioneering study, was by St John (1975), who showed that in dislocation-free samples the BDT occurred over a very narrow temperature range ($\sim 1°C$), and that above the transition temperature, dislocations were profusely generated from the crack tip. Furthermore, as detailed in the following sections, the transition temperature was dependent on both the rate of straining and the doping concentration.

---

[6]The stress field in and around an indentation is extremely complex and there is, in addition, an intricate distribution of dislocations generated also. However, Eq. (26) embodies the essential physics involved, albeit for a very simple case, in discussing the different contributions to the crack resistance.

[7]The transition is not, at least as presently understood, a phase transition in the thermodynamic sense, but rather marks a change in macroscopic behavior. In most engineering materials the change in fracture behavior occurs over many degrees, usually tens of degrees.

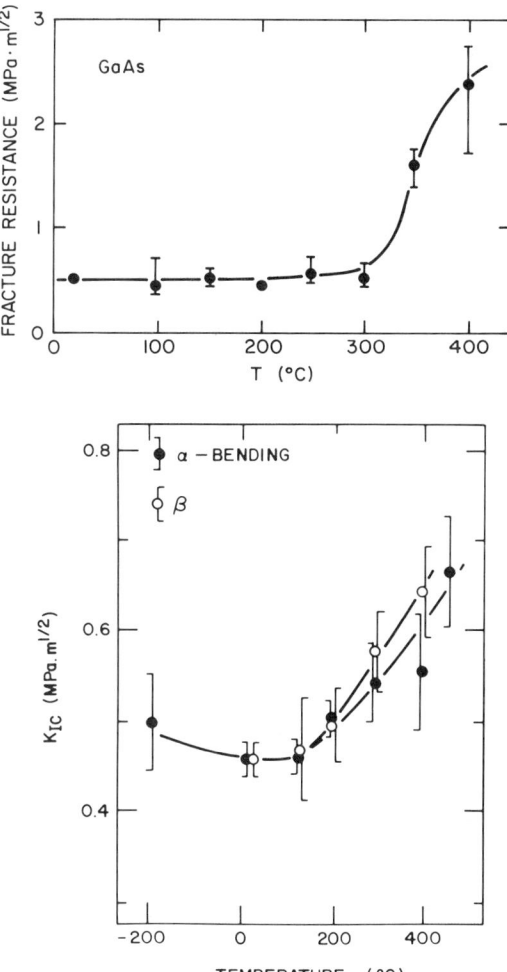

FIG. 14. Fracture resistance of GaAs as a function of temperature. (a) Measured in darkness on the Ga face by the indentation technique. Redrawn after Fujita. (b) Fracture resistance measured using a four-point bending technique. Reproduced courtesy of Yasutake et al. (1988).

St John's use of a fracture mechanics, single-crack test configuration combined with x-ray topographic observations facilitated the examination of the processes accompanying crack extension. It also provided a direct, quantifiable, and generally accepted measure of the fracture resistance of the material. Such studies have not yet been performed on other semiconductors as a function of temperature. The data that do exist have been obtained from either less direct measurements or from tests in which the fracture resistance can be less confidently calculated. For instance, Fig. 14 is the fracture toughness of undoped GaAs as a function of temperature reported by Yasutake *et al.* (1988). The fracture toughness has been calculated from the size of the stable crack formed during fracture. This is generally regarded as being satisfactory in the absence of plasticity, but probably is not at higher temperatures when dislocation plasticity occurs. Nevertheless, the data of Fig. 14 reveal that GaAs can undergo a transition to ductile behavior at temperatures in excess of $\sim 150°C$. This is substantiated by the transmission electron microscopy observations of crack tips in GaAs made at room temperature and at 400°C (Yasutake *et al.*, 1988). The former reveal no dislocations at the crack tip, whereas the latter do.

In addition to the preceding, there are a number of deformation experiments at different temperatures in which a semiconductor has been reported to deform or fracture. For instance, the deformation response of single crystals of indium phosphide to uniaxial compression up to 1000°C has been examined (Brown *et al.*, 1980). The authors noted that above 950°C the crystals deformed by dislocation slip, whereas up to $\sim 460°C$ they fractured. Between these two temperatures, deformation would be accommodated by a combination of slip and twinning. The change from fracture to slip and twinning exhibited some strain rate sensitivity, but the data are insufficient to allow any apparent activation energies to be deduced. In other experiments, single crystals of indium antimonide are reported to be deformable above about 200°C, being able to be bent through large angles without breaking (Allen, 1957). No details were given as to the straining rate or the impurity levels, but the crystals did contain grown-in dislocations as evidenced by etch-pitting. Only brittle fracture was reported at lower temperatures suggesting that the BDT temperature is approximately 200°C. Lastly, on the basis of their fracture surfaces, it has been reported that the BDT in silicon carbide is in the vicinity of 1700°C (Campbell *et al.*, 1989). As this is considerably higher than any processing temperature, it is unlikely that dislocation generation from cracks will pose a problem in creating defect states in silicon carbide devices.

16. STRAIN RATE EFFECT

The original work demonstrating a pronounced strain rate dependence of the BDT temperature was that of St John (1975). He showed that over a two-order-of-magnitude range of strain rate, the BDT temperature of a boron-doped silicon shifted by about 250°C, as reproduced in Fig. 15. In all cases the transition was abrupt, occurring within one to three degrees, and immediately above the transition the load–deflection curves changed from being straight (indicating purely elastic behavior) to showing curvature above a

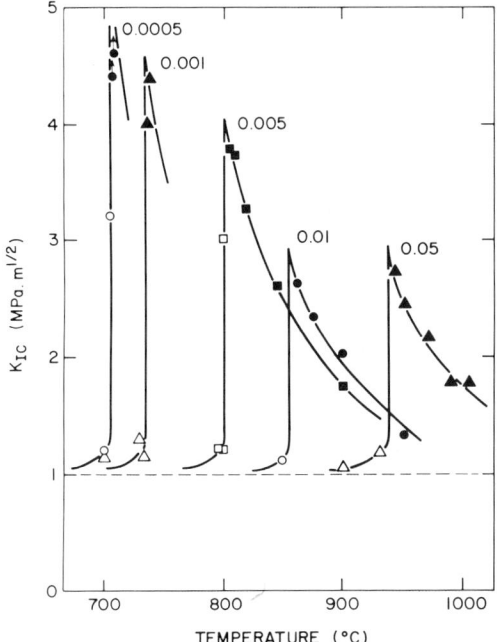

FIG. 15. Strain rate dependence of the fracture resistance of silicon at different temperatures. The numbers adjacent to the curves indicate the strain rates used. The open symbols represent brittle fractures, whereas the filled symbols indicate ductile fractures. Reproduced from St John (1975).

yield point. Also, above the transition the samples ceased to fail by fracture, but rather deformed plastically. Indeed, there are comments to the effect that well above the transition, the samples deform like butter. From the strain rate dependence, St John demonstrated that over the temperature range he studied, the BDT was a thermally activated process with an activation energy of ~1.9 eV. By noting the similarity with the then-reported activation energy for the glide of screw and 60° dislocations in silicon in approximately the same temperature range, he also concluded that the BDT was controlled by the velocity at which dislocations glide away from the crack tip. This conclusion, and hence the conclusion that the transition is not dislocation nucleation controlled, has in large measure stood the test of time, although there continues to be debate about the details of the controlling steps in the dislocation plasticity process.

Subsequent to St John's investigations, the strain rate dependence of the BDT temperature has been substantiated in the doping studies of Michot et al. (Michot, 1988; Michot et al., 1980; Michot and George, 1986); Brede and Haasen (1988; Haasen *et al.*, 1989), and Samuels and Roberts (1989). These studies have also extended the strain-rate dependence to almost three orders of magnitude, by examining material having differing dopant levels.

17. Doping Effects on the Transition Temperature

St John's finding that the apparent activation energy of the BDT temperature was close to that for dislocation glide has prompted studies of the BDT in silicon with different doping levels. Two classes of study may be distinguished: one that explores possible elastic effects, and another that investigates possible electronic effects of doping.

Michot *et al.* (1980) and Michot and George (1986) have investigated the effect of oxygen on the BDT. They found that the transition temperature is higher for materials that contain a higher concentration of oxygen. Thus, Czochralski-grown silicon exhibits a higher transition temperature than float-zone–grown silicon over at least two orders of magnitude in strain rate (Fig. 16). For Czochralski silicon containing $\sim 3.5 \times 10^{17}$ cm^{-3} of oxygen, the transition is about 40° higher than that of the float-zone material, containing $< 10^{15}$ cm^{-3} of oxygen at a similar strain rate. Variations in oxygen content intermediate between Czochralski and float-zone silicon have not been systematically studied. The apparent activation energy for the

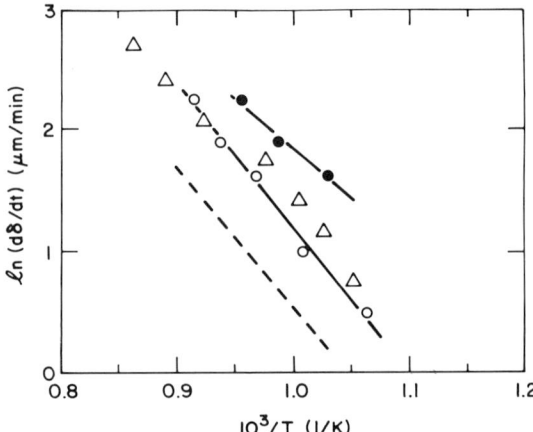

FIG. 16. Arrhenius plot of the data for oxygen and hydrogen doping of silicon. The circles and triangles are the data for the CZ and FZ silicon, containing 10^{17} and 10^{15} oxygen respectively, from Brede and Haasen (1988). The filled circles are the data on hydrogen charged silicon from the work of Zhang and Haasen (1989). For comparison, the dashed line is for the CZ silicon containing 3.5×10^{17} oxygen from the work of Michot and George.

transition is also greater for the Czochralski material, being about 2.4 eV as compared with about 2.0 eV for the float-zone silicon. In the silicon studied by Michot and co-workers, the oxygen is believed to be either in solution or in the form of extremely small silicon dioxide precipitates (<1 nm) and to decrease dislocation mobilities by solid-solution hardening, an elastic effect. In a more recent study, Behrensmeier et al. (1987) heat-treated oxygen-containing ($\sim 7 \times 10^{17}$ cm^{-3}), P-doped Czochralski-grown silicon in order to precipitate the oxygen deliberately in the form of silicon dioxide. Annealing at 750°C for 48 hours causes the silicon dioxide to form as platelets on {100}. Additional annealing, at 950°C for 48 hours not only coarsens the platelets, but also results in the formation of extrinsic {111} stacking faults being nucleated around the platelets. They found that the material having the largest precipitates had the lowest BDT temperature. The temperature was about 20° lower than that of the precipitated material at 750°C, and approximately 140° lower than that measured at the same strain rate by Michot et al. for float-zone silicon. All the measurements were made at a crack opening rate of 50 μm/min. The cause of the lower transition temperature in the precipitated material is not clear. Precipitates can be expected both to cause a decrease in the dislocation velocity and to serve as sources of dislocations in the stress field of the propagating crack. Behrensmeier et al. favor the argument that the decrease in transition is the result of there being a higher density of dislocation sources from the stacking faults around the precipitates.

The studies of the electronic doping effects on the BDT have sought to exploit the Patel effect, namely the enhanced dislocation velocities measured in n-type doped relative to those in p-type or intrinsic silicon (Patel et al., 1976; Hirsch, 1985). The results obtained to date are summarized in Fig. 17, expressed in terms of the applied rate of loading. The original data of St John, labelled S, is for lightly boron-doped material, but unfortunately the oxygen content was not reported. Similar activation energies were reported by the Oxford group (Samuels and Roberts, 1989) who employed a rather different fracture geometry to investigate the effect of n-type doping. Their data for intrinsic and n-type doped (2×10^{18} P cm^{-3}) are reproduced as the curves O_i and O_n, respectively. The BDT of the doped material was about 30°C lower than that of intrinsic silicon at the same strain rate with a correspondingly smaller activation energy of 1.6–1.7 eV. The most detailed study of doping has been carried out by Brede and Haasen (1988) using the same sample geometry as introduced by St John. Their measurements on four different doping levels, two of which are reproduced in Fig. 17, reveal a more complex situation in that the Arrhenius behavior changes at certain loading rates. At relatively low strain rates, comparable to those used by Samuels and Roberts (1989), the activation energy for the heavily phosphorus doped ($\sim 1.6 \times$

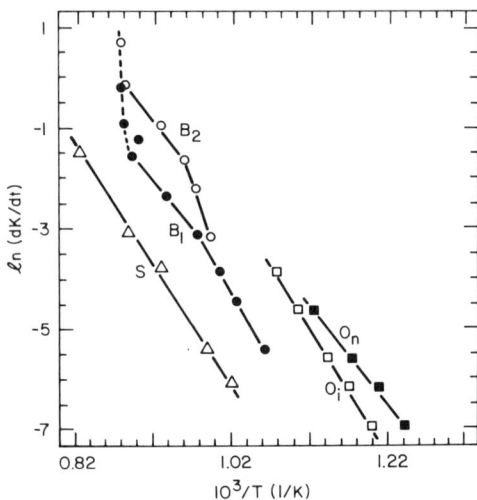

FIG. 17. Arrhenius plot of the displacement rate data for the BDT in silicon from Brede and Haasen (B), Samuels et al. (O), and St John (S) for silicon. Redrawn after Haasen et al. (1989).

10^{18} cm^{-3}) float-zone silicon, labelled B_1, was 2.3 ± 0.1 eV. At higher strain rates, where the BDT temperature was consequentially higher, the activation energy dropped to 1.6 ± 0.1 eV. The latter corresponds to the activation energy for dislocation velocity in such n-type material. In the higher-oxygen-content, Czochralski-grown silicon, doped with arsenic, curve B_2, the activation energy similarly had different values at different strain rates. At the higher temperatures, the value was expected for dislocation glide (1.7 ± 0.2 eV), but increasing to 3.8 ± 0.5 eV at the lower loading rates (i.e., higher temperatures). The explanation proposed by Brede and Haasen is that at the low strain rates, and corresponding low temperatures, the dislocation velocity is not simply proportional to the stress acting on them, as would be the case if there were simply an electronic interaction. Rather, the net stress acting on the dislocations is the difference between the local stress and that required to overcome any elastic interactions with impurity atoms, such as carbon and oxygen. Indeed, such a low-stress dependence on the activation energy for dislocation velocity has been noted by Imai and Sumino (1982) and by George and Champier (1979). However, whether this type of explanation is correct awaits a detailed dynamic crack-growth model that incorporates measured dislocation velocities and matches the results reported in Fig. 17. No explanation for the high strain rate data, shown dotted in Fig. 17, has been forthcoming.

An important point is that in none of the experiments performed to date has the degree of doping been found to affect the room-temperature fracture resistance. It can thus be concluded that the observed changes in the BDT temperature are the result of altering the dislocation mobility rather than the intrinsic fracture resistance. No experiments have, however, been performed on heavily *p*-type doped silicon.

18. X-Ray Topographic Studies

A number of workers have used x-ray topography techniques to observe the creation of dislocations at high temperatures in silicon. For instance, Tsunekawa and Weissman revealed the presence of a plastic zone around a notch in silicon deformed at high temperatures. Similarly, Nishino and Imura (1982) detected plastic zones forming at the edge of a silicon crystal stressed at 700°C. However, the first observations associated with the BDT were performed by St John (1975). Subsequently, by far the most detailed and comprehensive studies have been carried out by the Nancy group by Michot and his colleagues (Michot, 1988; Michot et al., 1980; Michot and George, 1976) using both laboratory x-ray sources and the the synchroton facilities in France. By recording x-ray topographs of their test samples during loading at temperature, both Michot *et al.* and St John showed that dislocations were generated in the vicinity of the crack tip at the BDT. Furthermore, at temperatures just above the transition, the dislocation density grew rapidly forming a distinct plastic zone around the crack tip. An example of x-ray topographs is shown in Fig. 18 for dislocation emission from a $\{110\}$ cleavage crack. From the characteristic contrast of the dislocations as a function of the diffraction conditions used by Michot *et al.*, they were able to deduce that the dislocations were glide dislocations with the usual Burgers vectors for dislocations in silicon of **a**/2 [110]. The studies of Michot *et al.* included measurements of crack extension, dislocation Burgers vector characterization, and size of the plastic zone around the crack tip as a function of temperature. From the size of the plastic zone and the dislocation velocities measured in simpler stressing configurations, Michot *et al.* were able to compute the crack driving forces for dislocation emission.

Although the majority of work has been with sharp cracks, both Tsunekawa and Weissmann (1974) and Nishino and Imura (1982) have also shown that dislocations are generated from notches when stressed above the BDT temperature. This is of importance because the stress distribution around notches is similar to that encountered around steps, suggesting that such sites would act as sources of dislocations when appropriately stressed, for instance, by thermal expansion mismatch stresses. Unfortunately, the manner by which the notches were introduced was not described by Nishino and Imura,

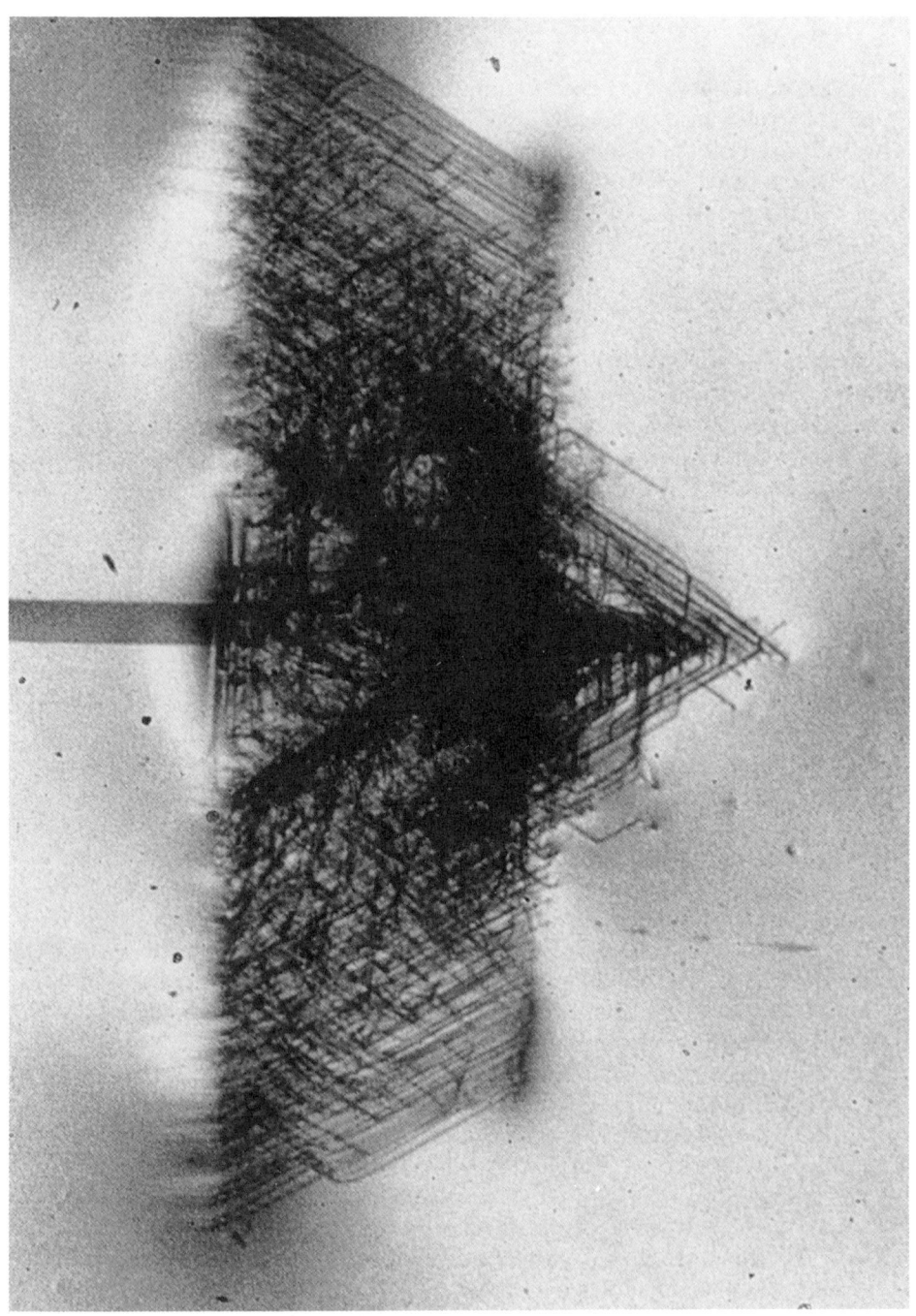

FIG. 18. X-ray topograph of the tip of a crack lying on the (110) plane of silicon recorded at the BDTT. The topograph was recorded using a $[11\bar{1}]$ diffracting vector. In this orientation the emitted dislocations, lying on all four of the {111} glide planes, are symmetrically arranged in a fan about the crack plane. Micrograph provided by G. Michot.

so it may well be that their formation was accompanied by the formation of small micro-cracks.

19. IN-SITU TEM STUDIES

The relatively poor spatial resolution ($\sim 1\ \mu m$) of the various x-ray topographic methods left unanswered a number of questions concerning the actual processes by which dislocations are generated at the BDT. In response to this limitation, higher-resolution *in-situ* studies have recently been reported as part of a transmission electron microscopy investigation of the BDT (Chiao and Clarke, 1989). In these experiments, short cleavage cracks (introduced by indentation) in thin foils of *n*-type silicon were stressed under biaxial tension in the microscope at elevated temperature and observations of crack growth and any dislocation activity were made in real time. Simultaneously recorded videotape provided a permanent record of the experiment, enabling subsequent analysis of the dislocation configurations and velocities.

After thinning but prior to deformation, transmission microscopy showed that the indentations were surrounded by punched-out dislocation rosettes, but that these did not extend as far as the crack tips. When the samples, and hence the cracks, were stressed at low temperatures, abrupt extension of the crack would occur without any evidence of dislocation emission at or near the crack tip. In these cases the cracks grew and the samples broke.

At temperatures associated with the brittle-to-ductile transition perfect dislocation loops having $\mathbf{b} = 1/2\langle 110 \rangle$ were seen to be emitted from the crack tip. The dislocations glided away from the crack-tip on the close-packed (111) planes having the highest resolved shear stress. In no instance were dislocations seen to migrate to the tip from sources behind the crack tip. Also, no dislocations were seen to be nucleated away from the crack tip or from regions in the immediate vicinity (~ 10 nm) of the tip (such as by the activation of some preexisting source in the crack-tip stress field). In the case of the (111) cracks loaded under near mode-I loading, the dislocations loops nucleated and rapidly grew to a distance from the crack and then abruptly stopped (Fig. 7).

Offsets in the moiré fringe pattern formed from the overlapping crack faces revealed the presence of 1/3 [111] high ledges, presumable left as blunting ledges after emission of those dislocation loops (Fig. 8).

In the sample in which the cracks were under a mode-III loading, the emitted dislocations adopted a coplanar arrangement ahead of the crack with an inverse spacing with distance. Such a configuration is the same as predicted from standard linear elastic models for dislocation interaction with a crack tip in mode III (Fig. 9) and used in shielding calculations.

The *in-situ* electron microscopy observations of the dislocation velocities and the positions of the dislocations have also enabled a calculation of the dislocation nucleation condition. From these experimental observations, the crack-tip stress intensity at which dislocations are emitted under mode-I and mode-III loadings was estimated to be $K_I = 0.94 \pm 0.06$ MPa m$^{1/2}$ and $K_{III} = 0.17 \pm 0.03$ MPa m$^{1/2}$. The former value is the same as that reported by a number of workers for the room-temperature fracture resistance of silicon. Unfortunately, no comparison can be made with the latter value since the brittle-to-ductile transition has not been investigated for mode-III cracks. It is noticeable that it is somewhat smaller than the values deduced for the emission of the first dislocations from the double-cantilever beam experiments of Brede and Haasen, who quote 0.24 and 0.36 MPa m$^{1/2}$ for the emission condition for float-zone and Czochralski silicon, respectively. These values were obtained under nominally mode-I loadings by loading above the BDTT, cooling to room temperature, and then measuring the fracture resistance. It is conceivable that under these conditions the first dislocations are emitted under mode III.

The estimates for the stress intensity factors for dislocation emission also enabled a calculation to be made of the energy barrier to nucleation, following the energy analysis introduced by Rice and Thomson (1974). The details are beyond the scope of this contribution, but the calculation leads to the conclusion that for the emission observed in the electron microscope (Fig. 7), the nucleation barrier was less than $k_B T$. Thus, for the emission events observed in the transmission electron microscope at the BDT dislocation nucleation is essentially spontaneous. This is consistent with the conclusion that dislocation glide is the rate-controlling process in the BDT, rather than the nucleation process.

Although *in-situ* TEM studies of the BDT have not been performed on other semiconductors, a number of observations of crack tips in material after deformation and subsequent thinning for TEM have been reported. These observations serve to substantiate the fact that dislocations are emitted from crack tips when the semiconductor is deformed above the BDT. An example is shown in Fig. 19, reproduced from the studies by Yasutake *et al.* (1988) of the fracture of GaAs wafers. Both images are of crack tips in indentation-precracked samples of undoped GaAs deformed at 400°C. In Fig. 19a, the loading configuration was an α-bending orientation, and in Fig. 19b, a β-bending orientation. The dislocations emitted lie on the (111) plane perpendicular to the crack surface and have a Burgers vector that would cause blunting of the tip. The observations of Figs. 19a and b are consistent with the difference in fracture resistance of GaAs under α and β bending at high temperature shown in Fig. 14. α dislocations are known to have a higher mobility than β dislocations and so will glide further from the crack tip, thereby providing a smaller shielding of the crack tip from the elastic interaction between the dislocations and the tip.

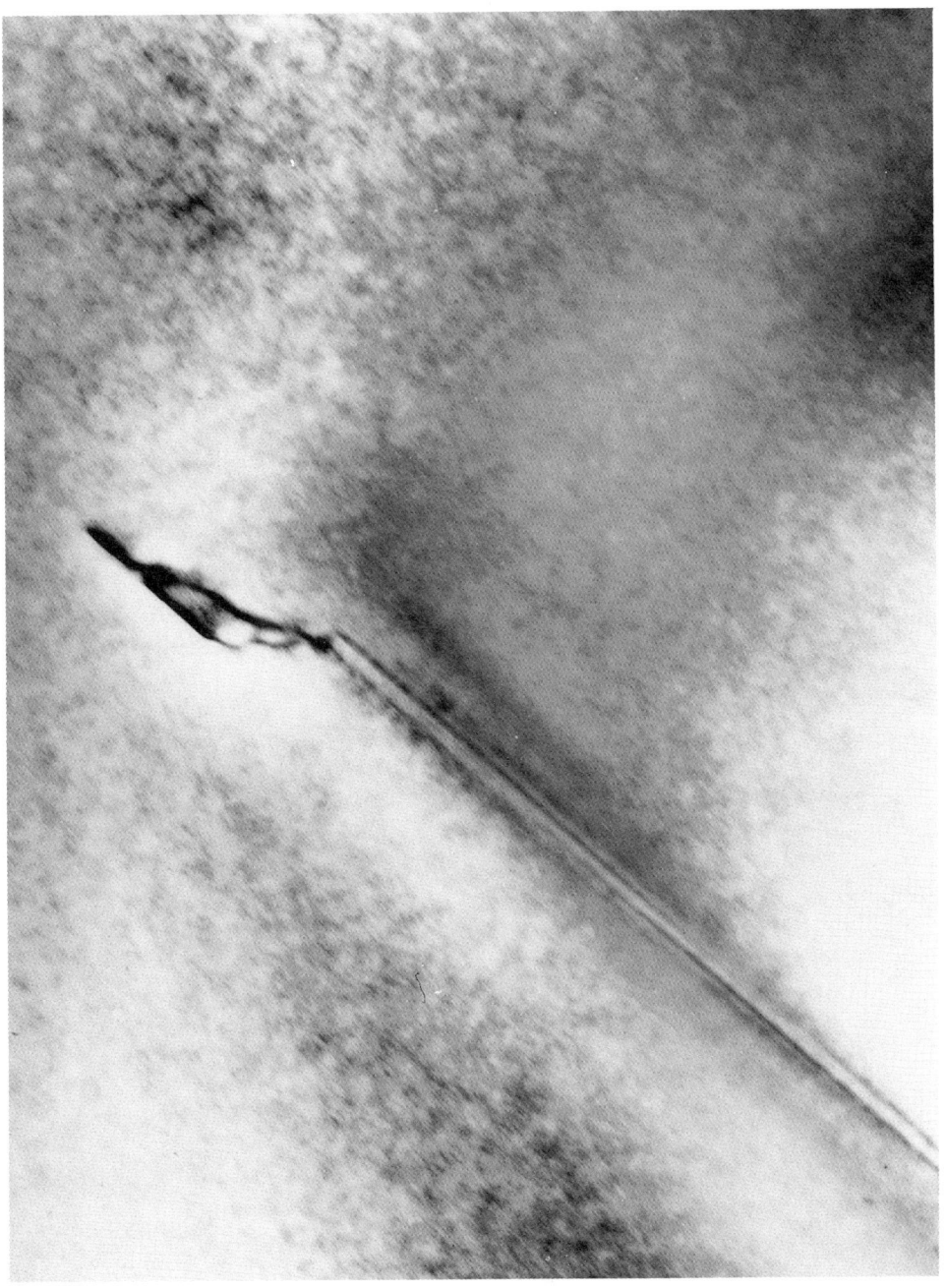

FIG. 19. Transmission electron micrographs of the dislocations emitted from crack-tips in loaded, pre-cracked undoped samples of GaAs stressed at 400°C. (a) α bending orientation and (b) β bending orientation. Photographs reproduced courtesy of Yasutake *et al.* (1988).

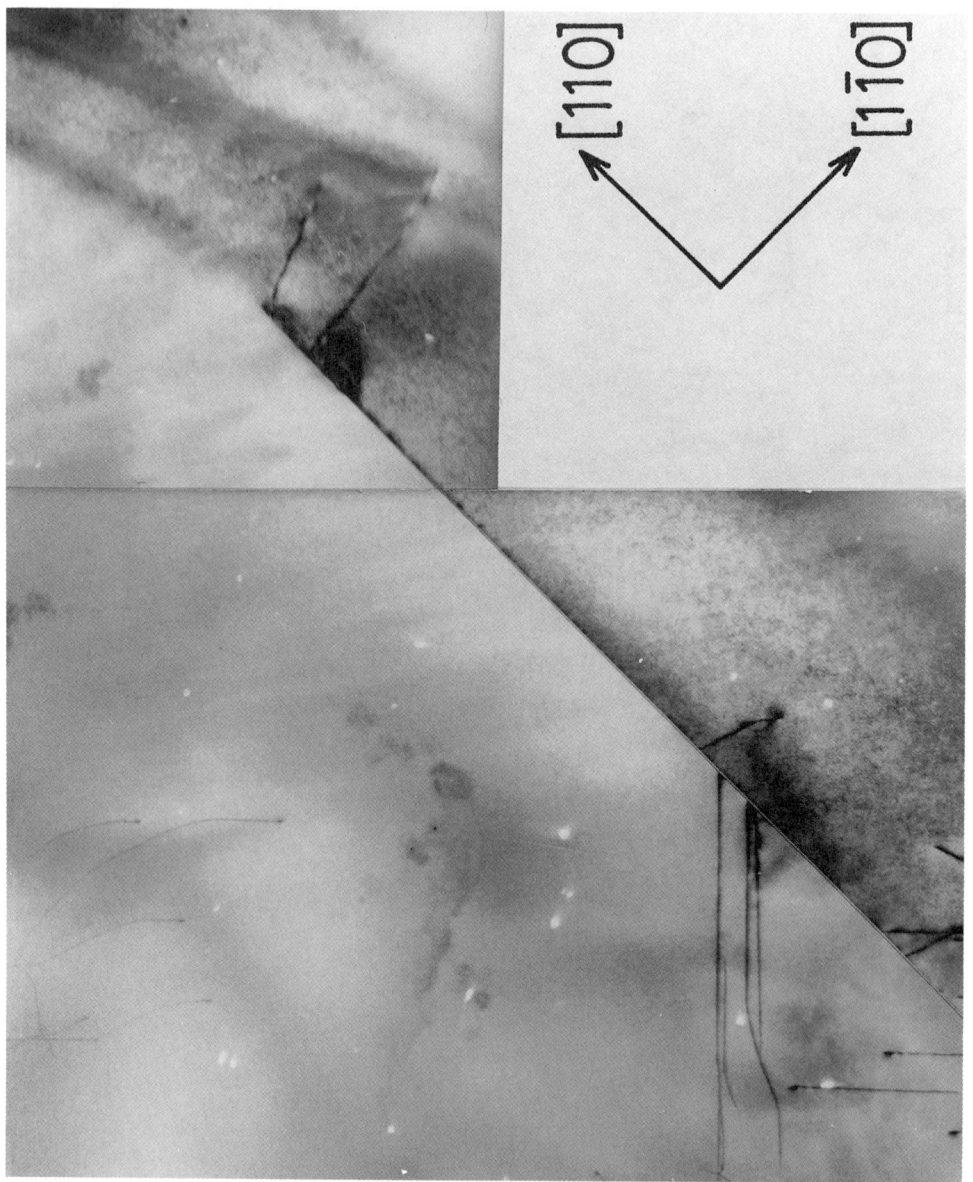

FIG. 19(b).

20. MODELLING THE BRITTLE-TO-DUCTILE TRANSITION

No adequate model for describing the doping and strain rate sensitivity of the brittle-to-ductile transition yet exists. For this reason space is not given here to describing the details of the various models proposed. Those that have been presented are almost all based on a consideration of the elastic interaction between individual dislocations and a crack tip. As outlined in Section III.9, a rigorous linear elastic solution exists for the interaction of a screw dislocation and a coplanar crack loaded under a pure shear, i.e., a mode III crack. However, rigorous solutions do not exist for an edge dislocation and a mode I crack, the case most directly comparable to the cracks studied experimentally. Whilst there have been significant advances (Anderson and Rice, 1987) in the formidable mathematical problem of describing the interaction between a general dislocation and a crack tip at an arbitrary angle, complete solutions are not yet available. In addition, although the interaction between a single dislocation and a crack, such as that illustrated in Fig. 20, may be tractable, the interaction between many dislocations may not be. As the x-ray topographs of Fig. 18 make plain, it is necessary, even at the transition, to consider the interaction of hundreds of dislocations on different slip planes. Furthermore, the nonlinear dependence of the dislocation velocity on stress, especially in the presence of a spatial distribution of dopants that interact elastically with the dislocations, complicates the dislocation dynamics and hence the calculation of the shielding contribution. An equally important shortcoming of the majority of the models is that the crack is held stationary, rather than being allowed to move in response to the effective driving force it experiences from the applied stress and the shielding.

Despite these difficulties, two models have been developed in some detail and contain much of the essential physics, although they use rather different criteria for the transition. The underlying assumption of each is that the transition arises as a result of the competition between cleavage and dislocation glide from the crack tip, shielding the tip and keeping the crack driving force below that for cleavage propagation. They thus incorporate not only the elastic interaction between a crack and a dislocation, but also the dislocation velocity dependence on the local stress. In a series of papers, Haasen and colleagues (Brede and Haasen, 1988; Haasen *et al.*, 1989; Behrensmeier *et al.*, 1987) argue that the controlling parameter is the velocity of the plastic zone developed around the crack tip, and that the distribution of dislocations in the plastic zone is in static equilibrium at all times. In essence, the transition occurs when the velocity of the leading dislocation of the plastic zone moves faster than the crack tip. In contrast, the assumption of the model developed by Hirsch and colleagues (1989) is that the dislocation array evolves continuously, with the transition temperature being

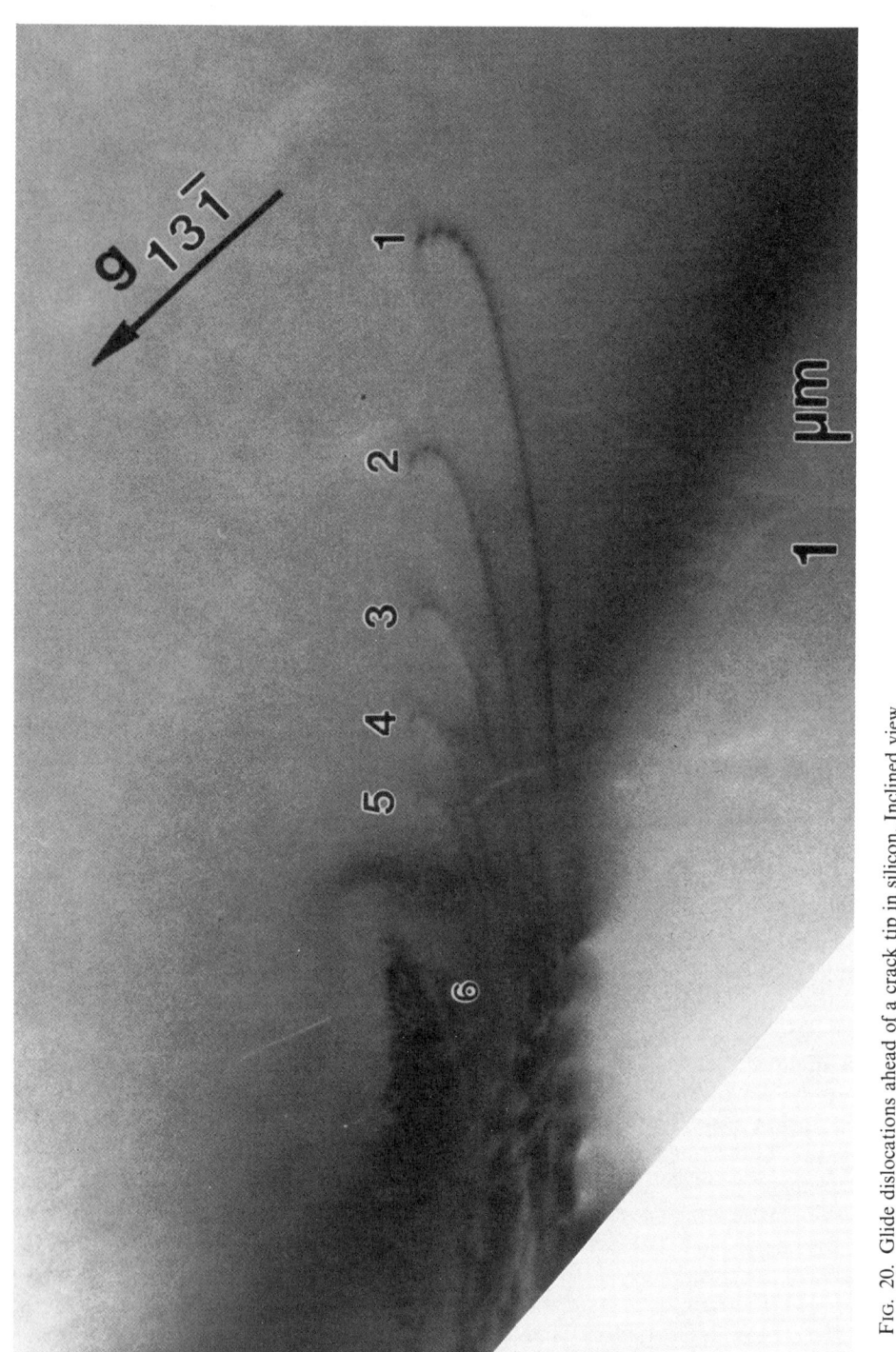

FIG. 20. Glide dislocations ahead of a crack tip in silicon. Inclined view.

primarily controlled by the time it takes for an existing dislocation to reach the crack tip to form dislocation sources. However, to fit the model predictions to their data, they have to make a somewhat arbitrary geometrical assumption, not supported by observation, of the nucleation of the dislocations and how they shield the crack tip. They are, however, able to fit some of the Brede and Haasen data (B_1 and B_2 in Fig. 17 at high strain rates), provided they assume a critical distance for the spacing of nucleation sites along the crack front.

The dislocation analyses of the brittle-to-ductile transition proposed by Brede and Haasen, Hirsch, Roberts and Samuels, Michot and George, and recently Maeda and Fujita (1989) have all been formulated so as to calculate a stress intensity for the crack, and then relate that to a critical stress intensity factor. The implicit assumption of the works being that when the critical stress intensity factor is attained, fracture occurs. However, exceeding the critical stress intensity factor represents only a necessary nonequilibrium condition for fracture as described in Section II.2.

The importance of including the dynamics of the crack has been pointed out by Clarke and Cook (1989). The emission of dislocations from the propagating crack tip affects the crack driving force and hence the crack velocity. It can even cause the crack to balk and thereby propagate in a discontinuous manner. Furthermore, dislocation nucleation at the crack tip and subsequent emission is a discrete process and can lead to abrupt changes in the dislocation velocities, and by extension, the extent of shielding. A general formulation of the brittle-to-ductile transition, appropriate to any cubic crystal, has been presented by Argon (1987). It attempts, for the first time, to incorporate both the dislocation mechanics and the mechanics of crack extension in describing the transition. As yet it has not been specifically applied to the BDT in covalent solids. From this work comes a way of portraying (Fig. 21) the fracture behavior of an intrinsically brittle solid that incorporates a number of the features of the brittle-ductile behavior. Two regimes are identified, one where cleavage dominates and another where ductile behavior dominates. The temperature is normalized by the activation energy, ΔG^*, of the particular controlling mechanism responsible for ductile fracture. As we have seen, this corresponds to the activation energy for dislocation mobility, at least for the semiconductors investigated to date. The horizontal axis is the crack velocity normalized to the phonon velocity. Under steady-state conditions, the velocity is proportional to the crack driving force, so the line separating the two regimes indicates how the BDT temperature varies with the rate of loading and the crack-driving force. Thus, as an example of the use of the figure, consider a crack propagating at a velocity represented by the vertical line and at a temperature given by the point. Under such conditions the crack will propagate in a brittle manner and cleavage fracture will occur. However, if the crack velocity is decreased or the temperature is raised until the boundary line is reached, the crack will arrest.

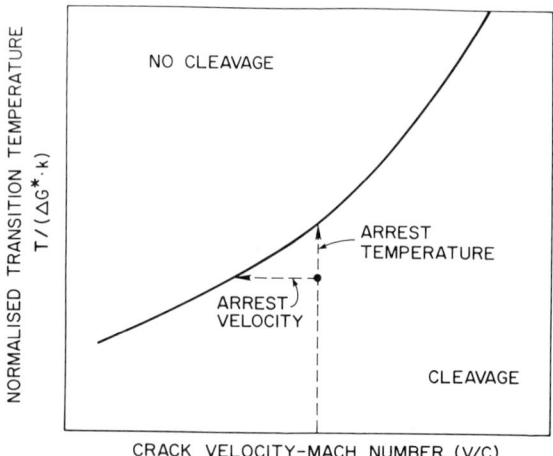

FIG. 21. Schematic for representing the transition from brittle, cleavage fracture behavior to ductile fracture as a function of normalized crack velocity and normalized temperature. Redrawn after Argon (1987).

21. Summary

It is probably fair to say that all the recent work on the brittle-to-ductile transition in silicon has served to substantiate the main conclusions reached by St John and to provide a clarification of the underlying mechanisms. There is now very little doubt that the transition is mediated by dislocations emitted from the tip of a brittle crack, but that the nucleation event is not rate-controlling. The similarity between the apparent activation energies for the strain rate dependence of the transition and those for dislocation glide suggest that the transition is controlled directly, or indirectly, by dislocation mobility away from the crack tip. There is also agreement that the crack-tip stress intensity factors can be calculated in terms of standard dislocation models. There also remains an uncertainty about the detailed mechanism of dislocation nucleation in the stress field at the tip of the crack. However, the central issue remains how to relate the observed dislocation emission process to the macroscopic brittle-to-ductile transition. This is of importance not only to an understanding of the brittle-to-ductile transition in semiconductors, but also as a test of our general understanding of toughening due to crack-tip process zones. Toughening by the emission of dislocations represents possibly the simplest example of a toughening phenomenon that is amenable to a complete analysis such that an entire crack resistance curve (R-curve) can be computed.

An important advance that has been made in the last few years, and one consistent with observations, is that unstable crack propagation (brittle fracture) can occur even if dislocation emission occurs during crack propagation, provided that the overall energy can be reduced. This is in accord with the observation of dislocation etch pits on the fracture surfaces immediately before brittle fracture by Brede and Haasen (1988) and the etch pit observations by Samuels and Roberts (1989) of a "transition" specimen (their Fig. 16). It is also consistent with our *in-situ* TEM observations.

A general point, pertinent for future experimentation, is that the loading configuration used can be expected to influence the possibility of observing dislocations emitted prior to unstable crack growth. For instance, it is notable that Michot and Brede and Haasen use a double cantilever beam sample so that while the loading rate is controlled the crack extends under a constant applied stress intensity. They report that dislocations are emitted at temperatures significantly below the brittle-to-ductile transition. Chiao and Clarke observed dislocation emission in samples that were loaded under essentially displacement control. In contrast, the samples of Samuels and Roberts were loaded at constant strain rates but in a nonstabilizing stress field so that emitted dislocations may not be seen being overtaken by the propagating crack.

VI. Environmental effects

One of the most striking aspects of the fracture of semiconductors studied to date is the *insensitivity* of the fracture resistance to the environment in which the fracture occurs. This is in marked contrast to the fracture of silica-based glasses and most crystalline ceramics, which do show subcritical crack growth, i.e., crack extension at low velocities before unstable or catastrophic fracture occurs (Thomson, 1986; Wiederhorn, 1968). The effect of water on crack growth in glasses has been extensively investigated, and from these studies a rather comprehensive picture has emerged that is believed to be applicable to any material. Even in the presence of the environment, there exists a threshold mechanical energy release rate, G_{th}, below which crack extension will not occur. This value is given by Eq. (20) with the appropriate surface energy for fracture in that environment. Above the threshold, the crack will propagate at a velocity dependent on the crack driving force and given by Eq. (23). In plotting crack velocity data in this regime, the data are often found to fit a power-law dependence, an approximation to the hyperbolic sine function of Eq. (23). At high driving forces, dependent on the particular combination of material and environment, the crack velocity is limited by the transport of the environment to the crack tip, and over a limited range of mechanical energy release rates the velocity is constant. At still higher driving forces, well away from equilibrium, no steady-state crack extension occurs, and the crack accelerates. This mechanical release rate corresponds to that which would be measured in vacuum. At the present time there have been, to the author's knowledge, no substantiated reports of subcritical crack growth at room temperature in semiconductors. There is, however, no reason to believe that in appropriate environments, ones that markedly lower the surface energy, semiconductors will also not exhibit environmentally dependent crack growth.

One such environment that has been posited to promote subcritical crack growth in silicon is hydrogen. There are a number of cryptic remarks in the literature to the effect that *p*-type silicon wafers are more brittle after exposure to hydrogen, but the effect has been neither substantiated nor studied in any detail. More recently, as part of a computational study into the behavior of hydrogen in silicon under tension and compression, it was found that the stable site for hydrogen depends critically on both the pressure and the hydrogen charge state (Nichols et al., 1989). Under ambient conditions, hydrogen is predicted to occur as H^+ in the bond center (BC) position in *p*-type silicon and as H^- in the tetrahedral site in *n*-type silicon. As the silicon lattice is expanded under tension, no change in *n*-type material is predicted. However, in *p*-type material it is predicted that H^+ would preferably adopt a BC position, i.e., bridging between two adjacent silicon atoms. It was therefore conjectured that *p*-type and *n*-type silicon should exhibit a difference in fracture resistance. In addition, the presence of hydrogen causes a softening of the lattice (i.e., a decrease in bulk modulus). At this time few experimental data appear to exist. An investigation by Zhang and Haasen (1989) of the fracture resistance of *n*-type silicon in the presence of silicon has shown no difference from that in the absence of hydrogen. The fracture resistance of *p*-type silicon in hydrogen has yet to be measured.

Hydrogen does, however, appear to have some effect on the BDT in *n*-type silicon (CZ grown with a doping of 1.2×10^{18} P cm^{-3} and an oxygen content of about 10^{17} cm^{-3}). Using the DCB sample geometry and testing at a constant strain rate, Zhang and Haasen (1989) report that ionized hydrogen lowered the BDTT by about 60 K. At the same time the apparent activation energy was lowered by 0.8 eV from 2.4 ± 0.6 eV in the uncharged material. Consistent with this decrease, they observed that hydrogen caused a marked decrease in flow stress (a softening) when the hydrogen plasma was introduced during deformation, indicating that the dislocation mobility was affected by the presence of hydrogen. No effect of the hydrogen was noted, within experimental uncertainty, on the fracture surface energy. The data are shown in Fig. 17 by the filled circles. The explanation proposed for these observations is the same as for earlier doping studies, namely that hydrogen increases the mobility of dislocations, thereby shifting the BDT but not affecting the intrinsic fracture resistance. The mechanism by which hydrogen increases dislocation mobility remains to be established, primarily because of a lack of knowledge of the binding of hydrogen to the core of dislocations. One possibility is that hydrogen acts as an electronic dopant, decreasing the formation energy of kinks along the dislocation line and hence lowering the activation energy for dislocation glide. A second possibility is that there exists an elastic interaction between hydrogen and a dislocation, with hydrogen decreasing the elastic self-energy of the dislocation. Yet a third possibility is that hydrogen reduces the oxygen content of the silicon, and thereby has an indirect effect of increasing the dislocation mobility.

VII. Closing Remarks

At low temperatures, below their individual BDT, all the semiconductors studied to date fracture in a brittle manner by cleavage. The cleavage planes, in bulk material at least, correspond to the lowest packed planes and are $\{111\}$ in silicon and germanium, and $\{110\}$ in the III-V compound semiconductors. The fracture resistance is independent of the dopant level for the semiconductors studied and is expected to be so for all others, unless the surface energy is radically altered by, for instance, extreme levels of doping. Above the BDT silicon will not fracture, but rather will deform plastically by dislocation glide (and climb). It is expected that the other semiconductors will behave in a similar manner, although the transition temperatures, the doping dependencies, the Burgers vectors of the dislocations involved, and the dislocation mechanisms will be different.

Although there is generally little possibility of altering doping of semiconductors to affect their mechanical properties, the effects of doping can nevertheless be important. For instance, doping GaAs with indium is known to decrease dislocation mobility and has been used to lower the dislocation density created on cooling. Such indium doping is now known to have no effect on the room-temperature fracture resistance, as shown in Fig. 22 (Michot *et al.*, 1988). However, the decrease in dislocation mobility can be expected to shift the BDT to higher temperatures than in the undoped material.

An aspect of the fracture of semiconductors that remains relatively unexplored is the effect of environments on crack growth rates. So far, fracture has been studied in rather benign environments, such as water, where protective oxides can be expected to form. Crack studies in more aggressive

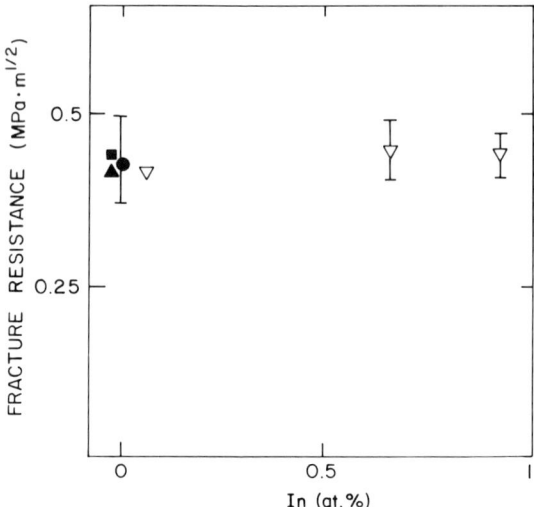

FIG. 22. Fracture resistance of GaAs as a function of indium doping. The symbols ▽, ▲, ■, ● refer to tests performed in air, helium, forming gas, and argon respectively. Redrawn with permission from *Scripta Met.* **22**, 1043–1048, "Fracture Toughness of Pure and In-doped GaAs." © 1988, Pergamon Press plc.

environments, such as those increasingly being used in semiconductor processing, need to be carried out.

Acknowledgments

This work was supported in part by the U.S. Office of Naval Research under contract number N00014-88-C-0176.

References

Ahlquist, C. N, and Carlsson, L. (1973), "The Influence of Light on Crack Propagation in Cadmium Telluride," *Philosophical Magazine* **30**, 733–738.

Allen, J. W. (1957). "On the Mechanical Properties of Indium Antimonide," *Philosophical Magazine* **2**, 1475–1481.

American Society for Testing Materials (1965), "Fracture Toughness Testing and Its Application," publication STP 381.

Anderson, P. M., and Rice, J. R. (1986). "Dislocation Emission From Cracks in Crystals or Along Crystal Interfaces," *Scripta Metallurgica* **20**, 1467–1472.

Anderson, P. M., and Rice, J. R. (1987), "The Stress Field and Energy of a Three Dimensional Dislocation Loop at a Crack Tip," *International Journal of Mechanics and Physics of Solids* **35**(6), 743–769.

Argon, A. S. (1987). "Brittle to Ductile Transition in Cleavage Fracture," *Acta Metallurgica* **35**(1), 185–196.

Barenblatt, G. I. (1962). "Mathematical Theory of Equilibrium Cracks in Brittle Solids," *Adv. Appl. Mechanics* **7**, 55.

Behrensmeier, R., Brede, M., and Haasen, P. (1987). "The Influence of Precipitated Oxygen on the Brittle–Ductile Transition of Silicon," *Scripta Metallurgica* **21**, 1581–1585.

Bilby, B., and Eshelby, J. D. (1968). *In* "Fracture" (Liebowitz, ed.), Vol. 1, p. 99. Academic Press, New York.

Bleloch, A. L., Howie, A., Milne, R. H., and Walls, M. G. (1989). "Elastic and Inelastic Scattering Effects in Reflection Electron Microscopy," *Ultramicroscopy* **29**, 175–182.

Brede, M., and Haasen, P. (1988). "The Brittle-to-Ductile Transition in Doped Silicon as a Model Substance," *Acta Metallurgica* **36**, (8), 2003–2018.

Broek, D. (1982). "Elementary Engineering Fracture Mechanics." Martinus Nijhoff, Boston.

Brown, G., Cockayne, B., and Macewan, W. (1980). "Deformation Behaviour of Single Crystals of InP in Uniaxial Compression," *Journal of Materials Science* **15**, 1469–1477.

Campbell, G. H., Dalgleish, B. J., and Evans, A. G. (1989). "Brittle-to-Ductile Transition in Silicon Carbide," *Journal of the American Ceramic Society* **72**, 1402–1408.

Chadi, D. J. (1983). "Semiconductor Surface Reconstructions," *Vacuum* **33**, 613.

Chan, D. Y. C., Hughes, B. D., and White, L. R. (1987). *J. Coll. Interface Science* **115**(1), 240–259.

Chiao, Y. H., and Clarke, D. R. (1989). "Direct Observation of Dislocation Emission from Crack Tips in Silicon at High Temperatures," *Acta Metallurgica* **37**(1), 203–219.

Chudnovsky, A., and Moet, A. (1985). "Thermodynamics of Translational Crack Layer Propagation," *J. Materials Science* **20**, 630–635.

Clarke, D. R. (1987). "On the Equilibrium Thickness of Intergranular Glass Phases in Ceramic Materials," *J. Am. Ceram. Soc.* **70**, 15.

Clarke, D. R., and Cook, R. F. (1989). "Dislocation Emission from Crack Tips and the Brittle-to-Ductile Transition," *in* "Structure And Properties of Dislocations In Semiconductors" (S. G. Roberts, D. B. Holt, and P. R. Wilshire, eds.). Institute of Physics.

Clarke, D. R., and Faber, K. T. (1987). "Fracture of Ceramics and Glasses," *Journal of Physics and Chemistry of Solids* **48**, 1115–1151.

Clarke, D. R., Lawn, B. R., and Roach, D. H. (1986). "The Role of Surface Forces in Fracture," *Fracture Mechanics of Ceramics* **8**, 341–350.

Cook, R. F. (1986). "Crack Propagation Thresholds: A Measure of Surface Energy,"*Journal of Materials Research* **1**, 852.

Cook, R. F. (1989). "Influence of Crack Velocity Thresholds on Stabilized Non-Equilibrium Fracture," *Journal of Applied Physics* **65**(5), 1902–1910.

Cotton, F. A., and Wilkinson, G. (1988). Advanced Inorganic Chemistry. Wiley Interscience, New York.

de Gennes, P. G. (1985). "Wetting: Statics and Dynamics," *Reviews of Modern Physics* **57**, 827.

Elliott, H. (1947). "An Analysis of the Conditions For Rupture Due to Griffith Cracks," *Proc. Phys. Soc. London* **59**, 208.

Evans, A. G. (1974). "Fracture Mechanics Determinations," *International Journal of Fracture* **1**, 17–48.

Fujita, S., Maeda, K., and Hyodo, S. (1988). "Radiation Enhanced Dislocation Glide Effect on Indentation-Induced Fracture of GaAs Single Crystals," *Phys. Stat. Sol.* **109**, 383–393.

Gehlen, P. C., and Kanninen, M. F. (1970). *In* "Inelastic Behavior of Solids" (Kanninen, Adler, Rosenfield, and Jaffee, eds.), p. 587. McGraw-Hill, New York.

Gehlen, P. C., Hahn, G. T., and Kanninen, M. F. (1972). *Scripta Metall.* **6**, 1087.
George, A., and Champier, G. (1979). "Velocities of Screw and 60° Dislocations in n and p-Type Silicon," *Phys. Stat. Sol.* **53**, 529–539.
Gilman, J. J., and Tong, H. C. (1971). "Quantum Tunneling as an Elementary Fracture Process," *Journal of Applied Physics* **42**, 3479–3486.
Green, D. J., Nicholson, P. S., and Embury, J. D. (1976). "Crack Particle Interactions in Brittle Particulate Composites," *in* "Ceramic Microstructures '76," (J. A. Pask and R. M. Fulrath, eds.).
Griffith, A. A. (1920). "The Phenomena of Rupture and Flow in Solids," *Phil. Trans. Roy. Soc.* **A221**, 163–198.
Haasen, P., Brede, M., and Zhang, T. "The Effect of Oxygen and Hydrogen on the Brittle-to-Ductile Transition of Silicon," *in* "Structure and Properties of Dislocations in Semiconductors" (S. G. Roberts, D. B. Holt, and P. R. Wilshaw, eds.). Institute of Physics.
Haneman, D. (1987) "Surfaces of Silicon," *Reports on Progress in Physics* **50**, 1045.
Haneman, D., and Lagally, M. G. (1989). "Three Bond Scission and the Structure of the Cleaved Si(111) Surface."
Harkins, W. D. (1942). "Energy Relations of the Surfaces of Solids," *Journal of Chemical Physics* **10**, 268.
Hirsch, P. B. (1985). "Dislocations in Semiconductors," *Materials Science and Technology* **1**, 666–677.
Hirsch, P. B., Pirouz, P., Roberts, S. G., and Warren, P. D. (1985). "Indentation Plasticity and Polarity of Hardness on {111} Faces of GaAs," *Philosophical Magazine B* **52**(3), 759–784.
Hirsch, P. B., Roberts, S. C., and Samuels, J. (1989). "The Brittle-to-Ductile Transition in Silicon. II. Interpretation," *Proceedings of the Royal Society* **A421**, 25–53.
Hockey, B. J., and Lawn, B. R. (1975). "Electron Microscopic Observations of Microcracking about Indentations in Aluminum Oxide and Silicon Carbide," *J. Mater. Sci.* **10**, 1275.
Hu, S. M. (1982). "Critical Stress in Silicon Brittle Fracture, and Effect of Ion Implantation and Other Surface Treatments," *Journal of Applied Physics* **53**(5), 3576–3580.
Huang, Y. M., Spence, J. C. H., Sankey, O. F., *et al.* (1991). "The Influence of Internal Surfaces on the (2 × 1) Shuffle and Glide Cleavage Reconstructions for Si (111)," *Surface Science*, **256**, 344.
Imai, M., and Sumino, K. (1982), "*In Situ* X-Ray Topographic Study of the Dislocation Mobility in High-Purity and Impurity-Doped Silicon Crystals," *Philosophical Magazine A* **47**(4), 599–621.
Irwin, G. R. (1958) "Handbuch der Physik," Vol. VI (S. Flugge, ed.). Springer-Verlag, Berlin.
Johansson, S., Schweitz, J.-A., Tenerz, L., and Tiren, J. (1988). "Fracture Testing of Silicon Microelements *In Situ* in A scanning Electron Microscope," *Journal of Applied Physics*.
Kelly, A., and Macmillan, N. H. (1986). "Strong Solids." Clarendon Press, Oxford.
Kelly, A., Tyson, W. R., and Cottrell, A. H. (1967). "Ductile and Brittle Crystals," *Philos. Mag.* **15**, 567–586.
Kerkhof, F. (1970). "Bruchvorgange in Glasern." Verlag der Deutche Glastech. ges., Frankfurt.
Küsters, K. H., and Alexander, H. (1982). "Photoplastic Effect in Silicon," *Physica* **116**, 594–599.
Lawn, B. R. (1975). "An Atomistic Model of Kinetic Crack Growth in Brittle Solids," *Journal of Materials Science* **10**, 469.
Lawn, B. R., and Sinclair, J. E. (1972). "An Atomistic Study of Cracks in Diamond-Structure Crystals," *Proceedings of the Royal Society* **A329**, 83–103.
Lawn, B. R., Wiederhorn, S. M., and Hockey, B. J. (1980). "Atomically Sharp Cracks in Brittle Solids: An Electron Microscopy Study," *Journal of Materials Science* **15**, 1207–1223.
Lawn, B. R., and Wilshire, T. R. (1975). "Fracture of Brittle Solids." Cambridge University Press.
Maeda, K., and Fujita, S. (1989). "Ductile-to-Brittle Transition Caused by Dynamical Work Hardening at a Crack-Tip," *Scripta Metallurgica* **23**, 383–388.

Maeda, K., and Takeuchi, S. (1983). "Recombination Enhanced Mobility of Dislocations in III-V compounds," *Journal de Physique* **C4**, 375–385.

Maeda, K., Ueda, O., Murayama, Y., and Sakamoto, K. (1977). "Mechanical Properties and Photomechanical Effect in GaP Single Crystals," *Journal of Physical Chemical Solids* **38**, 1173–1179.

Maeda, K., Sato, M., Kubo, A., and Takeuchi, S. (1983). "Quantitative Measurements of Recombination Enhanced Dislocation Glide in Gallium Arsenide," *Journal of Applied Physics* **54**(1), 161–168.

Maugis, D. (1985). "Review: Subcritical Crack Growth, Surface Energy, Fracture Toughness, Stick-Slip and Embrittlement," *J. Mater. Sci.* **20**, 3041.

McLaughlin, J. C., and Willoughby, A. F. W. (1987). "Fracture of Silicon Wafers," *Journal of Crystal Growth* **85**, 83–90.

Michot, G. (1988), "Fundamentals of Silicon Fracture," *Crystal Properties & Preparation* **17 & 18**, 55–98.

Michot, G., and George, A. (1986). "Dislocation Emission from Cracks — Observations by X-ray Topography in Silicon," *Scripta Met.* **20**, 1495–1500.

Michot, G., Badawi, K., Abd el Halim, A. R., and George, A. (1980). "Observation par topographie aux rayons X des configuration de dislocations developpées a l'extrémité d'une fissure dans le silicium," *Philosophical Magazine A* **42**(2), 195–215.

Michot, G., George, A., Chabli-Brenac, A., and Molva, E. (1988). "Fracture Toughness of Pure and In-Doped GaAs," *Scripta Met.* **22**, 1043–1048.

Nichols, C. S., Clarke, D. R., and Van de Walle, C. G. (1989). "Properties of Hydrogen in Crystalline Silicon under Compression and Tension," *Physical Review Letters* **63**(10), 1090–1093.

Nishino, Y., and Imura, T. (1982). "Dislocation Generation in the Initiation of Fractures in Silicon Crystals," *Japanese Journal of Applied Physics* **21**(9), 1283–1286.

Orowan, E. (1949). "Fracture and Strength of Solids," *Reports in Progress in Physics* **12**, 185.

Patel, J., Testardi, L., and Freeland, P. (1976). "Electronic Effects on Dislocation Velocities in Heavily Doped Silicon," *Physical Review B* **13**(8), 3548–3557.

Pearson, G. L., Read, W. T., Jr., and Feldmann, W. L. (1957). "Deformation and Fracture of Small Silicon Crystals," *Acta Metallurgica* **5**, 181–191.

Petersen, K. E. (1982), "Silicon as a Mechanical Material," *Proceedings of the IEEE* **70**(5), 420–457.

"Physics and Chemistry of Fracture" (1987), Special issue, *Journal of Physics and Chemistry of Solids* **48**(11), 965–1157.

Rice, J. R. (1968). "A Path Independent Integral and the Approximate Analysis of Strain Concentration by Notches and Cracks," *J. Appl. Mech.* **35**, 379.

Rice, J. R. (1978). "Thermodynamics of the Quasi-Static Growth of Griffith Cracks," *J. Mech. Phys. Solids* **26**, 61–78.

Rice, J. R. (1985). "Three Dimensional Elastic Crack Tip Interactions with Transformation Strains and Dislocations," *International Journal of Solids and Structures* **21**, 781.

Rice, J. R., and Anderson, P. M. (1987). *International Journal of Mechanics and Physics of Solids* **35**, 743.

Rice, J. R., and Thomson, R. M. (1974). "Ductile Versus Brittle Behaviour of Crystals," *Philos. Mag.* **29**, 73–97.

Samuels, J., and Roberts, S. G. (1989). "The Brittle–Ductile Transition in Silicon. I. Experiments," *Proceedings of the Royal Society* **A421**, 1–23.

Sinclair, J. E. (1974). "The Influence of the Interatomic Force Law and of Kinks on the Propagation of Brittle Cracks," *Proceedings of the Royal Society*, 647–671.

St John, C. (1975). "The Brittle-to-Ductile Transition in Pre-Cleaved Silicon Single Crystals," *Philosophical Magazine* **32**, 1193–1212.

Thomson, R. M. (1980), "Theory of Chemically Assisted Fracture," *Journal of Materials Science* **15**, 1014–1026.
Thomson, R. M. (1986). "Physics of Fracture," *in* "Solid State Physics," Vol. 39 (H. Ehrenreich and D. Turnbull, eds.), pp. 1–129. Academic Press, New York.
Tsunekawa, Y., and Weissmann, S. (1974). "Importance of Microplasticity in the Fracture of Silicon," *Metallurgical Transactions* **5**, 1585–1593.
Weiderhorn, S. M. (1968). "Mechanisms of Sub-critical Crack Growth in Glass," *International Journal Fracture Mechanics* **4**, 549–580.
Willis, J. R. (1967). "A Comparison of the Fracture Criterion of Griffith and Barenblatt," *J. Mech. Phys. Solids* **15**, 151–162.
Wolf, D. (1984). "Effect of Interatomic Potential on the Calculated Energy and Structure of High Angle Coincident Site Grain Boundaries," *Acta Metall.* **32**, 245.
Wong, B. and Holbrook, R. J. (1987). "Microindentation for Fracture and Stress Corrosion Cracking Studies in Single Crystal Silicon," *Journal of the Electrochemical Society* **134**, 2254–2256.
Yasutake, K., Iwata, M., Yoshi, K., Umeno, M., and Kawabe, H. (1986). "Crack Healing and Fracture Strength of Silicon Crystals," *Journal of Materials Science* **21**, 2185–2192.
Yasutake, K., Konishi, Y., Adachi, K., Yoshi, K., Umeno, M., and Kawabe, H. (1988). "Fracture of GaAs Wafers," *Japanese Journal of Applied Physics* **27**(12), 2238–2246.
Zhang, T., and Haasen, P. (1989). "The Influence of Ionized Hydrogen on the Brittle-to-Ductile Transition in Silicon," *Philosophical Magazine A* **60**(1), 15–38.

Appendix: Fracture Resistance Measurement by Indentation

The methods prescribed by the various standards organizations for measuring the fracture resistance of a material require large samples, usually having sizes unobtainable in most semiconductors other than silicon and, more recently, gallium arsenide. A similar predicament was faced by those wishing to measure the fracture resistance of many ceramic materials. In response, a method based on the cracking produced by a microhardness indentation has been developed that is suited for small samples of semiconductors. The application of the technique is to load a pyramid-shaped diamond onto the surface of brittle material and measure the length of the cracks (Fig. 1) created as a function of the applied load.

The basis of the technique is as follows. On indenting a polished surface, the diamond leaves a permanent impression whose size is proportional to the applied load. The material displaced in forming the impression is accommodated in the surrounding local region by dislocations creating a volume under residual stress. Some of the dislocations created on indentation are revealed on etching and can be seen in weak contrast in Fig. 1, together with the distance they have propagated, marked R. At sufficiently high loads, the accommodation stresses can be large enough to propagate (semicircular) radial cracks from the indentation. They then extend until their crack driving force falls to the value of the fracture resistance of the material.

2. FRACTURE OF SILICON AND OTHER SEMICONDUCTORS

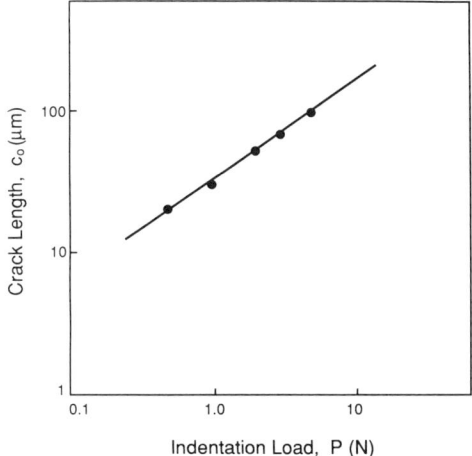

FIG. A1. An example of crack length data from the indentation of GaAs.

The equilibrium crack length, c_0, generated by an indentation load P is related to the fracture resistance of the material by the relationship

$$K_c = 1.42\xi(E/H)^{1/2} P/c_0^{3/2}, \qquad (A1)$$

where $\xi = 0.016$ is a material-independent, geometrical constant, E is Young's modulus, and H is the material hardness. The hardness can also be obtained from the indentation test, being given by the equation

$$H = P/2a^2, \qquad (A2)$$

where a is the hardness impression semidiagonal. In carrying out an experiment, a series of indentations at successively higher loads are made with the diamond pyramid oriented so that the diagonals of the square impression are parallel to the traces of the cleavage planes in the material. The fracture resistance can then be determined from the slope of a logarithmic plot of indentation load P against crack length c_0. An example of crack length data from the indentation of GaAs is shown in Fig. A1.

Whilst the simple indentation test technique described above is particularly suited to very small crystals (1 mm² or smaller), the difficulty in identifying the tip of the crack often makes measuring the true crack length uncertain, introducing considerable uncertainty into the determination of the fracture resistance. A more precise indentation technique, suitable for wafer samples, is to measure the strength of thin disks of material containing deliberately introduced indentation cracks. The basis of the technique is that when a material containing a preexisting crack is stressed, the crack will extend under equilibrium conditions until, with increasing load, a critical stress is reached at which the crack becomes unstable and the material breaks (see Section II.2). The analysis is as follows. When a crack, such as that in Fig. 1, is created on indentation it is held open, after removal of the indenter, by the plastic strain mismatch between the indented volume and the surrounding elastic matrix caused by the dislocations introduced in the indented volume. The mechanics of the crack can then be described, by analogy, as a crack subjected to a center loaded point force. Such a crack has a crack driving force, K_r, of

$$K_r = \chi P/c^{3/2}, \tag{A3}$$

where the subscript, r, is used to denote the residual crack driving force after the removal of the indentation load. The parameter χ is a constant related to materials parameters (compare as in Eq. (A1)). (In the absence of any applied stress, the indentation cracks are stable, i.e., they do not grow, since $dK_r/dc < 0$.) When an external stress is applied on loading the material, the indentation crack is subject to the combined effects of the applied stress and its own residual stress field. Making use of the additive properties of stress intensity factors, the net driving force on the crack may be written as

$$K = K_a + K_r = \chi \sigma_a c^{1/2} + \chi P/c^{3/2}, \tag{A4}$$

where χ is a constant and σ_a is the applied stress. As the load on the material is increased, the crack driving force increases, the crack grows incrementally until the system just becomes unstable ($dK/dc > 0$), and the crack accelerates, causing the material to break. Applying the equilibrium condition $dK/dc = 0$ to Eq. (A4) leads to an expression for the equilibrium failure strength, σ_m, in terms of the equilibrium fracture resistance, K_c, and the indentation load, P, viz.,

$$\sigma_m = A K_c^{4/3} P^{-1/3}, \tag{A5}$$

where A is constant. To ascertain the fracture resistance, a typical experiment would be to form a crack in a wafer using an indentation load, P, and then measure the failure strength on loading, for instance in biaxial bending. A plot of strength against indentation load then gives a straight line and provides a value for the fracture resistance.

CHAPTER 3

The Plasticity of Elemental and Compound Semiconductors

Hans Siethoff

PHYSICAL INSTITUTE
UNIVERSITY OF WÜRZBURG
WÜRZBURG, GERMANY

I.	INTRODUCTION	143
II.	GENERAL	145
	1. *Material Properties Relevant for Plasticity*	145
	2. *The Stress–Strain Curve of Semiconductors*	147
	3. *Strain-Hardening Parameters*	150
	4. *Equivalence of Stress–Strain and Creep Curve*	153
III.	LOWER YIELD POINT AND CREEP AT INFLECTION POINT	154
	5. *Intrinsic Effects*	156
	6. *Surface Effects and Influence of Predeformation*	161
IV.	DYNAMICAL RECOVERY	164
	7. *First Recovery Stage*	164
	8. *Second Recovery Stage*	169
	9. *Steady-State Creep*	171
	10. *Discussion*	173
V.	EFFECT OF HIGH DOPING	175
	11. *Dragging of Solute Atmospheres*	176
	12. *Breaking Away of Solute Atmospheres*	177
	13. *Nonlocal Interaction*	181
IV.	ADDENDUM	182
	ACKNOWLEDGMENTS	184
	REFERENCES	185

I. Introduction

The plasticity of semiconductors has been intensively studied for more than three decades. The growing interest in this field of scientific research has two main causes. One aim of the investigations, which is especially important for technological applications, is to understand the mechanisms of generation and multiplication of dislocations in the processes of crystal growth and device fabrication. Other work is more occupied with the fundamental question of plasticity, how macroscopic deformation behavior can be derived from microscopic dislocation properties. In this domain, the semiconductors offer a unique possibility to study the behavior of individual dislocations by micromechanical methods such as measurements of the dislocation velocity, multiplication and interaction, and by electrical and spectroscopic methods such as Hall effect and electron paramagnetic spin resonance (EPR), which are not available for other materials in such a variety.

Alexander and Haasen (1968) summarized the work performed so far on Si, Ge, and InSb in a first comprehensive review. Their main goal was to show that the macroscopic plastic behavior at the beginning of deformation, i.e., of the yield point in the stress–strain curve and of the inflection point in the sigmoidal creep curve, could be traced back to dynamical properties of individual dislocations. In the following years the activities changed to a certain extent, and problems connected with the structure of the dislocation core, with the dislocation mobility, and with the influence of doping and other point defects were intensively studied. These topics have been reviewed in some recent publications (Hirsch, 1985; Alexander, 1986; George and Rabier, 1987). One of the most remarkable results, in this context, was that dislocations in elemental and compound semiconductors move by the emission or absorption of various point defects. In the same period of time, however, the knowledge of macroscopic plasticity also was enlarged by some new findings: alloy properties of highly doped Si and Ge such as microcreep, Lüders-band spreading, and the Portevin–LeChatelier effect (Siethoff, 1970; Brion et al., 1971); new deformation stages in the stress–strain curves of Si, Ge, and InSb (Brion et al., 1981); and the explanation of the unusual steady-state creep properties of Si and Ge (Siethoff, 1983). It is one aim of the present paper to give a comprehensive review on these phenomena.

In the last few years, the technological importance of electronic materials such as GaAs, InP, and GaP has led to an increasing number of publications on the plasticity of these III–V compounds. Most of the investigations were devoted to the yield point in the stress–strain curve, although, quite recently, higher deformation degrees were also studied (Guruswamy et al., 1987; Siethoff et al., 1988; Rai et al., 1989; Siethoff and Behrensmeier, 1990). Concerning the lower yield point, the experiments indicate that GaAs, InP, and GaP behave similarly to Si, Ge, and InSb, but that a new deformation regime emerges at high temperatures and low strain rates, which has not been observed in the latter materials. Above 700°C the experiments have to be performed under liquid encapsulation to avoid the evaporation of the volatile elements phosphorus and arsenic, which affects the surface of the crystals. It is an open question for the moment whether the new deformation regime found in InP and GaAs has to do with impurities or precipitates (Lee et al., 1989), or has to be ascribed to a change in rate-controlling mechanism, as has been suggested by Rai *et al.* (1989). Current work is devoted to these problems; therefore, in the present review only the actual state of art can be summarized. It is finally noted that this new deformation regime is typical for the beginning of deformation, because similar deviations have not been observed in the later stages of the stress-strain curves.

The plasticity of II-VI compounds has attracted attention since the discovery of a pronounced photoplastic effect, which means a change of the flow stress of the crystals under illumination (Osip'yan *et al.*, 1986). Because the results on the initial flow stress are somewhat contradictory and higher deformation degrees have not been systematically investigated (just a few stress-strain curves of CdTe have been recorded by Hall and Vander Sande [1978]), these materials will be only marginally treated.

II. General

From the various methods of plastic deformation customary in current research, only two have been applied to semiconducting materials to a greater extent: the deformation at constant strain rate, where the flow stress τ is recorded as a function of strain ε, yielding the stress-strain curve $\tau(\varepsilon)$; and the deformation at constant stress, where the strain is obtained as a function of time t yielding the so-called creep curve $\varepsilon(t)$. Only for special problems has torsional deformation (Siethoff and Alexander, 1964; Wagner *et al.*, 1971) been used. In practically all measurements, single crystals were investigated. For constant strain-rate deformation, a specimen orientation such as $\langle 123 \rangle$, which is suitable for single slip, has been preferred, since for this orientation the different hardening and recovery stages of the stress-strain curves are best revealed. The symmetrical $\langle 111 \rangle$, $\langle 110 \rangle$, and $\langle 001 \rangle$ orientations lead to more or less structureless stress-strain curves of parabolic shape, and to multiple slip from the beginning of deformation. For creep experiments, however, the $\langle 111 \rangle$ orientation has been proved to be advantageous for reasons that will be outlined in Section II.4.

1. MATERIAL PROPERTIES RELEVANT FOR PLASTICITY

The elemental semiconductors crystallizing in the diamond cubic lattice are characterized by a strong covalent bonding. This, via a high Peierls potential, leads to a dislocation mobility exponentially decreasing with decreasing temperature and is, finally, responsible for the low-temperature brittleness of Si and Ge. In principle, for the III–V compounds, which crystallize in the sphalerite (cubic zincblende) structure, the same is true; the chemical bonding, however, is not only of covalent but also of ionic character (Gottschalk et al., 1978). To take into account the different bonding strengths, it has become customary (Frost and Ashby, 1982) to normalize flow stresses and strain-hardening parameters by the elastic constants that, for a cubic lattice, are suitably represented by the isotropic shear modulus $G = [(c_{11} - c_{12})c_{44}/2]^{1/2}$. Both the elemental and compound semiconductors have {111} glide planes and ⟨110⟩ glide directions. Their Burgers vectors are rather large (Table I); this will have consequences for the cross-slip behavior as discussed in Section IV.8.

In Table I, besides the magnitude b of the Burgers vectors, the melting points T_m and the stacking-fault energies γ, the parameters of self-diffusion are indicated, which are important for the interpretation of dynamical recovery. The diamond cubic lattice is characterized by a rather low concentration of native point defects, and vacancies and interstitials are assumed to be mainly involved. For Ge a vacancy-controlled diffusion mechanism has been confirmed (Vogel et al., 1983). For Si, there is increasing evidence in the literature that both vacancies and interstitials contribute to self-diffusion (Frank et al., 1984; Fahey et al., 1989). In Table I, the vacancy parameters are given. For GaAs, GaP, and InP, there are further complications, because they exhibit an extended homogeneity region at elevated temperatures (Morozov and Bublik, 1986). This leads to high concentrations of native point defects such as vacancies and antisite defects, which can be easily influenced by changing the composition of the melt during crystal growth, by thermal treatment of the bulk material, or by doping (Corbel et

TABLE I

MATERIAL PARAMETERS

	T_m/K^a	$b/10^{-10}$ m^a	γ/mJ m^{-2}	Q^{SD}/eV	D_0/m^2 s^{-1}
Si	1,685	3.84	58[b]	4.03[d]	6×10^{-5} [d]
Ge	1,210	4.0	75[b]	3.14[e]	2.5×10^{-3} [e]
InSb	800	4.58	38[c]	—	—
InP	1,335	4.15	18[c]	—	—
GaAs	1,513	4.0	55[c]	—	—
GaP	1,740	3.85	41[c]	—	—

[a] Madelung (1982); b from lattice constant.
[b] Gottschalk (1979).
[c] Gottschalk et al. (1978).
[d] Tan and Gösele (1985).
[e] Vogel et al. (1983).

al., 1988; Dlubek et al., 1988; Parsey et al., 1988; Dannefaer et al., 1989). As a consequence, measurements of self-diffusion are difficult, and relevant publications are scarce and controversial (Tuck, 1988). The intrinsic defects may react with deformation-induced point defects (Omling et al., 1986; Vignaud and Farvacque, 1989), eventually leading to various point-defect complexes. The whole subject is intensely studied in GaAs and far from being completely understood (Bourgoin et al., 1988).

2. THE STRESS-STRAIN CURVE OF SEMICONDUCTORS

A set of stress–strain curves typical for a $\langle 123 \rangle$ orientation is shown in Fig. 1 for Ge. At low temperatures, three deformation stages following the pronounced yield point are observed: a stage of low work hardening (I), a stage of strong work hardening (II), and a stage of increasing softening (III), where recovery processes operate in the crystals exposed to an external stress (dynamical recovery). At high temperatures, two further stages appear: a new hardening stage (IV) and a second recovery stage (V). This deformation behavior is also found in Si and III–V compounds; for the latter, only the yield point may be suppressed at high temperatures. It has become customary

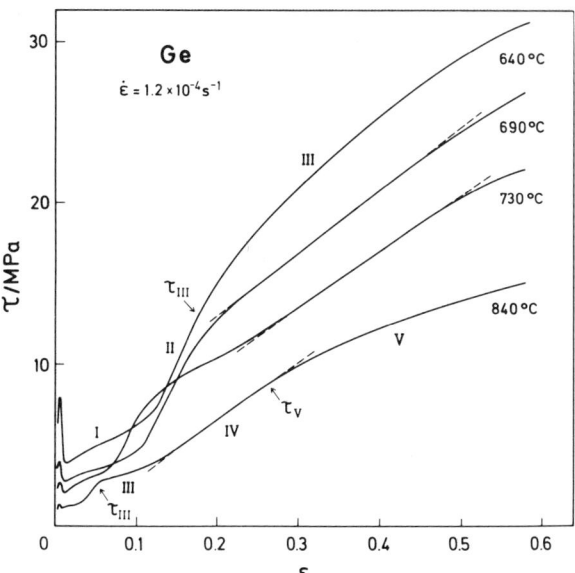

FIG. 1. Stress–strain curves of Ge with a $\langle 123 \rangle$ orientation, obtained in compression at different temperatures. From Brion et al. (1981).

to denote the characteristic points in the $\tau(\varepsilon)$ curve by their appropriate stresses: the stress at the lower yield point τ_{1y}, the stresses τ_{II}, τ_{III}, and τ_V at the beginning of the respective stages, and the saturation stress τ_s in stage V; they are indicated in the schematical plot of Fig. 2a and, partly, in Fig. 1. If, at low temperatures, the stages IV and V do not emerge, τ_s is reached in stage III. The strong temperature dependence of these stresses (and of the respective strains) is particularly noted, since it leads to a remarkable shrinkage of the curves with increasing temperature. The strain to reach τ_{III}, for instance, varies typically between 0.05 and 0.7, while the strain at the beginning of stage V lies between 0.2 and 0.75. The stages of strong work hardening, II and IV, are usually characterized by their slopes θ_{II} and θ_{IV}, respectively.

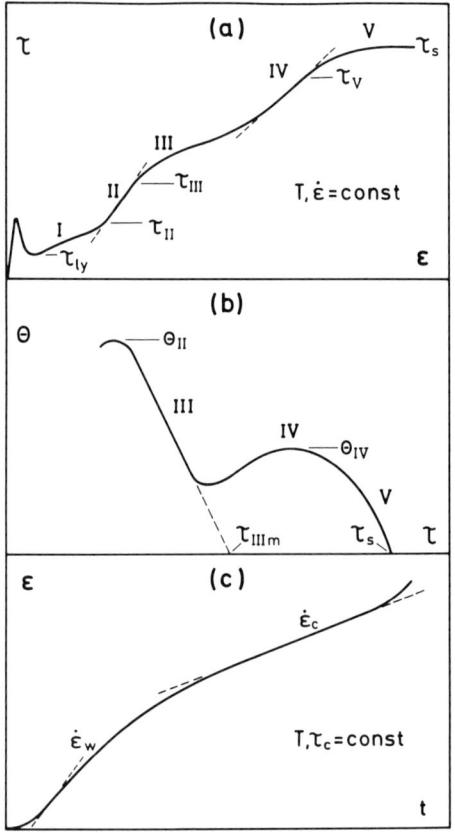

FIG. 2. Schematical plot (a) of the stress–strain curve of semiconductors with a ⟨123⟩ orientation, (b) of the work-hardening coefficient as a function of stress, and (c) of the creep curve; the characteristic deformation parameters are indicated.

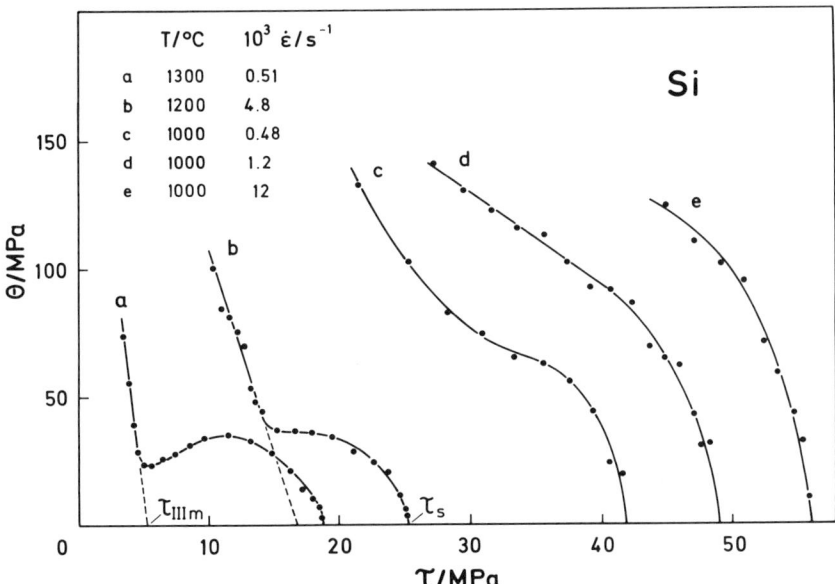

FIG. 3. Work-hardening coefficient as a function of stress for Si with a $\langle 123 \rangle$ orientation, starting at the onset of stage III. From Siethoff and Schröter (1983).

For some purposes, it has been proved to be advantageous to use the first derivative of the stress–strain curves, i.e., the so-called work-hardening coefficient $\theta = d\tau/d\varepsilon$, and to plot it as a function of stress, as it is shown schematically in Fig. 2b. This procedure is especially useful for crystals with a $\langle 111 \rangle$ orientation, since it reveals the different deformation stages, although the stress–strain curves are more or less parabolic (Siethoff et al., 1986). Furthermore, it has been found experimentally (Siethoff and Schröter, 1983) that within stage III, θ varies linearly with stress, as shown in Fig. 3 for Si. The extrapolation $\theta \to 0$ defines a stress τ_{IIIm}, which can be used as an alternative measure of stage III. For stage V, the relation $\theta(\tau)$ is nonlinear. The clear separation between the two recovery stages is lost at lower temperatures (and larger strain rates), when τ_v approaches τ_{III}. Such a behavior is typical for all semiconductors studied so far, as is demonstrated in Fig. 4 for GaAs.

Figure 1 also indicates that the stages IV and V are not the mere result of the distortions that the specimens undergo during compressive deformation, because the stages are more pronounced at high temperatures, where the strains to reach the different stages become increasingly smaller. Moreover, these stages appear in crystals with a symmetrical $\langle 110 \rangle$ or $\langle 111 \rangle$ orientation, although the specimen distortion is quite different from that of the $\langle 123 \rangle$ orientation (Siethoff et al., 1986). To verify the predominance of the

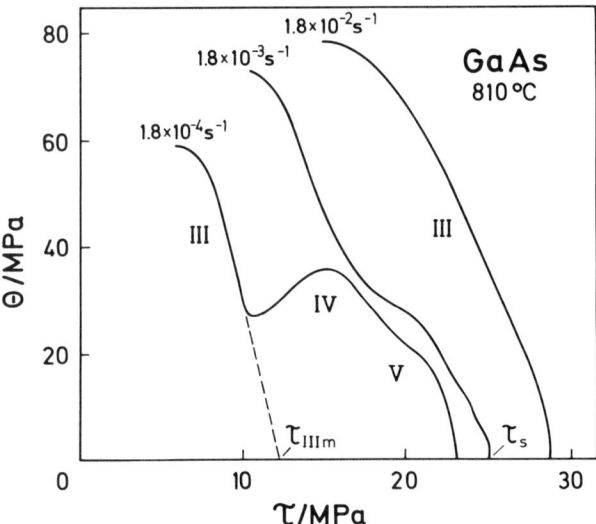

FIG. 4. Work-hardening coefficient as a function of stress for GaAs with a ⟨123⟩ orientation, deformed under B_2O_3 encapsulation; the curves start at the onset of stage III. From Siethoff and Behrensmeier (1990).

main slip system in stage IV and stage V for the ⟨123⟩ orientation, specimens were deformed into these stages, the slip lines were polished away, and the deformation was continued by a few per cent of strain. It was found that the new slip lines belonged to the main glide system, also showing that the specimen distortions are not responsible for the observed effects (Brion et al., 1981). It is finally noted that two recovery stages also emerge in Si crystals deformed in tension (Gross et al., 1989).

3. STRAIN-HARDENING PARAMETERS

In the hardening stages II and IV, the stress–strain curves exhibit a portion of constant slope, which, however, degenerates to an extended inflection point at higher temperatures. The temperature dependence of the respective strain-hardening coefficients θ_{II} and θ_{IV} (normalized to the shear modulus G) is shown in Fig. 5 for different semiconductors deformed along ⟨123⟩. At low temperatures, θ_{II}/G remains constant with the values given in Table II, and then decreases linearly with increasing temperature. In InSb, the scatter is rather large and seems to obliterate a possible temperature dependence. The shift to lower temperatures of the θ_{II} (T) curve of GaAs is noted. The

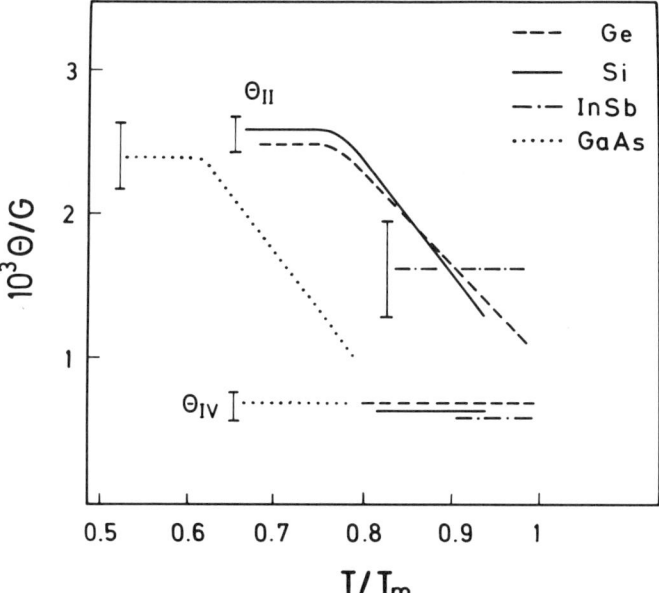

FIG. 5. Work-hardening coefficients of stage II and stage IV, corrected by the shear modulus, as a function of the temperature, which is normalized to the melting point. From Siethoff and Schröter (1984) and Siethoff and Behrensmeier (1990).

TABLE II

STRAIN-HARDENING PARAMETERS FOR UNDOPED SEMICONDUCTORS

	$10^3\Theta_{II}/G^f$	$10^3\Theta_{IV}/G^f$	v	Q_{II}/eV
Si	2.6[a]	0.65[a]	3.2[e]	0.95[e]
Ge	2.4[a]	0.7[a]	3.2[e]	0.71[e]
	2.6[b]	—	3.3[b]	0.68[b]
InSb	1.7[a]	0.6[a]	—	—
InP	2.7[c]	—	3.6[e]	0.49[e]
GaAs	2.4[d]	0.7[d]	4.1[d]	0.45[d]

[a] Siethoff and Schröter (1984).
[b] Kojima and Sumino (1971).
[c] Siethoff et al. (1988).
[d] Siethoff and Behrensmeier (1990).
[e] Siethoff (1989)
[f] Shear moduli G quoted in the references.

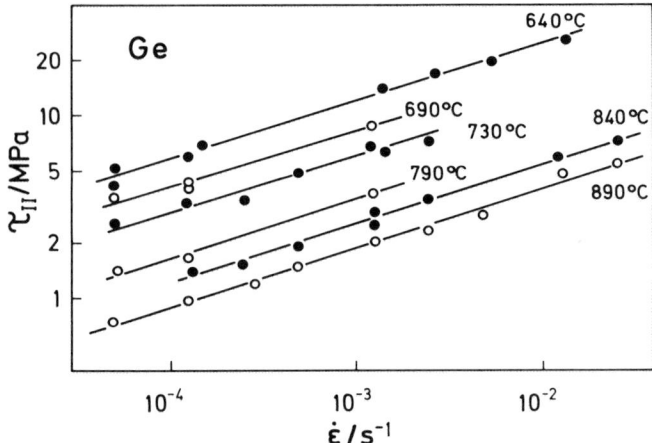

FIG. 6. Strain-rate dependence of the stress at the transition from stage I to stage II for Ge ⟨123⟩, in a log–log plot.

parameter θ_{IV}/G, on the other hand, does not depend on temperature and is nearly the same for the materials indicated in Fig. 5. For InP, θ_{IV} has not yet been measured. A few θ_{II} values obtained by other authors have been collected elsewhere (Siethoff and Schröter, 1984).

Kojima and Sumino (1971) were the first to note that the stress τ_{II} at the transition from stage I to stage II (see Fig. 2a) depends on strain rate by a power law, as shown in Fig. 6 for Ge. The distance of the curves, i.e., the temperature dependence of τ_{II}, obeys an Arrhenius law. Therefore the τ_{II} ($\dot{\varepsilon}$, T) relationship may be suitably described by

$$\tau_{II} = C_{II}\dot{\varepsilon}^{1/\nu} \exp\left(\frac{Q_{II}}{kT}\right), \qquad (1)$$

where C_{II} is a constant independent of strain rate $\dot{\varepsilon}$ and temperature T, k is Boltzmann's constant, $1/\nu$ is the strain-rate exponent, and Q_{II} is an appropriate activation energy. Such a behavior is typical for all the semiconductors discussed here (for InP and GaAs at least at low temperatures); the parameters ν and Q_{II} are listed in Table II. A tentative interpretation of these results and of the temperature dependence of θ_{II} has been given elsewhere (Siethoff and Behrensmeier, 1990).

4. Equivalence of Stress–Strain and Creep Curve

There are some particular points in the stress–strain curve that need special attention: the lower yield point and the deformation at τ_s in stage V (or, at low temperatures, in stage III). They are characterized by the fact that the work-hardening coefficient becomes zero. Such a situation is usually called steady-state (or stationary) deformation if the stress–strain curve reaches a stress plateau, as is often the case for τ_s, or it is called quasi-stationary deformation if the stress–strain curve exhibits an extremum, as it is the case for the lower yield point. Alexander and Haasen (1968) have shown that the lower yield point may be regarded as a state of plasticity, i.e., that it can be described by a mechanical equation of state with the state variables stress τ_{1y}, strain rate $\dot{\varepsilon}$, and temperature T. They have further shown that this state is equivalent to that occurring in the creep curve at the inflection point (see Fig. 2c). This means that the relationship $\dot{\varepsilon}_w(\tau_c, T)$, where $\dot{\varepsilon}_w$ is the creep rate at the inflection point and τ_c the creep stress, is governed by the same physical processes as the relationship $\tau_{1y}(\dot{\varepsilon}, T)$ at the lower yield point, and that the parameters obtained in both deformation experiments are the same. The reason for this behavior is that, at the inflection point and at the lower yield point, there is an equilibrium between the processes of dislocation generation and dislocation interaction. Alexander and Haasen (1968) have given the theoretical background for this situation, which will be outlined in Section III.

For τ_s, in principle, the same conditions are expected to hold. The stationary relationship $\tau_s(\dot{\varepsilon}, T)$ describes a state of plasticity, which should be equivalent to that obtained in the steady-state regime of the creep curve, which is characterized (see Fig. 2c) by the relationship $\dot{\varepsilon}_c(\tau_c, T)$. In this case, the steady state results from an equilibrium between hardening and recovery processes within the deforming crystals. Although in the literature for metals and other materials, the equivalence of the steady state of constant strain-rate and creep deformation is established by numerous investigations, for the semiconductors reviewed here there are some problems. In different creep experiments on Si and Ge it was found that the steady-state creep rates did not coincide, although the orientation of the samples was identical. The dependencies on stress and temperature and the resulting parameters, however, were more or less the same. Furthermore, for the $\langle 111 \rangle$ orientation, the strain to reach the steady state is appreciably smaller than that necessary for the $\langle 110 \rangle$ and the $\langle 123 \rangle$ orientation (Demenet et al., 1983). Chiang and Kohlstedt (1985) have pointed out that the $\langle 111 \rangle$ orientation especially favors cross-slip, while the $\langle 123 \rangle$ and $\langle 111 \rangle$ orientations do not (Rai et al., 1989). This is the recovery mechanism that has been used to explain the steady state of the semiconductors in creep and in the second recovery stage of stress–strain curves. Rai *et al.* (1989), however, have proposed a different mechanism for the second recovery stage, as will be discussed in Section IV. In stage III of the stress–strain curve, if it is followed by stage V (see Section II.2), a steady state is not attained, because the hardening stage IV sets in before. Therefore, the first recovery stage is not visible in the creep curve, where only stationary deformation stages are revealed. Concerning stage III, however, the steady state can be established by the extrapolation method introduced in Section II.2, which defines a stress τ_{IIIm}, the temperature and strain-rate dependence of which will be discussed in Section IV.7.

Sometimes, in the literature, the upper yield point has also been investigated. Although the respective stress depends on strain rate and temperature in a way similar to the lower yield stress, an equilibrium dislocation density is not attained, and experimental details such as the initial dislocation density and the hardness of the deformation machine play an important role. This indicates that the upper yield point does not result from quasi-stationary deformation conditions. Alexander (1986) has reviewed the present state of art.

III. Lower Yield Point and Creep at Inflection Point

As has been discussed in the preceding section, the lower yield stress τ_{1y} and the creep rate $\dot{\varepsilon}_w$ at the inflection point obey the same mechanical equation of state $f(\dot{\varepsilon}, \tau, T) = 0$. For this situation, Alexander and Haasen (1968) have derived the following equations:

$$\tau_{1y} = C_{1y} \dot{\varepsilon}^{1/(2+m)} \exp\left[\frac{U}{(2+m)kT}\right], \qquad (2)$$

$$\dot{\varepsilon}_w = C_w \tau^{2+m} \exp\left(-\frac{U}{kT}\right), \qquad (3)$$

with

$$C_{1y} = C_w^{-1/(2+m)}.$$

Here U and m are the activation energy and the stress exponent, respectively, of the dislocation mobility v according to

$$v = B\tau_{\text{eff}}^m \exp\left(-\frac{U}{kT}\right). \tag{4}$$

The underlying physical process is assumed to be the motion of dislocations in a Peierls potential via nucleation and migration of kinks. The effective stress τ_{eff} takes into account the interaction between the dislocations and is the difference between the applied and the internal stress:

$$\tau_{\text{eff}} = \tau - A\sqrt{N}. \tag{5}$$

In these equations C_{1y}, C_w, A, and B are constants, N is the dislocation density, and k is Boltzmann's constant. It is one remarkable result of the theory that, in Eqs. (2) and (3), the applied stress appears as a variable (which can be easily measured), although the effective stress has been explicitly taken into account in their derivation, and that the effective stress at the lower yield point turns out to be $\tau_{\text{eff}} = \tau_{ly}\, m/(2 + m)$, i.e., is a constant portion of the lower yield stress (Alexander and Haasen, 1968). There are competitive ideas in the literature concerning the problem of the effective stress; here the reader may be referred to the review of Alexander (1986). Recently, Oueldennaoua et al. (1988) and Allem et al. (1989), from an analysis of the dislocation structure of Si in the yield point region, concluded that the effective stress should be nearly equal to the lower yield stress. Their experiments, however, were performed on predeformed specimens, which, according to the findings presented in Section III.6, show properties that are different from those of as-grown crystals.

Equation (4) has been experimentally proven to be a correct description of the mobility of dislocations in Si and Ge at high stresses, with similar parameters for 60° and screw dislocations; at low stresses and temperatures below about 0.75 T_m, however, the activation energy was found to depend on stress, while the stress exponent became temperature-dependent. Alexander (1986) has argued that these peculiarities might be a special property of the dislocation half-loops used in the mobility measurements, and that straight dislocations penetrating the whole crystal behave normally, i.e., in accordance with Eq. (4). In fact, macroscopic deformation tests of the lower yield stress and of the creep rate at the inflection point do not reflect the peculiarities mentioned above. This, in principle, is also true for III-V compounds. Although in these materials three types of dislocations (screw, α, and β dislocations) have to be taken into account, parameters obtained by analyzing the lower yield point (via Eq. (2)) and dislocation mobility (via Eq. (4)) agree within the expected frame (Siethoff *et al.*, 1987, 1990). Appreciably higher activation energies published in the literature (Rabier and George, 1987) are typical for predeformed material, as will be discussed in Section III.6.

5. INTRINSIC EFFECTS

Figure 7 shows the strain-rate dependence of the lower yield stress of undoped GaAs. Between 415 and 660°C the deformation was carried out under argon as a protective atmosphere, while at higher temperatures the samples were encapsulated with liquid B_2O_3. The chain line will be discussed in Section III.6. The full lines represent a fit of Eq. (2) to the appropriate data, with the parameters given in Table III (Siethoff and Behrensmeier, 1990). Apparently, in this regime, there is no remarkable difference between specimens deformed under argon and liquid encapsulation. There are, however, deviations at high temperatures and low strain rates (dashed portions in Fig. 7), which are not compatible with Eq. (2). Rai *et al.* (1989) argue that the rate-controlling mechanism changes from kink nucleation at low temperatures to dragging of jogs on screw dislocations above 700°C. It is noted that similar deviations are observed for the stress τ_{II} at the transition from stage I to stage II (Siethoff and Behrensmeier, 1990), but not for the recovery stages, as will be shown in Section IV.

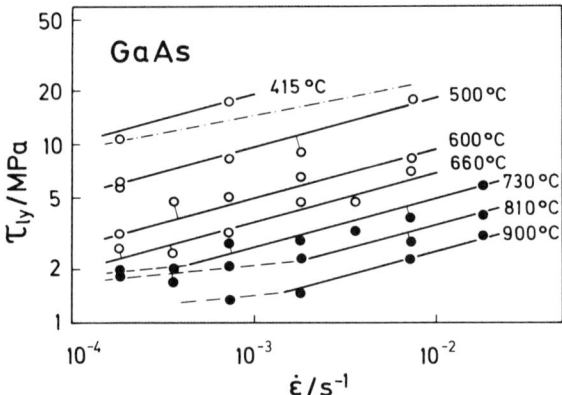

FIG. 7. Lower yield stress as a function of strain rate in a log-log plot for GaAs with a $\langle 123 \rangle$ orientation; open and full points are obtained from deformation under argon and liquid B_2O_3, respectively; the solid lines represent a fit of Eq. (2) to the data, and the chain line is from repeatedly deformed specimens. From Siethoff and Behrensmeier (1990).

TABLE III
PARAMETERS OF THE LOWER YIELD POINT AND OF CREEP AT THE INFLECTION POINT FOR UNDOPED SEMICONDUCTORS, FROM ANALYSIS OF EQS. (2) AND (3)

Material	Method	T/T_m	$2+m$	U/eV^a	U/kT_m	References
Si	τ_{1y}	0.61–0.93	2.9	<u>2.32</u>	16.0	Schröter et al. (1983)
Ge	τ_{1y}	0.68–0.96	$m(T)$	$U(\tau)$	—	Schröter et al. (1983)
InSb	τ_{1y}	0.62–0.96	3.1	<u>0.96</u>	13.9	Schäfer et al. (1964)
	$\dot{\varepsilon}_w$	0.68–0.83	3.3	0.88	12.8	Peissker et al. (1962)
	τ_{1y}	0.63–0.87	3.1	0.99	14.4	Shimizu and Sumino (1975)
InP	$\dot{\varepsilon}_w$	0.70–0.75	3.4	0.8	7.0	Völkl et al. (1987)
	τ_{1y}	0.61–0.75	2.9	<u>1.43</u>	12.4	Siethoff et al. (1990)
	τ_{1y}	0.65–0.90	3.5	1.1	9.6	Reppich et al. (1990)
GaAs	$\dot{\varepsilon}_w$	0.41–0.58	3.2	1.45	11.1	Osvenskii et al. (1969)
	τ_{1y}	0.45–0.51	3.1	1.4	10.7	Yonenaga et al. (1987)
	τ_{1y}	0.45–0.78	3.6	<u>1.37</u>	10.5	Siethoff and Behrensmeier (1990)
GaP	τ_{1y}	0.50–0.62	2.9	1.42	9.5	Yonenaga and Sumino (1989b)
	τ_{1y}	0.45–0.57	3.1	1.67	11.2	Paufler and Retzlaff (1989)

^{a}The underlined values are further used in Table IV.

The applicability of Eqs. (2) and (3) to the lower yield point and to creep at the inflection point, respectively, has been confirmed for other semiconducting materials and by other authors; relevant parameters are listed in Table III (some earlier measurements on Ge and Si have been collected elsewhere [Schröter et al., 1983]). For Ge, however, the formalism represented by Eqs. (2) and (3) does not work without further modifications. This is demonstrated in Fig. 8, where the lower yield stress (solid lines) is plotted as a function of the strain rate for a large temperature range (Schröter et al., 1983). Obviously, the slope of the straight lines increases with rising temperature. In a similar way, the activation energy U becomes stress-dependent. The same behavior has been observed by Weiss (1975) in measurements of the creep rate at the inflection point (dashed lines in Fig. 8). This author tried to reconcile the measurements with Eq. (3) by introducing a temperature-dependent stress exponent $m(T) = m_0 + U_1/kT$. As has been shown by Schröter et al. (1983), this concept inevitably leads to a curvature in the Arrhenius plot of the lower yield stress. According to Fig. 9, this is not the case. Further improvements are needed to regain the applicability of Eqs. (2) and (3) to the data for Ge. For the other materials considered here, the model of Alexander and Haasen (1968) is compatible with the measurements. A reason for the deviating behavior of Ge is not yet known.

Castaing et al. (1981) investigated the lower yield point of Si at temperatures between 300 and 600°C and had to apply a hydrostatic pressure to

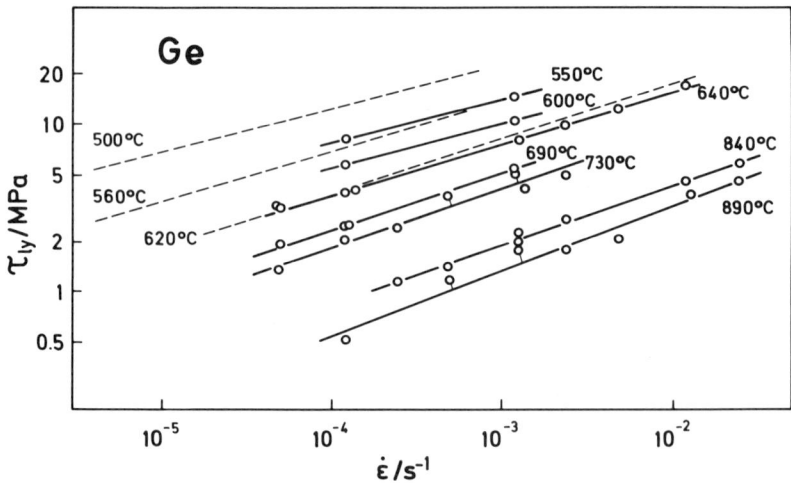

FIG. 8. Lower yield stress as a function of strain rate in a log–log plot for Ge with a ⟨123⟩ orientation; the dashed lines indicate measurements of the creep rate at the inflection point (Weiss, 1975). From Schröter et al. (1983).

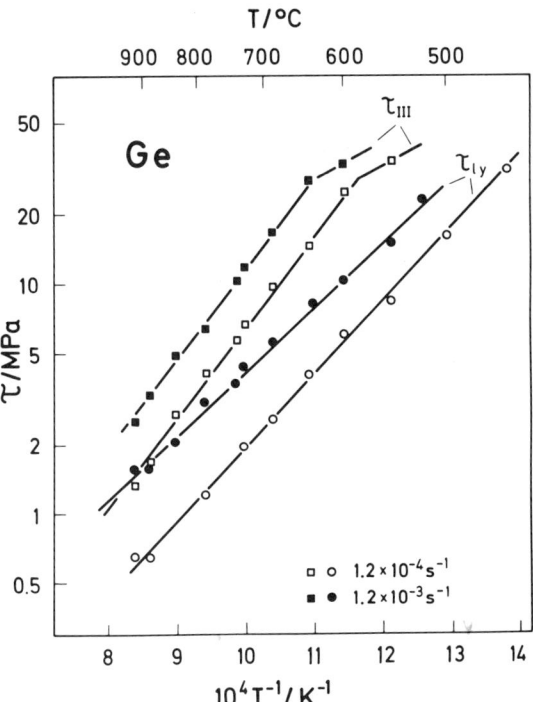

FIG. 9. Arrhenius plot of the lower yield stress τ_{ly} and of the stress τ_{III} at the beginning of stage III at two different strain rates for Ge $\langle 123 \rangle$. From Schröter et al. (1983) and Brion et al. (1979).

avoid fracture of the samples. Above about 450°C, the measurements were shown (Alexander, 1986) to be compatible with those reported above, while at lower temperatures deviations occurred, whose origin, however, was not elucidated. In this regime, the lower yield stress could be described either by a power law such as Eq. (2) — with other parameters, however, than in the high-temperature regime — or by a simple Arrhenius law such as $\dot{\varepsilon} \sim \exp(-\Delta G(\tau)/kT)$ (Rabier et al., 1983).

On account of limited information, it was formerly suggested (Siethoff, 1969a) that the relative strength of the semiconductors might be correlated with the ionicity of the bonds of the III-V compounds. Figure 10 shows that this is apparently not the case. The ionicities of InSb and GaAs, for example, are of similar magnitude, and are distinctly smaller than that of InP (Gottschalk et al., 1978). As can be seen from Table III, the quantity U/kT_m also cannot be easily correlated with ionicity. It is argued that another

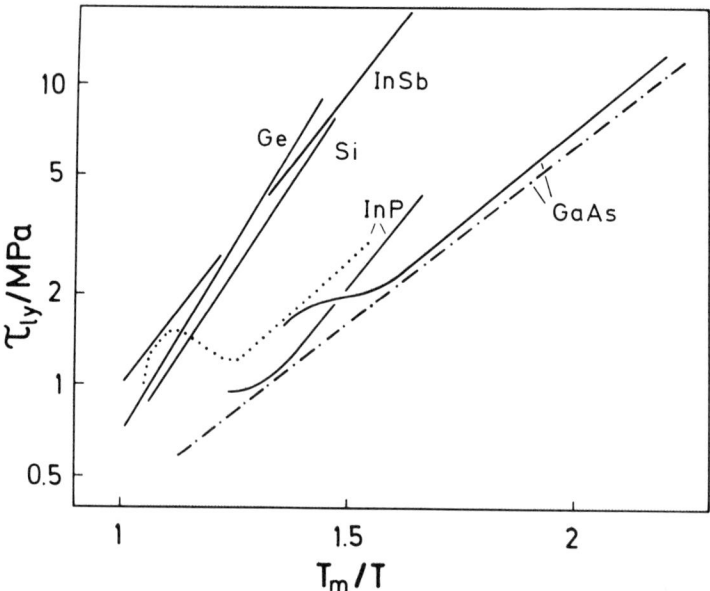

FIG. 10. Lower yield stress of different semiconductors with $\langle 123 \rangle$ orientation in an Arrhenius plot, where the temperature is normalized to the melting point. Data: Si and Ge, $\dot{\varepsilon} = 2 \times 10^{-4}\,\mathrm{s}^{-1}$ (Schröter et al., 1983); InSb, $\dot{\varepsilon} = 2.75 \times 10^{-4}\,\mathrm{s}^{-1}$ (Schäfer et al., 1964); InP, solid line, $\dot{\varepsilon} = 2 \times 10^{-4}\,\mathrm{s}^{-1}$ (Siethoff et al., 1987); InP, dotted line, $\dot{\varepsilon} = 10^{-4}\,\mathrm{s}^{-1}$ (Reppich et al., 1990); GaAs, chain line, $\dot{\varepsilon} = 10^{-4}\,\mathrm{s}^{-1}$ (Bourret et al., 1987); GaAs, solid line, $\dot{\varepsilon} = 2 \times 10^{-4}\,\mathrm{s}^{-1}$. From Siethoff and Behrensmeier (1990).

material parameter might be involved to set up an order for the strength of the semiconductors. This problem will be further discussed in Section IV.10.

If one compares the lower yield stress obtained by different authors, it is found that for Si and Ge there is a rather good agreement in the literature (Schröter et al., 1983). For the III–V compounds, and especially for InP, however, there are larger discrepancies in the absolute stress values and the deformation parameters involved, as can be seen from Fig. 10 and Table III. The reason for these differences has to be elucidated by future work; for the moment, it may be assumed that different amounts of native point defects, impurities, and precipitates (Lee et al., 1989) originating from crystal growth and different heating procedures prior to deformation may be responsible for the observed discrepancies.

Such effects are well-known and more pronounced in the II–VI compounds (Osip'yan et al., 1986). These materials, which crystallize either in the cubic zincblende or the hexagonal wurtzite structure, are characterized by a ionicity higher than that of the III–V compounds. The glide system of the cubic modification is well established to be {111} ⟨110⟩, as is the case for the other semiconductors regarded here. The flow stress of the II–VI compounds has been found to vary from crystal to crystal. This behavior has been correlated to the fact that the dislocations in these materials carry a large charge, which can be influenced by doping, stoichiometry and illumination (photoplastic effect). Activation energies U/kT_m derived from the temperature dependence of the stress at the beginning of deformation (a yield point is rarely observed) typically vary between 3 and 8, with a discrepancy up to a factor of two for the same compound. These values are appreciably smaller than those listed in Table III; recently, however, a pronounced yield point and $U/kT_m = 17$ has been reported in the literature for a $Cd_xHg_{1-x}Te$ alloy (Barbot et al., 1986). Creep experiments and deformation to higher strains are rather scarce in II–VI compounds (Hall and Vander Sande, 1978). Measurements of steady-state creep on CdTe are under way (Behrensmeier, 1990).

6. Surface Effects and Influence of Predeformation

In this section, phenomena will be dealt with that deviate from the intrinsic behavior described in the foregoing section. Figure 11 shows the lower yield stress of Si (as a function of the strain rate) obtained from experiments that were performed under different protecting atmospheres. The dashed lines and

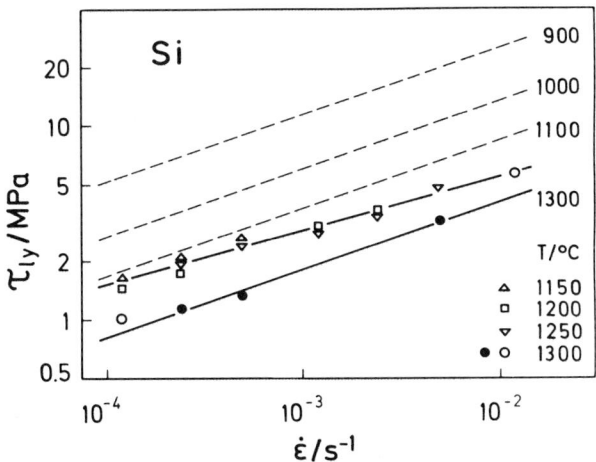

FIG. 11. Strain-rate dependence of the lower yield stress of Si ⟨123⟩ in a log–log plot; data obtained under forming gas and under argon as protecting atmospheres are represented by the open and solid symbols, respectively. From Siethoff (1988).

the open symbols were derived from measurements under forming gas (90% N_2, 10% H_2), while the full points were obtained under argon (Siethoff, 1988). The results described by the dashed lines and by the full points (at 1,300°C) are compatible; the data stretch on a straight line in the Arrhenius plot of Fig. 10 (with the parameters given in Table III), the yield points show the usual shape, and the samples were clean after deformation. The data characterized by the open symbols exhibit a deviating behavior: In the log τ_{1y} (log $\dot{\varepsilon}$) plot of Fig. 11, they practically fall on the same line (with a slope different from that of the other curves) in the temperature range between 1,150 and 1,300°C, thus leading to a temperature plateau; furthermore, the yield points are diminished to a certain extent, and the surface of the samples is coated by a white film. According to Denisova et al. (1975), this surface film probably consists of Si_3N_4. Because the diffusion depth of nitrogen into Si is expected to be only a few microns during the deformation experiments, it was concluded (Siethoff, 1988) that the deviations observed in specimens deformed under forming gas at temperatures above 1,100°C have to be ascribed to a surface effect. Omri et al. (1987) reported on the same behavior for deformation under forming gas (although they did not arrive at the same conclusions); in their work, the yield points even vanished, and the temperature plateau degenerated to a maximum in the log τ_{1y} $(1/T)$ plot at temperatures above 1,000°C.

If InP, GaAs, and GaP are deformed above about 700°C in a gas atmosphere, the surface of the crystals is increasingly disturbed by the evaporation of phosphorus and arsenic (Siethoff et al., 1987, 1988; Paufler and Retzlaff, 1989). Simultaneously, a temperature plateau of the lower yield stress is observed, as is indicated in Fig. 10 for InP (solid line). The influence of a coated or disturbed surface on plasticity may be twofold: either by impeding the operation of surface dislocation sources, or by piling up dislocations generated in the bulk. In the semiconductors considered here, the underlying mechanisms have not yet been investigated. In metals the influence of oxide coatings or of thin surface alloy layers on flow stress and other mechanical properties has been studied in some detail, but a unifying concept to account for all the phenomena observed has not been arrived at so far (see, for example, Nabarro, 1977; Wilsdorf and Ruddle, 1977). Charging the surface of Ni single crystals with hydrogen led to prominent serrations in the stress–strain curves (Latanision and Staehle, 1968). Serrations are also observed in GaP deformed at 900°C under argon (Yonenaga and Sumino, 1989b). It is an open question for the moment whether these are caused by a large amount of impurities in the bulk of the material (as assumed by the authors), i.e., have to be ascribed to a Portevin–LeChatelier effect (see section V.12), or are the result of the evaporation of phosphorus from the specimen surface that, according to Paufler and Retzlaff (1989), increasingly takes place above about 700°C.

It has to be noted in this context that a plateaulike temperature dependence is also observed in GaAs (solid line in Fig. 10) deformed under liquid encapsulation in the regime that, in Fig. 7, is indicated by the dashed lines. Because in these experiments the specimen surfaces do not look disturbed or coated after deformation, a surface effect does not seem to be a suitable explanation for the observed temperature dependence. Therefore, in Section III.5, a different rate-controlling process was assumed to be involved in this regime. However, an influence of impurities, complexes, or precipitates (Lee et al., 1989) cannot be totally excluded for the moment. Bourret et al. (1987) approximate their measurements on GaAs (chain line in Fig. 10) by a straight line. Closer inspection, however, reveals that a plateau between 800 and 1,000°C may better fit the data. Djemel et al. (1988), on the other hand, report on an extended plateau in GaAs between 500 and 900°C, while Guruswamy et al. (1987) observe such an effect between 700 and 1,100°C. The dotted curve in Fig. 10 demonstrates that InP deformed under liquid B_2O_3 behaves similarly to GaAs and that even a maximum emerges in the temperature dependence (Reppich et al., 1990). Moreover, the authors report on a decrease in weight of the deformed crystals and on precipitation of metallic indium on the specimen surfaces at temperatures close to the melting point, thus revealing a surface reaction even under encapsulation. Apparently, this leads to a complicated temperature dependence of the lower yield point.

It is finally important to note that deviations like those described above for the lower yield point of Si, GaAs, and InP do not emerge in the recovery stages of the stress–strain curves, which will be dealt with in Section IV.

Omri et al. (1987) reported on a study of the yield point of Si crystals, which were predeformed by a strain $\varepsilon = 0.07$. Adopting the formalism of Eq. (2), as was done by Omri (1981) in his thesis, for the lower yield point, an activation energy $U/(2 + m) = 0.65$ eV is obtained, which is independent of strain rate, and a parameter $2 + m$, which increases from 2.7 at 600°C to 3.8 at 750°C, thus establishing a behavior different from that of material that has not been predeformed. Demenet et al. (1984) even measured $2 + m = 4.8$ and $U = 3.8$ eV. It is not clear whether the formalism of Eq. (2) can be applied to such data without modifications. The influence of predeformation is well documented also in InP and GaAs. The chain line in Fig. 7 represents measurements that were carried out by repeated deformation of the same specimens at different strain rates (Siethoff et al., 1987). By this procedure, a parameter $2 + m = 5$ was obtained, which is appreciably higher than that of not-predeformed GaAs. The increase of the parameter $2 + m$ seems to be correlated with activation energies nearly twice as high as those listed in Table III (Astié et al., 1986; Gall et al., 1987; Rabier and George, 1987). Such values are not compatible with activation energies drawn from measurements of the dislocation mobility, which have been collected elsewhere (Siethoff et al., 1987, 1990). The origin of the different behavior of predeformed material has not yet been systematically studied. It is finally noted that the influence of predeformation is more or less suppressed in highly doped Si (Demenet et al., 1984).

IV. Dynamical Recovery

A general description of the different stages of dynamical recovery as observed in creep and constant strain-rate experiments, and a discussion of some fundamental problems involved, have been given in Sections II.2 and II.4. One conclusion was that, for principle causes, the first recovery stage can be only observed in stress–strain curves.

7. First Recovery Stage

The strain-rate dependences of the stress τ_{III} at the beginning of the first recovery stage in case of GaAs, and of the steady-state flow stress τ_{IIIm} in the case of Si, are shown in Figs. 12 and 13, respectively. Quite similar relationships are observed for the other semiconductors reviewed here. For

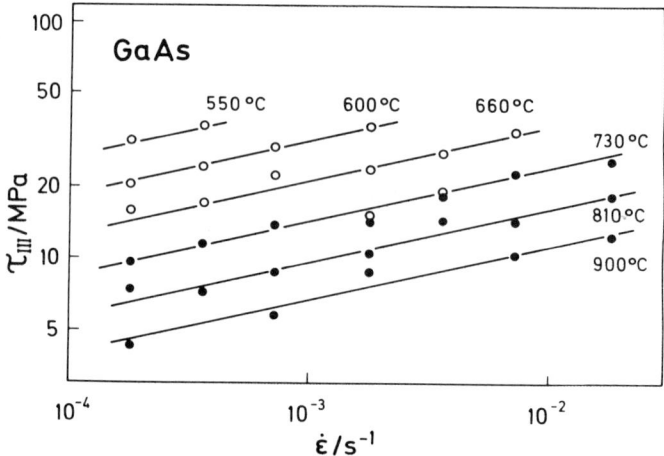

Fig. 12. Stress at the beginning of stage III as a function of strain rate for GaAs $\langle 123 \rangle$ in a log–log plot; open and solid points are obtained from deformation under argon and liquid B_2O_3, respectively; the lines represent a fit of Eq. (6) to all data. From Siethoff and Behrensmeier (1990).

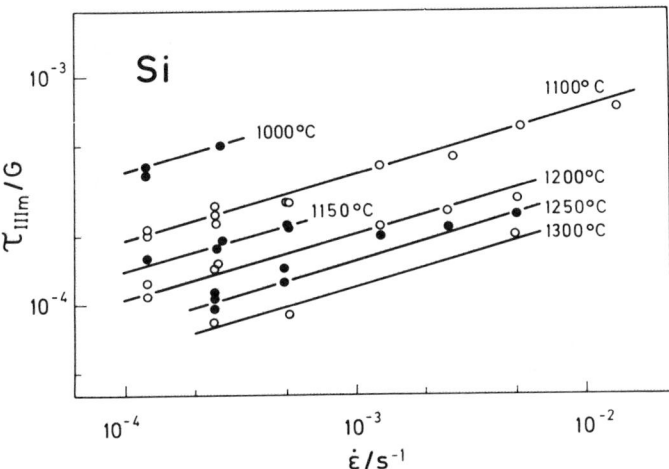

FIG. 13. Steady-state flow stress of stage III, normalized by the shear modulus, as a function of strain rate for Si ⟨123⟩ in a log–log plot; the lines represent a fit of Eq. (6) to the data. From Siethoff and Schröter (1983).

τ_{III} of GaAs, there is no remarkable difference between crystals deformed under argon (open points) and those deformed under liquid B_2O_3 (full points). Straight parallel lines in the log–log plots indicate that stresses and strain rates might be correlated by a power law. Simultaneously, the temperature dependence of the stresses obeys an Arrhenius law, as is shown in Fig. 9 for τ_{III} of Ge. (The deviations at high stresses will be explained in Section IV.10.) It was formerly proposed (Alexander and Haasen, 1961) that the elementary recovery process working in stage III might be diffusion-controlled climb: Edge dislocations surmount glide barriers in their slip plane by climb and are annihilated, thereby releasing internal stresses. The climb process involves emission or absorption of point defects; if local equilibrium of these point defects at the dislocations and transport of them through the bulk is assumed, climb is controlled by self-diffusion. Such a mechanism was originally proposed for the interpretation of high-temperature creep of metals (Weertman, 1955), and related models can be found in the literature (see, for example, Siethoff and Schröter, 1984). For the description of the first recovery stage of the semiconductors, Siethoff and Schröter (1978) adopted the formalism of Mohamed and Langdon (1974):

$$A_{ML}\left(\frac{\tau}{G}\right)^n = \frac{kT}{D_0 Gb}\left(\frac{Gb}{\gamma}\right)^3 \dot{\varepsilon} \exp\left(\frac{Q^{SD}}{kT}\right). \tag{6}$$

Here τ is either τ_{IIIm} or τ_{III}; A_{ML} is a model constant; n is the appropriate stress exponent; and Q^{SD} is the activation energy of self-diffusion as obtained from steady-state deformation experiments. In Eq. (6) the applied stress appears as a variable, though the dislocation processes in the interior of the crystals occur under the action of the effective stress, which is the difference between applied and internal stresses. This problem is not taken into account in the semi-empirical derivation of Eq. (6). There is, however, experimental evidence in the literature on metals (e.g., Blum and Finkel, 1982; Blum *et al.*, 1984) that the effective stress is a constant portion of the applied stress (as is the case for the yield-point theory of Alexander and Haasen, 1968; see Section III). This may justify the use of Eq. (6) to a certain extent. Other justification arises from the excellent simultaneous fits to the experimental data sets, as is demonstrated in Figs. 12 and 13. The appropriate parameters are listed in Table IV, together with the results obtained for the other semiconductors. It is worth noting that the values for n and Q^{SD} derived from the onset of stage III, i.e., from the analysis of τ_{III} ($\dot{\varepsilon}$, T), and those resulting from the steady state, i.e., from τ_{IIIm} ($\dot{\varepsilon}$, T), are rather similar. Measurements of τ_{III} performed by other authors are compatible with those reported above and have been discussed in a former review (Siethoff and Schröter, 1984).

The activation energies collected in Table IV have to be compared to those obtained by diffusional methods (Table I). For Ge, the Q^{SD} value derived from τ_{IIIm} ($\dot{\varepsilon}$, T) is not much lower than that derived from tracer self-diffusion; the value drawn from τ_{III} ($\dot{\varepsilon}$, T) is somewhat smaller. It is noted that the analysis of τ_{IIIm} of Ge crystals with a $\langle 111 \rangle$ orientation yielded an activation energy of 3.07 eV, in close agreement with Q^{SD} from Table I (Siethoff *et al.*, 1986). For Si, there is a similar agreement, if a vacancy mechanism is assumed to work; it is, however, an open question why there is no indication of a mechanism mediated by interstitials in the deformation experiments. For III–V compounds, tracer-diffusion measurements are scarce and controversial. In GaAs, activation energies of Ga self-diffusion of 2.6 eV (Palfrey *et al.*, 1983) and 6 eV (Tan and Gösele, 1988) have been published. Only the former value is consistent with the activation energy derived from the first recovery stage. If such a high value as 6 eV will turn out to be relevant, the diffusion mechanisms and the point defects involved are expected to be not the same in the tracer measurements and in the deformation experiments. Indications of rather different activation energies for various point defects have been recently reported on for GaAs (Yu *et al.*, 1989; Iguchi, 1989). Widely differing activation energies for self-diffusion are also known for InSb (Eisen and Birchenall, 1957; Kendall and Huggins, 1969). The activation energy derived from the first recovery stage is compatible with the measurements of Eisen and Birchenall (1957). These discrepancies are explained in the Addendum. It is finally noted that the quantity Q^{SD}/kT_m in Table IV depends on material in a way similar to that already found for U/kT_m drawn from the lower yield

TABLE IV
PARAMETERS OF THE FIRST STAGE OF DYNAMICAL RECOVERY FOR UNDOPED SEMICONDUCTORS WITH A $\langle 123 \rangle$ ORIENTATION, FROM ANALYSIS OF EQ. (6)

	Analysis of τ_{III} ($\dot{\varepsilon}$, T)				Analysis of τ_{IIIm} ($\dot{\varepsilon}$, T)			
	n	Q^{SD}/eV	Q^{SD}/kT_m	Q^{SD}/U^d	n	Q^{SD}/eV	Q^{SD}/kT_m	Q^{SD}/U^d
Si[a]	3.5	3.5	24.1	1.51	3.4	3.7	25.5	1.59
Ge[a]	3.7	2.8	26.9	—	3.4	2.9	27.8	—
InSb[a]	3.8	1.5	21.8	1.56	—	—	—	—
InP[b]	3.3	2.3	20.0	1.61	—	—	—	—
GaAs[c]	4.4	2.0	15.3	1.46	4.2	2.3	17.6	1.68

[a] Siethoff and Schröter (1984).
[b] Siethoff et al. (1988).
[c] Siethoff and Behrensmeier (1990).
[d] U taken from Table III (underlined values).

point (see Section III.5). These findings will be dealt with more thoroughly in Section IV.10.

In recent works (Siethoff et al, 1988; Siethoff and Behrensmeier, 1990), a method was described that allows one to estimate the pre-exponential factor D_0 of self-diffusion from the deformation data by taking into account the formalism of Eq. (6). This method requires only the knowledge of D_0 of one of the materials involved. If the value for Ge is taken from Table I, one obtains $D_0 = 10^{-4}$ m^2 s^{-1} for Si and $D_0 = 9 \times 10^{-7}$ m^2 s^{-1} for InSb. The former quantity is compatible with that of Table I, while the latter is not far from that derived from the diffusion measurements of Eisen and Birchenall (1957). For InP, finally, $D_0 = 2.5 \times 10^{-6}$ m^2 s^{-1}, and for GaAs a value of 10^{-7} to 10^{-8} m^2 s^{-1} can be estimated from the first recovery stage; both quantities are consistent with a vacancy mechanism. These results corroborate the view that the first stage of dynamical recovery is compatible with a diffusion-controlled process. It has eventually to be noted that climb as rate-controlling mechanism is not unquestioned in the literature for principal reasons; this problem has been discussed recently by Patu et al. (1986) and Weertman and Weertman (1987). Siethoff and Schröter (1984) have shown that the first recovery stage of the semiconductors can be also described by the jog-dragging model of Barrett and Nix (1965), which, for low stresses, is characterized by a power-law stress dependence and the activation energy of self-diffusion, as is the case for the climb model.

Alexander (1986) has argued that dynamical recovery in stage III might be better explained by a combined climb and cross-glide process, during which nonequilibrium point defects produced in the earlier deformation stages are absorbed; the activation energy of this process is assumed to be the sum of the migration energy of divacancies and of the activation energy of dislocation glide. Apart from other arguments, these ideas are not easy to reconcile with Fig. 9: In Section III.5 it was found that the activation energy U of the dislocation mobility of Ge as measured by the lower yield point depends on stress; this, in the Arrhenius plot of Fig. 9, manifests itself by straight lines of different slope for τ_{1y}. The activation energy of the first recovery stage, on the other hand, is independent of stress, as demonstrated by the straight parallel lines for τ_{III} in Fig. 9. If this activation energy would be the sum of U and the migration energy of divacancies, the latter quantity has to fully compensate the stress dependence of U, i.e., might be itself stress-dependent. A stress-dependent migration energy of divacancies, however, does not seem to be a promising idea at the high temperatures (and low stresses) necessary to reach the first recovery stage.

Haasen et al. (1987) put forward a model for the work-hardening coefficient θ in the deformation stages III and IV, and explicitly took into account the substructure that is produced in the deforming crystals by annihilation of edge dislocations by climb (Brion and Haasen, 1985). The following relation was derived:

$$\theta = \theta_{\text{II}} + A_{\text{H}}\sqrt{\tau} - B_{\text{H}}\frac{D}{\dot{\varepsilon}}\sqrt{\tau^5} \qquad (7)$$

Here A_{H} and B_{H} are model constants, and D is the coefficient of vacancy self-diffusion. In Eq. (7), the square-root term describes $\theta(\tau)$ in the hardening stage IV (see Fig. 2b), while the term proportional to $\tau^{2.5}$ represents the decrease of $\theta(\tau)$ in stage III. In the experiments, however, θ in stage III has been always found to vary linearly with stress (see, for example, Siethoff and Schröter, 1983). Furthermore, a fit of Eq. (7) to the appropriate data of Si yielded parameters that were different from the expected values to some extent (Haasen et al., 1987). Such problems may arise from difficulties in quantifying the properties of the dislocation substructure. Nevertheless, models involving the substructure are needed, if both recovery and work-hardening behavior are to be understood.

8. SECOND RECOVERY STAGE

Figure 14 shows the saturation stress in stage V for Ge with a $\langle 111 \rangle$ orientation (Siethoff et al., 1986). In the log–log plot, the data do not stretch on parallel lines; therefore, a law like Eq. (6) is not compatible with the measurements. Siethoff (1983) has proposed that cross-slip might be the recovery process underlying stage V and that a theory put forward by Escaig (1968) may be adopted to describe the second stage of dynamical recovery. In

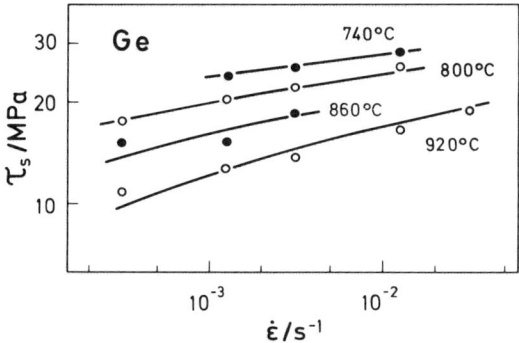

FIG. 14. Saturation stress of stage V as a function of strain rate for Ge $\langle 111 \rangle$ in a log–log plot; the lines represent a fit of Eq. (8) to the data. From Siethoff et al. (1986).

Escaig's model it is assumed that in extended screw dislocations, constrictions exist from which the dislocations dissociate on the cross-slip plane by a stress-assisted, thermally activated process. In this way, glide barriers may be overcome and internal stresses released. Escaig finally arrived at the relation

$$\tau = \tau_0\left(1 + \frac{kT}{E_Q^0} \ln \frac{\dot\varepsilon}{\dot\varepsilon_0}\right). \tag{8}$$

Here τ is either τ_v or τ_s, and $\dot\varepsilon_0$ is a rate constant with $\ln(\dot\varepsilon_0 \times 1\text{s}) = 24 \pm 2$. The effective activation energy of the process is $E_Q^0 (1 - \tau/\tau_0)$; the maximum cross-slip energy E_Q^0 and the parameter τ_0 are derived as

$$E_Q^0 = \frac{G^2 b^4}{1875\gamma} \left[\ln\left(\frac{Gb}{14.5\gamma}\right)\right]^{1/2}; \quad \tau_0 = \gamma/3b. \tag{9}$$

Both quantities can be calculated from the shear modulus G, the magnitude of the Burgers vector, b, and the stacking-fault energy γ (see Section II.1); they are listed in Table V in the first line for each material for which measurements are available. It is emphasized that the high values of E_Q^0 are a direct consequence of the large Burgers vectors (b^4-dependence of E_Q^0) and are, finally, responsible for the occurrence of the second recovery stage in the semiconductors at high temperatures. In metals, for example, E_Q^0 is smaller than the activation energy of self-diffusion, and dynamical recovery by cross-slip occurs at low temperatures (Schröter and Siethoff, 1985). Möller (1978)

TABLE V

PARAMETERS OF THE SECOND STAGE OF DYNAMICAL RECOVERY AND OF STEADY-STATE CREEP FOR UNDOPED SEMICONDUCTORS, FROM ANALYSIS OF EQ. (8)

	Method	E_Q^0/eV	τ_0/MPa	$\ln(\dot\varepsilon_0 \times 1\text{s})$	Data source
Ge	Eq. (9)	4.3	63	24 ± 2	—
	$\tau_v(\dot\varepsilon, T) \langle 123 \rangle$	3.8	112	27.2	Siethoff (1983)
	$\tau_s(\dot\varepsilon, T) \langle 111 \rangle$	4.4	89	30.2	Siethoff et al. (1986)
	$\dot\varepsilon_c(\tau_c, T) \langle 111 \rangle$	3.95	460	—	Betekhtin and Bakhtibaev (1970)
	$\dot\varepsilon_c(\tau_c, T) \langle 111 \rangle$	4.5	349	25.3	Myshlyaev and Khodos (1971)
	$\dot\varepsilon_c(\tau_c, T) \langle 111 \rangle$	4.5	260	—	Demenet et al. (1983)
Si	Eq. (9)	7.4	50	24 ± 2	—
	$\tau_v(\dot\varepsilon, T) \langle 123 \rangle$	4.4	123	21.9	Siethoff (1983)
	$\dot\varepsilon_c(\tau_c, T) \langle 111 \rangle$	5.6	338	25.3	Myshlyaev et al. (1969)
	$\dot\varepsilon_c(\tau_c, T) \langle 111 \rangle$	4.9	296	—	Betekhtin and Bakhtibaev (1970)
GaAs	Eq. (9)	5.3	41	24 ± 2	—
	$\dot\varepsilon_c(\tau_c, T) \langle 111 \rangle$	5.1	55	28	Behrensmeier (1990)

has shown that Escaig's formalism is also a suitable means to describe the cross-glide behavior of individual dislocations in Ge. Note that Eq. (8) is only valid for high stresses; at low stresses back-jumps (of the elementary cross-slip process) have to be taken into account.

The solid lines in Fig. 14 represent a fit of Eq. (8) to the data; the resulting parameters are given in the third line of Table V. The agreement with E_Q^0 and τ_0 as calculated from Eq. (9) is rather good. Also, the stress at the beginning of the second recovery stage of crystals with $\langle 123 \rangle$ orientation obeys Eq. (8) (Siethoff, 1983). The analysis of τ_v ($\dot{\varepsilon}$, T), however, yields a cross-slip energy that is somewhat smaller than that obtained under steady-state conditions (Table V). Such a behavior is known from the first recovery stage, where the stress at its beginning turned out to be controlled by a smaller activation energy than the steady-state flow stress (see Section IV.7). Recently, Rai et al. (1989) have proposed that stage V may be controlled by a diffusional process rather than by cross-slip. They refer to the work of Brion et al. (1981), where the onset of stage V was interpreted by a power law with an activation energy close to that of self-diffusion. For Si, however, problems arose with such an interpretation (Siethoff, 1983). Moreover, if a power law is applied to the data shown in Fig. 14 in the temperature range between 800 and 920°C, an activation energy of 3.5 eV and a stress exponent of about 9 is derived; if a model like that of Barrett and Nix (1965) is used, which takes into account the temperature dependence of the slope of the curves in Fig. 14, an activation energy of 3.9 eV results. Such quantities are not very convincing for a diffusional process in Ge at temperatures close to the melting point. For $\langle 123 \rangle$ GaAs, however, the situation may be different to some extent: The activation energy deduced from the saturation stress of the second recovery stage is estimated as 2.3 eV (Siethoff, 1989); such a value is compatible with a diffusion-controlled mechanism rather than with cross-slip (see Section IV.7). On the other hand, steady-state creep of $\langle 111 \rangle$ GaAs can be nicely described by Escaig's cross-slip model (see Section IV.9 and Table V). The whole subject will be further discussed in Section IV.10).

9. STEADY-STATE CREEP

In Section II.4 it was pointed out that the steady states attained in stress–strain curves and those in creep curves obey the same recovery process, although for the semiconductors some problems arose (see also Section IV.8). Nevertheless, high-temperature creep of Si, Ge, and GaAs is governed by Eq. (8) (which may be used in inverted form for creep experiments), and not by a relation such as Eq. (6), which was found to be compatible with power-law creep in metals and various other materials (and, of course, with the first recovery stage of semiconductors). Equation (8) is characterized by straight

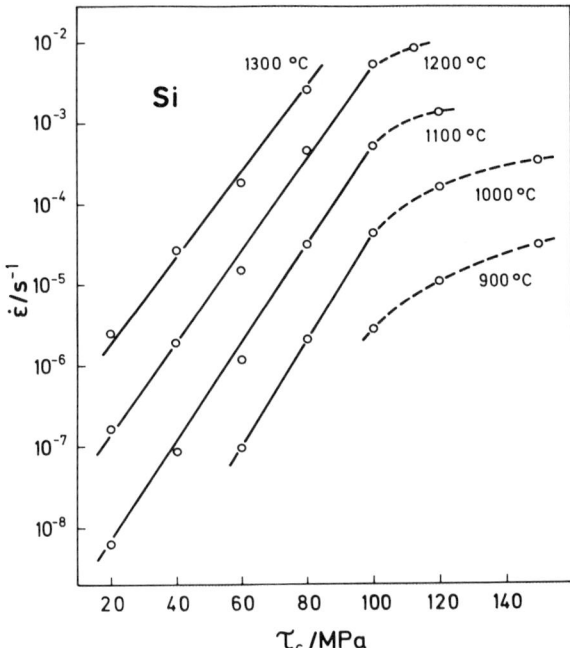

FIG. 15. Steady-state creep of Si with a $\langle 111 \rangle$ orientation, in a semilogarithmic plot of creep rate versus stress (data from Myshlyaev et al., 1969); the solid lines represent a fit of the (inverted) Eq. (8) to the data. From Schröter and Siethoff (1985).

lines in a semilogarithmic plot of strain rate versus stress. Such a behavior is shown in Fig. 15 for the case of Si by the solid lines; the deviations at high stresses (broken lines) will be explained in the following section. The appropriate parameters are indicated in Table V, together with those drawn from other creep experiments performed on Ge, Si, and GaAs, which also obey Eq. (8) (Siethoff, 1983; Behrensmeier, 1990).

Closer inspection of Table V shows that, at least for Ge and GaAs, the calculated and measured cross-slip energies E_Q^0 nicely agree (the somewhat lower values from the τ_v-measurements have been commented on in the preceding section). This obviously is not the case for the parameter τ_0. Similarly, the absolute stress levels attained in the various creep experiments do not coincide, although in most of the works the orientation of the crystals was the same. The origin of these differences is not fully understood, but there are some arguments: In Eq. (8), the stress τ is the applied stress, and not the effective stress, acting on the dislocations that are involved in the recovery process. Contrary to the lower yield point and to diffusion-controlled recovery, for which arguments were presented in Section III and Section IV.7, respectively, that effective and applied stress might be correlated in some way, such a dependence is not known for the cross-slip mechanism. Furthermore, it is not clear in some publications whether normal stresses or resolved shear stresses have been used. For example, the saturation stress of stress–strain curves of ⟨111⟩ Ge is lower than the steady-state creep stress by a factor of 3 to 5, if the two are compared at the same temperature and strain rate (Siethoff *et al.*, 1986), and differences among the creep experiments amount to a factor of 1.5 to 2. Therefore, if data from different works are to be compared, corrections have to be used.

10. Discussion

The preceding analysis has shown that the deformation of the semiconductors, if carried to higher strain, is characterized by different stages of dynamical recovery. Since the temperature (and strain-rate) dependence of the data in these stages obeys different laws, a temperature is expected to exist where the regimes intersect (Siethoff *et al.*, 1984). Figure 16 demonstrates that for Si and Ge, the point of intersection of two recovery stages lies within the

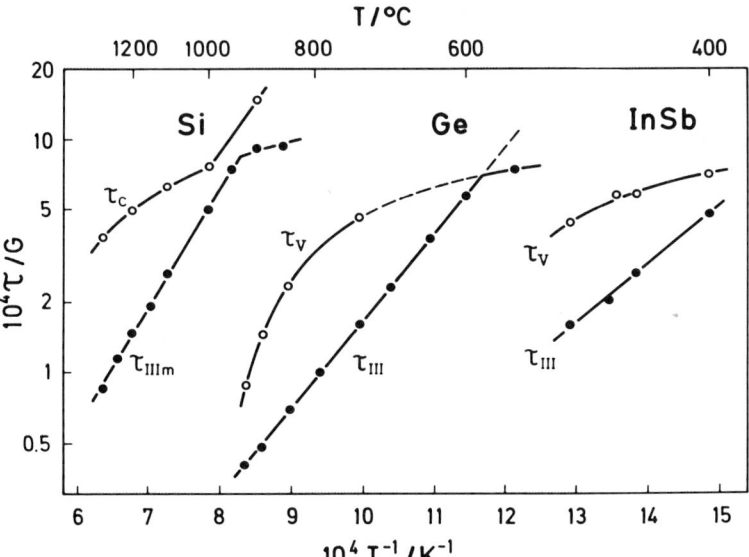

Fig. 16. The stages of dynamical recovery for Si, Ge, and InSb with a ⟨123⟩ orientation in an Arrhenius plot of the modulus-corrected flow stress at a strain rate $\dot{\varepsilon} = 3 \times 10^{-4}\,\mathrm{s}^{-1}$ (InSb: $\dot{\varepsilon} = 1.3 \times 10^{-3}\,\mathrm{s}^{-1}$). Data from Siethoff *et al.* (1986, 1988) and from Siethoff (1989).

temperature range accessible to measurements. GaAs shows a quite similar behavior (Siethoff and Behrensmeier, 1990); for InSb, the measurements have to be extended to lower temperatures. In this way, the deviations observed at high stresses for τ_{III} of Ge (Fig. 9) and for τ_c of Si (Fig. 15) are explained as the manifestation of a crossover of the recovery mechanisms involved.

Figure 16, however, also indicates that there remains a problem: The character of the break is different for steady-state creep of Si $\langle 111 \rangle$ (open points) and for the first recovery stage of Si and Ge with $\langle 123 \rangle$ orientation (closed points). In the former case, mechanisms work that, on either side of the break, lead to the *highest* flow stress. Such a behavior is typical for two thermally activated processes that are not independent on one another (sequential processes), as is known for recovery- and glide-controlled mechanisms. For Si $\langle 111 \rangle$, these are cross-slip at low and could be jog-dragging at high stresses. For the first recovery stage the situation is different: On both sides of the break the *lowest* flow stress is attained. This is typical for two independent (parallel) recovery processes that, for τ_{III} and τ_{IIIm} of crystals with a $\langle 123 \rangle$ orientation, are assumed to be climb at low and cross-slip at high stresses. Consequently, the high-temperature branch of τ_c $\langle 111 \rangle$ and the low-temperature branch of τ_{IIIm} $\langle 123 \rangle$ are the manifestation of the same mechanism (as may be the case for the respective branches of τ_v and τ_{III}), while for the low-temperature regime of τ_c and the high-temperature regime of τ_{IIIm}, different mechanisms have to be assumed. From all this follows that it cannot be excluded that three mechanisms of steady-state deformation are involved in the high-temperature plasticity of the semiconductors. These admittedly complicated problems are not completely understood. It is also an open question whether specimens with different orientations behave in the same way, or whether the observed phenomena depend on the individual material. It may be remembered in this context that the second recovery stage of GaAs $\langle 123 \rangle$ is perhaps controlled not by cross-slip but by a different mechanism (see Section IV.8). Stress–strain curves of Ge $\langle 111 \rangle$, on the other hand, show a recovery behavior similar to that of steady-state creep of Si $\langle 111 \rangle$ (Siethoff et al., 1986). It is finally noted that, at temperatures below the break, only one stationary regime has been observed in the same deformation curve in all known cases. On these questions, a unique answer cannot be given for the moment; further investigations have to be performed to arrive at more definite conclusions.

There is another point of interest that needs further consideration. In Sections III.5 and IV.7 it was shown that the quantities U/kT_m (Table III) as well as Q^{SD}/kT_m (Table IV) are different for the various semiconductors, and that the ionicity of the chemical bond is not a suitable material property to set up an order among these materials. Therefore, it was argued that another parameter might be involved here (Siethoff et al., 1990). In Table IV the ratio

Q^{SD}/U is indicated; apparently this quantity is independent of material, although the individual (normalized) activation energies are material-dependent. This, eventually, means that the activation energy of the first stage of dynamical recovery (which is assumed to be governed by self-diffusion; see Section IV.7) and that of dislocation mobility may be correlated. It is not a new idea that dislocation mobility and self-diffusion might be connected in some way. Patel and Chaudhuri (1966) noted a similarity in the doping behavior of both phenomena in highly doped Ge; this view, however, has been given up in the meantime. More recently, spectroscopic methods, which allow determination of the energy levels introduced in the energy gap by plastic deformation, have clearly revealed that dislocations in Si and Ge move by the emission or absorption of point defects at low temperatures (e.g., Alexander, 1986; Kisielowski-Kemmerich and Alexander, 1989). Recently, this effect has been corroborated for high-temperature deformation of Si (Thibault-Desseaux *et al.*, 1989). Also, in GaAs the creation of point-defect–related defects by plastic deformation has been reported in the literature (Omling *et al.*, 1986; Vignaud and Farvacque, 1989). It is one significant result of the deformation experiments described so far that, probably, a quantitative relation exists between the activation energies of dislocation mobility and those of the first recovery stage. These findings, if they are corroborated by further investigations, do not *a priori* rule out the kink mechanism, which has been found to underlie the motion of dislocations in semiconductors. The question, however, must be answered how kink mechanism and point defect production can be reconciled, i.e., which elementary process is eventually rate-controlling.

V. Effect of High Doping

In this part the effects of a strong interaction between electrically active impurity atoms and oppositely charged dislocations are described. Such phenomena occur at doping concentrations of 10^{19}–10^{20}/cm^3 ($c = 2 \times 10^{-4}$–2×10^{-3}) for substitutionally or interstitially dissolved, n- and p-type impurities in a temperature range where the conductivity of the material is still extrinsic. Although the stress–strain curves of the heavily doped crystals exhibit five deformation stages, as is the case for the undoped material, only the lower yield point has been analyzed to a greater extent. The influence of doping manifests itself in three different regimes, A, B, and C, which are best documented for highly As-doped Ge, as shown in Fig. 17 in a plot of the lower yield stress versus strain rate.

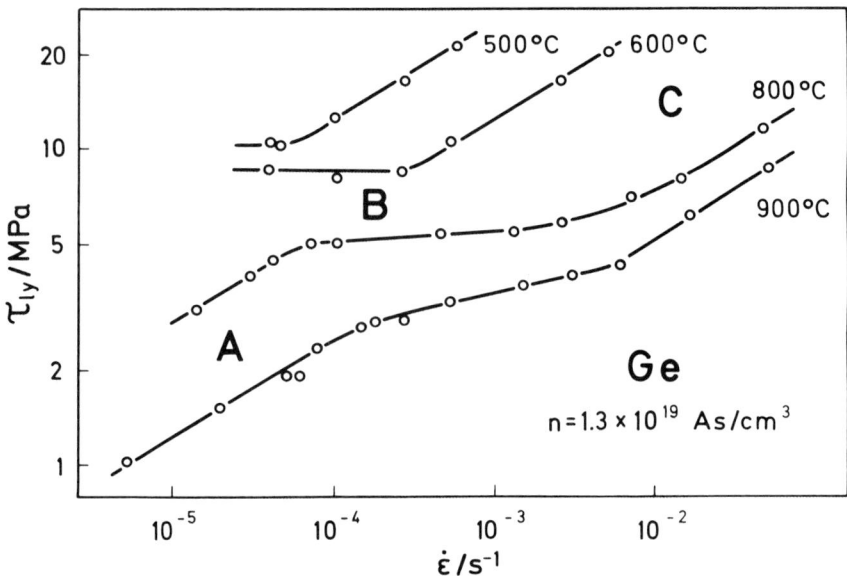

FIG. 17. The three regimes of plastic deformation occurring in highly doped semiconductors, for the case of n-type Ge ($n = 1.3 \times 10^{19}$ As/cm^3), in a log–log plot of lower yield stress versus strain rate. Regime A: dragging of impurity clouds by dislocations; regime B: breaking away of dislocations from impurity clouds; regime C: free dislocation motion. From Brion et al. (1971).

11. Dragging of Solute Atmospheres

For regime A it is assumed that solute atmospheres are formed around and dragged along with moving dislocations on account of a strong interaction (which will be discussed in detail in Section V.12), and that the rate-controlling process is the diffusion of the solute atoms. Brion et al. (1971) put forward a model to describe the temperature, strain-rate and concentration dependence of the lower yield stress, which may be written as

$$\tau_{ly}^A = \left(\frac{9\pi A^2 U_0^2}{4b^3 D_0 kT}\right)^{1/3} \dot{\varepsilon}^{1/3} c^{1/3} \exp\left(\frac{U_0 + U_D}{3kT}\right). \tag{10}$$

Here A is a constant, U_0 is the binding energy between dislocation and solute atom, c is the solute concentration, and U_D and D_0 are the activation energy and the pre-exponential factor, respectively, of impurity diffusion. Measurements on Ge doped with 1.3×10^{19} As/cm^3 (see Fig. 17) were found to agree with Eq. (10) concerning the temperature and strain-rate dependence of τ_{ly}^A. The equivalence of constant strain-rate and creep deformation (see

Section II.4) was also established in this case. For U_D a value of 2.8 eV was deduced (Brion et al., 1971), which has to be compared to 2.5 eV obtained from diffusion experiments (Seeger and Chik, 1968); D_0 was estimated from Eq. (10) as 6.5×10^{-3} m^2/s.

Probably, regime A has been also observed in InP doped with 2×10^{18} Zn/cm^3 (Völkl et al., 1987). The activation energy U_D was deduced as 2.2 eV, which is higher than the value of 1.4 eV for Zn-diffusion in InP (Van Gurp et al., 1987, and literature quoted therein). On the other hand, the $\dot{\varepsilon}^{1/3}$ and the $c^{1/3}$ dependence of τ_{1y}^A could be well established.

12. Breaking Away of Solute Atmospheres

Regime B has been intensely studied in Si and Ge (Siethoff, 1970; Brion et al., 1971). At not too high temperatures, the lower yield stress does not depend on strain rate, as shown in Fig. 17 for the data at 600°C. It is assumed that, at a certain stress, the dislocations break away from their impurity clouds. For the breakaway stress, the following relation has been derived (Brion et al., 1971):

$$\tau_{1y}^B = \frac{kT}{b^3} c \exp\left(\frac{U_0}{kT}\right). \tag{11}$$

Equation (11) describes a process that is not thermally activated and, therefore, is independent of strain rate. The temperature dependence of τ_{1y}^B enters Eq. (11) via the equilibrium concentration of the impurities around the dislocations, which depends exponentially on the binding energy U_0. It has been shown (Siethoff, 1980) that Eq. (11) is compatible with a model discussed by Mohamed (1979), which is characterized by an impurity cloud highly segregated at the dislocation core for $U_0 > kT$. The linear c-dependence of τ_{1y}^B is excellently fulfilled in the experiments (Siethoff, 1970). The binding energy U_0 is deduced from the experiments as 0.2 eV for As-doped (n-type) Ge and 0.36 eV for P-doped (n-type) Si. These values have to be compared to those estimated from the two types of interaction that are involved here (Siethoff, 1969b). There is, at first, an elastic interaction between the stress field of the dislocations and the strain around the solutes, if their size is different from that of the lattice atoms. The corresponding energies are typically 0.05 to 0.1 eV (and only for B-doped [p-type] Si 0.4 eV). Secondly, a short-range electrostatic interaction exists between the ionized dopants (extrinsic conductivity range) and the oppositely charged dislocations, with a binding energy of 0.4 and 0.6 eV for As in Ge and P in Si, respectively. These values are nearly as twice as high as those derived from the measurements. It is argued that the electrostatic binding energy may be overestimated to a certain extent, since the strong segregation of the impurities at the dislocation core was not taken into account.

It is well-known from metal alloys that breakaway of dislocations from impurity clouds is a collective phenomenon, in which a large ensemble of dislocations is involved. As a result, the deformation occurs inhomogeneously, leading to visible serrations in the stress–strain curves (jerky flow). Figure 18 demonstrates that such effects are also typical for heavily doped semiconductors. Two kinds of serrations may be distinguished: one emerging exclusively in the yield point region (curve a), which, however, is not observed in all cases (curve b), although deformation procedure and material properties are the same; the other occurring in the stages I and II of the stress–strain curves. The first kind is connected with a rather peculiar dislocation distribution across the specimens, as shown by crystal a in Fig. 19. The dislocation density rises steeply from zero to a high value. Such structures, which, during deformation, migrate along the specimens, are usually called

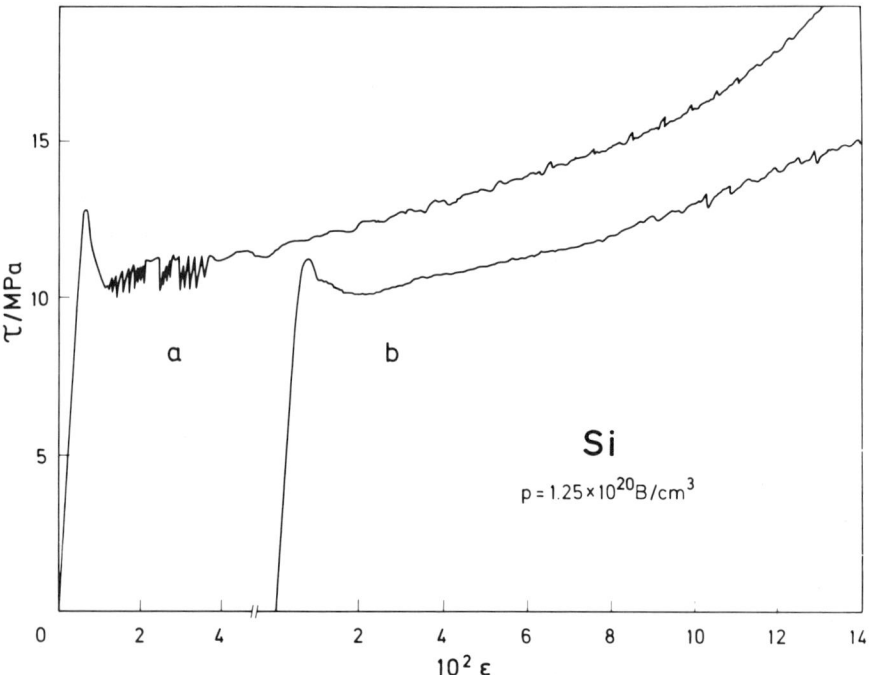

FIG. 18. Stress–strain curves in the yield region and stage I of highly B-doped Si ($p = 1.25 \times 10^{20}/\text{cm}^3$, $T = 1,100°C$, $\dot\varepsilon = 4.8 \times 10^{-4} \text{ s}^{-1}$). From Siethoff (1970).

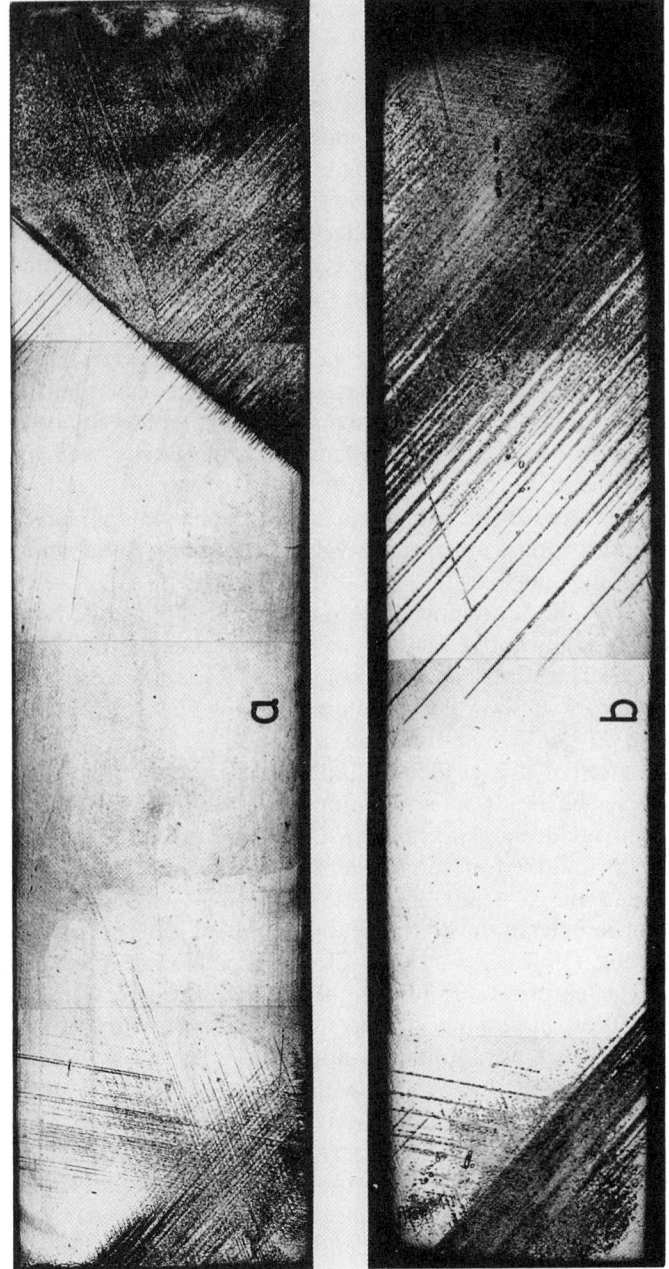

FIG. 19. Dislocation distribution at the lower yield point, revealed by etch pits on (111): (a) Type K Lüders band in P-doped Si ($n = 10^{20}/\text{cm}^3$, $T = 1{,}000°\text{C}$, $\dot{\varepsilon} = 4.8 \times 10^{-4}\,\text{s}^{-1}$); (b) type G Lüders band in B-doped Si ($p = 1.25 \times 10^{20}/\text{cm}^3$, $T = 1{,}100°\text{C}$, $\dot{\varepsilon} = 4.8 \times 10^{-4}\,\text{s}^{-1}$); specimen length 15 mm. From Siethoff (1969b, 1973).

Lüders bands. The band front is a kink band with a highly polygonized structure of pure edge dislocations of the same sign, which are arranged in walls perpendicular to the primary glide plane and which lead to a macroscopically visible kink in the specimens. These type-K Lüders bands move discontinuously and produce the stress fluctuations shown by curve a in Fig. 18 (Siethoff, 1969b, 1973). Occasionally, under the same doping and deformation conditions, other Lüders bands (type-G bands) develop, the fronts of which are more or less parallel to the main glide plane (crystal b in Fig. 19), and which move continuously in the yield point region, i.e., without serrations (curve b in Fig. 18). The lower yield stress, however, is practically the same for both types of bands. The stress fluctuations emerging in the later stages of the stress–strain curves have not been further investigated. They are ascribed to a Portevin–LeChatelier effect that is frequently observed in metal alloys (Haasen, 1965).

The regimes A and B (including Lüders-band phenomena) have also been observed in Czochralski-grown (oxygen-containing) Si crystals with an additional doping concentration of 2×10^{18} Sb/cm^3 (Gross et al., 1989). The authors attribute the hardening effect to oxygen precipitation in the course of deformation; a quantitative analysis of the lower yield stress, however, was not performed. The stress τ_{III} at the beginning of the first recovery stage, on the other hand, was shown to be compatible with the measurements on undoped and oxygen-free Si discussed in Section IV.7; only the overall stress level was somewhat higher. These findings are not in agreement with the behavior of τ_{III} of highly n- and p-doped Si, which is characterized by a weaker strain-rate dependence, i.e., by a higher stress exponent in Eq. (6), in comparison to undoped material (Siethoff, 1989). For further information, see the Addendum.

It has been claimed in the literature that regime B is observed in InP doped with 10^{19} S/cm^3 (Völkl et al., 1987). Closer inspection of the data, however, reveals that the features typical for highly doped Si and Ge (stress plateau at low temperatures, increasing strain-rate dependence with rising temperatures; see Fig. 17) are not found in S-doped InP. The stress seems to depend on strain rate rather by a power law with a stress exponent of about 7, and it is not easy to see how the resulting activation energy of 0.8 to 1 eV could be reconciled with the mechanism underlying Eq. (11). It is noted that this regime occurring in doped InP shows some similarity to the regime found in undoped material at high temperatures and low strain rates (Reppich et al., 1990), which has been discussed in Section III. More information, however, is needed to arrive at definite conclusions.

Concerning GaAs, serrated flow has been detected in crystals doped with 5×10^{19} to 4×10^{20} In/cm^3 at high temperatures (Djemel et al., 1988); the strain-rate dependence of the flow stress, however, was not studied. A strong hardening effect is somewhat surprising for an isovalent impurity such as In in GaAs, for which an electrostatic interaction with dislocations is not expected to occur. Adding a few percent of Ge to Si, for example, produces only a small hardening on account of a weak elastic interaction, and serrations and Lüders bands are not observed (Siethoff, 1969a). Ehrenreich and Hirth (1985) propose an $InAs_4$ complex as an effective obstacle to dislocation motion in In-doped GaAs. This has been criticized by Yonenaga and Sumino (1987), who found that isovalent impurities did not affect the motion of dislocations to a greater extent. Their measurements, however, were performed at temperatures lower than those used in the work of Djemel et al. (1988).

13. NONLOCAL INTERACTION

In regime C, finally, the dislocations are assumed to move freely. The dislocation mobility, however, is affected by doping via a nonlocal interaction between impurities and dislocations, because the Fermi level of heavily doped material is different from that of undoped crystals. This effect changes the charge on the dislocations and, probably, the density and mobility of the kinks. As a consequence, the activation energy U of the dislocation mobility is altered, but eventually, the formalism of Eqs. (2)–(4) can be still used for the interpretation of the data. For a comprehensive discussion of this subject, the reader is referred to reviews of Hirsch (1985) and Alexander (1986), where the theoretical background is also dealt with. Measurements of the lower yield stress (Brion et al., 1971) and of the creep rate at the inflection point (Shea and Heldt, 1967) in Ge yielded a decrease of U for n-doping (As) and an increase for p-doping (Ga). For P-doped (n-type) Si, a decrease of U was also found, while B-doping (p-type) did not influence U to a greater extent (Siethoff, 1970; Sumino et al., 1981). In all experiments, the parameter $2 + m$ was practically unchanged. These results are compatible with those obtained from measurements of the dislocation velocity, with the exception of B-doping in Si, where a slight decrease of U was found. It has, however, to be taken into account that the dislocation velocity has been investigated at lower temperatures than the lower yield point in most cases, and that the activation energies deduced from the mobility measurements may depend on stress (see, for example, Alexander, 1986).

Regime C has been also studied in highly doped III–V compounds. Concerning GaAs, satisfactory agreement of the different investigations can be noticed. Osvenskii et al. (1969) measured the creep at the inflection point and found a strong increase of the activation energy U (via Eq. (3)) for n-type material (Te-doping) with $U = 2.3$ eV, and a moderate decrease of U for p-type material (Zn-doping), with $U = 1.1$ eV in comparison to 1.45 eV for undoped crystals (see Table III); the parameter $2 + m$ was not appreciably changed. These results were fully corroborated by Steinhardt and Haasen (1978). More recently, the temperature dependence of the yield point was investigated (Bourret et al., 1987; Nakada and Imura, 1987), yielding information on the quantity $U/(2 + m)$. Assuming the parameter $2 + m$ to lie in the range of 3 to 3.5, activation energies U are obtained that confirm the doping dependence found in the creep experiments. Activation energies drawn from measurements of the dislocation mobility (Choi et al., 1977; Yonenaga and Sumino, 1989a) vary between 0.85 and 1.7 eV; their doping dependence is qualitatively the same as in the macroscopic investigations for β dislocations, while for α and screw dislocations, the sequence is reversed; the parameter $2 + m$ varies between 1.4 and 2.1. High doping with isoelectronic impurities such as In has practically no influence on the yield point at low temperatures (Hobgood et al., 1986) and on the dislocation mobility (Matsui and Yokoyama, 1985; Burle-Durbec et al., 1987; Yonenaga and Sumino, 1989a), although at high temperatures impurity effects such as serrated flow have been observed (see Section V.12). Probably, the defect structure in the crystals sensibly depends on temperature. In this context, the ability of dopant atoms to form complexes in a certain temperature range (Dannefaer et al., 1989) may be of importance.

For InP, the situation concerning regime C is less satisfactory. Reppich et al. (1990), from analysis of the lower yield point, found the activation energy U to increase to 2 eV for n-type material (S-doping), in some agreement with dislocation mobility measurements of George and Jacques (1987), who deduced values between 1.4 and 1.8 eV. Völkl et al. (1987), on the other hand, measured $U = 0.6$ eV in a creep experiment. These authors also investigated the influence of p-doping (Zn) and found $U = 1.7$ eV, i.e., an increase in comparison to undoped InP.

VI. Addendum

Since completion of the manuscript, new material has been published and shall be briefly summarized. The investigations on III-V compounds were extended to GaSb (Siethoff et al., 1991a). In this material, the lower yield stress (Eq. 2) is characterized by an activation energy $U = 1.2$ eV and a stress exponent $2 + m = 3$, while the stress τ_{III} of the first stage of dynamical recovery (Eq. 6) yields $Q^{SD} = 1.7$ eV and $n = 3.8$. The ratio $Q^{SD}/U = 1.42$ is of the same magnitude as that of the other semiconductors (see Table IV). This corroborates the view that a relation exists between diffusion-controlled dynamical recovery and the dislocation mobility, which has been discussed in Section IV.10. However, the activation energy of 1.7 eV is smaller than Q^{SD}

values obtained in tracer diffusion measurements (Weiler and Mehrer, 1984). This has been explained as follows: In elemental semiconductors, edge-type dislocations act as perfect sinks and sources for vacancies and interstitials, thereby allowing for an independent approach of each native point defect concentration to its equilibrium value, and the activation energy of dynamical recovery is close to that of self-diffusion (see Section IV.7). In III–V compounds, on the other hand, climb of dislocations simultaneously affects the point defect concentrations in the two sublattices and therefore only allows for a local equilibrium between them. This, eventually, leads to activation energies of diffusion-controlled dynamical recovery smaller than Q^{SD} not only for GaSb, but also for other III–V compounds (Siethoff et al., 1991a).

In the same work, comparison of the different semiconductors investigated so far established a linear relation between stacking-fault energy and the stress exponent n of the first recovery stage, which is shown in Fig. 20. Obviously, n approaches 3 for small values of γ/Gb. This indicates that the stress dependence of the strain rate in Eq. (6) can be separated in a term proportional to τ^3, which may be identified with the so-called "natural" creep law (Stocker and Ashby, 1973); and in a term that is probably related to the

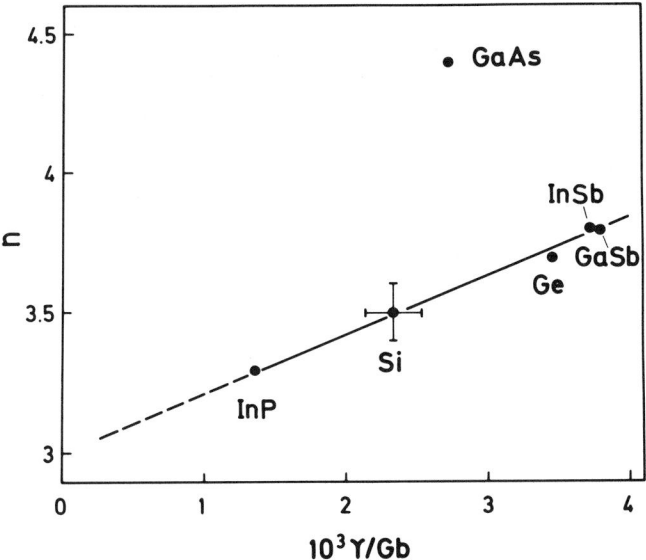

FIG. 20. Stress exponent n, characteristic of the first recovery stage of the various semiconductors (Eq. 6), as a function of the (normalized) stacking-fault energy γ; G: shear modulus (room-temperature values); b: Burgers vector. From Siethoff et al. (1991a).

evolution of the dislocation substructure during deformation and is expected to depend on the stacking-fault energy. GaAs, however, is an exception to this rule, which is not yet understood.

The investigations of dynamical recovery were also extended to heavily doped Si and Ge (Siethoff *et al.*, 1991b); the material was the same as that used for the measurements of the lower yield stress (see Section V). The most striking result was that, contrary to undoped crystals, the first recovery stage could not be interpreted by a diffusion-controlled mechanism. Rather a behavior typical of metal solid solutions at high stresses was found (Siethoff and Ahlborn, 1991). It is concluded that alloying not only affects the yield point of semiconductors, but also the higher degrees of deformation. It is interesting to note, however, that the second recovery stage, which is governed by a cross-slip mechanism (see Section IV.8), is not influenced by doping. The strong interaction between dislocations and solute atoms responsible for the above alloying effects was formerly assumed to be of elastic and electrostatic nature (Siethoff, 1969b). Recent theoretical work (Heggie *et al.*, 1989) has shown that, at least for phosphorus in Si, there may be a different type of interaction: If phosphorus is trapped in the dislocation core, an energetically favorable configuration occurs by bond reconstruction which, finally, leads to a strong local (electrically inactive) pinning center for the dislocation.

At very low temperatures, where deformation has to be performed under a confining pressure to avoid damage of the specimens, the yield stress shows deviations from the high-temperature behavior (see Section III.5). Recent investigations indicate that twinning (besides dislocation glide) is a predominant deformation mechanism in this regime (Androussi *et al.*, 1989; Boivin *et al.*, 1990; Pirouz and Hazzledine, 1991). The yield point is suppressed and the stress-strain curve is parabolic; the influence of doping is reversed. Rabier and Boivin (1990) interpret these results by cross-slip in terms of the model originally proposed by Escaig (1968).

Acknowledgments

The author is indebted to Professor P. Haasen for providing experimental facilities at his institute over a couple of years. Valuable comments on the manuscript by Professor W. Schröter are greatfully acknowledged. Some of the work described here has been supported by the Deutsche Forschungsgemeinschaft.

References

Alexander, H. (1986). *In* "Dislocations in Solids" (F. R. N. Nabarro, ed.), Vol. 7, p. 113. North-Holland, Amsterdam.

Alexander, H., and Haasen, P. (1961). *Acta Met.* **9**, 1001.
Alexander, H., and Haasen, P. (1968). *Solid State Phys.* **22**, 27.
Allem, R., Michel, J.-P., and George, A. (1989). *Phil. Mag. A* **59**, 273.
Androussi, Y., Vanderschaeve, G., and Lefebvre, A. (1989). *Phil. Mag. A* **59**, 1189.
Astié, P., Couderc, J. J., Chomel, P., Quelard, D., and Duseaux, M. (1986). *Phys. Stat. Sol. (a)* **96**, 225.
Barbot, J. F., Rivaud, G., and Desoyer, J. C. (1986). *Materials Science Forum* **10–12**, 809.
Barrett, C. R., and Nix, W. D. (1965). *Acta Met.* **13**, 1247.
Behrensmeier, R. (1990). Thesis, University of Göttingen.
Betekhtin, V. I., and Bakhtibaev, A. N. (1970). *Soviet Phys. Dokl.* **14**, 1007.
Blum, W., and Finkel, A. (1982). *Acta Met.* **30**, 1715.
Blum, W., Münch, H., and Portella, P. D. (1984). *In* "Creep and Fracture of Engineering Materials and Structures" (B. Wilshire and D. R. J. Owen, eds.), Vol. 1, p. 131. Pineridge Press, Swansea, UK.
Boivin, P., Rabier, J., and Garem, H. (1990), *Phil. Mag. A* **61**, 647.
Bourgoin, J. C., von Bardeleben, H. J., and Stiévenard, D. (1988). *J. Appl. Phys.* **64**, R65.
Bourret, E. D., Tabache, M. G., Beeman, J. W., Elliot, A. G., and Scott, M. (1987). *J. Cryst. Growth* **85**, 275.
Brion, H. G., and Haasen, P. (1985). *Phil. Mag. A* **51**, 879.
Brion, H. G., Haasen, P., and Siethoff, H. (1971). *Acta Met.* **19**, 283.
Brion, H. G., Schröter, W., and Siethoff, H. (1979). *Inst. Phys. Conf. Ser.* **46**, 508.
Brion, H. G., Siethoff, H., and Schröter, W. (1981). *Phil. Mag. A* **43**, 1505.
Burle-Durbec, N., Pichaud, B., and Minari, F. (1987). *Phil. Mag. Lett.* **56**, 173.
Castaing, J., Veyssière, P., Kubin, L. P., and Rabier, J. (1981). *Phil. Mag. A* **44**, 1407.
Chiang, S.-W., and Kohlstedt, D. L. (1985). *J. Mater. Sci.* **20**, 736.
Choi, S. K., Mihara, M., and Ninomiya, T. (1977). *Jap. J. Appl. Phys.* **16**, 737.
Corbel, C., Stucky, M., Hautojärvi, P., Saarinen, K., and Moser, P. (1988). *Phys. Rev. B* **38**, 8192.
Dannefaer, S., Mascher, P., and Kerr, D. (1989). *J. Phys.: Condens. Matter* **1**, 3213.
Demenet, J. L., Grosbras, P., and Desoyer, J. C. (1983). *Phys. Stat. Sol. (a)* **75**, K33.
Demenet, J. L., Desoyer, J. C., Rabier, J., and Veyssière, P. (1984). *Scripta Met.* **18**, 41.
Denisova, N. V., Zorin, E. I., Pavlov, P. V., Tetel'baum, D. I., and Khokhlov, A. F. (1975). *Izv. Akad. Nauk. SSSR, Neorg. Mater.* **11**, 2236.
Djmel, A., Castaing, J., and Duseaux, M. (1988). *Phil. Mag. A* **57**, 671.
Dlubek, G., Dlubek, A., Krause, R., Brümmer, O., Friedland, K., and Rentzsch, R. (1988). *Phys. Stat. Sol. (a)* **106**, 419.
Ehrenreich, H., and Hirth, J. P. (1985). *Appl. Phys. Lett.* **46**, 668.
Eisen, F. H., and Birchenall, C. E. (1957). *Acta Met.* **5**, 265.
Escaig, B. (1968). *J. Phys., Paris* **29**, 225.
Fahey, P. M., Griffin, P. B., and Plummer, J. D. (1989). *Rev. Mod. Phys.* **61**, 289.
Frank, W., Gösele, U., Mehrer, H., and Seeger, A. (1984). *In* "Diffusion in Crystalline Solids" (G. E. Murch and A. S. Nowick, eds.), p. 63. Academic Press, Orlando, Florida.
Frost, H. J., and Ashby, M. F. (1982). "Deformation-Mechanism Maps," p. 71. Pergamon Press, Oxford.
Gall, P., Peyrade, J. P., Coquillé, R., Renaud, F., Gabillet, S., and Albacete, A. (1987). *Acta Met.* **35**, 143.
George, A., and Jacques, A. (1987). *Izv. Akad. Nauk. SSSR, Ser. Phys.* **51**, 105.
George, A., and Rabier, J. (1987). *Revue Phys. Appl.* **22**, 941.
Gottschalk, H., (1979). *J. Phys., Paris* **40**, C6-127.
Gottschalk, H., Patzer, G., and Alexander, H. (1978). *Phys. Stat. Sol. (a)* **45**, 207.
Gross, T. S., Mathews, V. K., De Angelis, R. J., and Okazaki, K. (1989). *Mater. Sci. Eng. A* **117**, 75.

Guruswamy, S., Rai, R. S., Faber, K. T., and Hirth, J. P. (1987). *J. Appl. Phys.* **62**, 4130.
Haasen, P. (1965). In "Physical Metallurgy" (R. W. Cahn, ed.), p. 821. North-Holland, Amsterdam.
Haasen, P., Ahlborn, K., and Schröter, W. (1987). *Izv. Akad. Nauk. SSSR, Ser. Phys.* **51**, 749.
Hall, E. L., and Vander Sande, J. B. (1978). *J. Amer. Ceram. Soc.* **61**, 417.
Heggie, M., Jones, R., Lister, G. M. S., and Umerski, A. (1989). *Inst. Phys. Conf. Ser.* **104**, 43.
Hirsch, P. B. (1985). In "Dislocations and Properties of Real Materials," p. 333. The Institute of Metals, London.
Hobgood, H. M., McGuigan, S., Spitznagel, J. A., and Thomas, R. N. (1986). *Appl. Phys. Lett.* **48**, 1654.
Iguchi, H. (1989). *Jap. J. Appl. Phys.* **28**, L 2115.
Kendall, D. L., and Huggins, R. A. (1969). *J. Appl. Phys.* **40**, 2750.
Kisielowski-Kemmerich, C., and Alexander, H. (1989). *Sov. Phys. Sol. State* **31**, 864.
Kojima, K., and Sumino, K. (1971). *Cryst. Latt. Defects* **2**, 147.
Latanision, R. M., and Staehle, R. W. (1968). *Scripta Met.* **2**, 667.
Lee, B.-T., Gronsky, R., and Bourret, E. D. (1989). *J. Cryst. Growth* **96**, 333.
Madelung, O. (1982). "Landolt-Börnstein," Vol. III/17a. Springer-Verlag, Berlin.
Matsui, M., and Yokoyama, T. (1985). *Inst. Phys. Conf. Ser.* **79**, 13.
Mohamed, F. A. (1979). *Mater. Sci. Eng.* **38**, 73.
Mohamed, F. A., and Langdon, T. G. (1974). *J. Appl. Phys.* **45**, 1965.
Möller, H.-J. (1978). *Phil. Mag.* **37**, 41.
Morozov, A. N., and Bublik, V. T. (1986). *J. Cryst. Growth* **75**, 497.
Myshlyaev, M. M., and Khodos, I. I. (1971). *Phys. Stat. Sol. (b)* **43**, 83.
Myshlyaev, M. M., Nikitenko, V. I., and Nesterenko, V. I. (1969). *Phys. Stat. Sol.* **36**, 89.
Nabarro, F. R. N. (1977). In "Surface Effects in Crystal Plasticity" (R. M. Latanision and J. T. Fourie, eds.), p. 49. Nordhoff, Leyden.
Nakada, Y., and Imura, T. (1987). *Phys. Stat. Sol. (a)* **103**, 435.
Omling, P., Weber, E. R., and Samuelson, L. (1986). *Phys. Rev. B* **33**, 5880.
Omri, M. (1981). Thesis, L'Institut National Polytechnique de Lorraine, Nancy.
Omri, M., Tete, C., Michel, J.-P., and George, A. (1987). *Phil. Mag. A* **55**, 601.
Osip'yan, Y. A., Petrenko, V. F., Zaretskii, A. V., and Whitworth, R. W. (1986). *Advances in Physics* **35**, 115.
Osvenskii, V. B., Mil'vidskii, M. G., and Stolyarov, O. G. (1969). *Sov. Phys. Cryst.* **13**, 718.
Oueldennaoua, A., Allem, R., George, A., and Michel, J.-P. (1988). *Phil. Mag. A* **57**, 51.
Palfrey, H. D., Brown, M., and Willoughby, A. F. W. (1983). *J. Electron. Mater.* **12**, 863.
Parsey, J. M., Jr., Asom, M. T., Kimerling, L. C., Sauer, R., and Thiel, F. A. (1988). In "Defects in Electronic Materials" (M. Stavola, S. J. Pearton, and G. Davies, eds.), p. 429. Material Research Society, Pittsburgh.
Patel, J. R., and Chaudhuri, A. R. (1966). *Phys. Rev.* **143**, 601.
Patu, S., Arsenault, R. J., and Kramer, I. R. (1986). *Mater. Sci. Eng.* **78**, 145.
Paufler, P., and Retzlaff, U. (1989). *Cryst. Res. Technol.* **24**, 701.
Peissker, E., Haasen, P., and Alexander, H. (1962). *Phil. Mag.* **7**, 1279.
Pirouz, P., and Hazzledine, P. M. (1991). *Scripta Met.* **25**, 1167.
Rabier, J., and Boivin, P. (1990). *Phil. Mag. A* **61**, 673.
Rabier, J., and George, A. (1987). *Revue Phys. Appl.* **22**, 1327.
Rabier, J., Veyssière, P., and Demenet, J. L. (1983). *J. Phys., Paris* **44**, C4-243.
Rai, R. S., Guruswamy, S., Faber, K. T., and Hirth, J. P. (1989). *Phil. Mag. A* **60**, 339.
Reppich, B., Rieger, K., and Müller, G. (1990). *Z. Metallk.* **81**, 166.
Schäfer, S., Alexander, H., and Haasen, P. (1964). *Phys. Stat. Sol.* **5**, 247.
Schröter, W., and Siethoff, H. (1985). *Czech. J. Phys. B* **35**, 307.
Schröter, W., Brion, H. G., and Siethoff, H. (1983). *J. Appl. Phys.* **54**, 1816.

Seeger, A., and Chik, K. P. (1968). *Phys. Stat. Sol.* **29**, 455.
Shea, M. M., and Heldt, L. A. (1967). *Abstr. Bull. Inst. Met.* **2**, 33.
Shimizu, H., and Sumino, K. (1975). *Phil. Mag.* **32**, 123.
Siethoff, H. (1969a). *Mater. Sci. Eng.* **4**, 155.
Siethoff, H. (1969b). *Acta Met.* **17**, 793.
Siethoff, H. (1970). *Phys. Stat. Sol.* **40**, 153.
Siethoff, H. (1973). *Acta Met.* **21**, 1523.
Siethoff, H. (1980). *Scripta Met.* **14**, 601.
Siethoff, H. (1983). *Phil. Mag. A* **47**, 657.
Siethoff, H. (1988). *Phil. Mag. Letters* **58**, 129.
Siethoff, H. (1989). Unpublished results.
Siethoff, H., and Ahlborn, K. (1991). *Phys. Stat. Sol. (a)* **128**, 397.
Siethoff, H., and Alexander, H. (1964). *Phys. Stat. Sol.* **6**, K165.
Siethoff, H., and Behrensmeier, R. (1990). *J. Appl. Phys.* **67**, 3673.
Siethoff, H., and Schröter, W. (1978). *Phil. Mag. A* **37**, 711.
Siethoff, H., and Schröter, W. (1983). *Scripta Met.* **17**, 393.
Siethoff, H., and Schröter, W. (1984). *Z. Metallk.* **75**, 475.
Siethoff, H., Ahlborn, K., and Schröter, W. (1984). *Phil. Mag. A* **50**, L1.
Siethoff, H., Brion, H. G., Ahlborn, K., and Schröter, W. (1986). *Phys. Stat. Sol. (a)* **97**, 153.
Siethoff, H., Völkl, J., Gerthsen, D., and Brion, H. G. (1987). *Phys. Stat. Sol. (a)* **101**, K13.
Siethoff, H., Ahlborn, K., Brion, H. G., and Völkl, J. (1988). *Phil. Mag. A* **57**, 235.
Siethoff, H., Behrensmeier, R., Ahlborn, K., and Völkl, J. (1990). *Phil. Mag. A* **61**, 233.
Siethoff, H., Brion, H. G., and Schröter, W. (1991a). *Phys. Stat. Sol. (a)* **125**, 191.
Siethoff, H., Ahlborn, K., and Brion, H. G. (1991b). *Acta Met. Mater.* **39**, 1133.
Steinhardt, H., and Haasen, P. (1978). *Phys. Stat. Sol. (a)* **49**, 93.
Stocker, R. L., and Ashby, M. F. (1973). *Scripta Met.* **7**, 11.
Sumino, K., Yonenaga, I., Harada, H., and Imai, M. (1981). In "Dislocation Modelling of Physical Systems" (M. F. Ashby, R. Bullough, C. S. Hartley, and J. P. Hirth, eds.), p. 212. Pergamon Press, New York.
Tan, T. Y., and Gösele, U. (1985). *Appl. Phys. A* **37**, 1.
Tan, T. Y., and Gösele, U. (1988). *Mater. Sci. Eng. B* **1**, 47.
Thibault-Desseaux, J., Kirchner, H. O. K., and Putaux, J. L. (1989). *Phil. Mag. A* **60**, 385.
Tuck, B. (1988). "Atomic Diffusion in III-V Semiconductors," Adam Hilger, Bristol.
Van Gurp, G. J., Boudewijn, P. R., Kempeners, M. N. C., and Tjaden, D. L. A. (1987). *J. Appl. Phys.* **61**, 1846.
Vignaud, D., and Farvacque, J. L. (1989). *J. Appl. Phys.* **65**, 1516.
Vogel, G., Hettich, G., and Mehrer, H. (1983). *J. Phys. C: Solid State Phys.* **16**, 6197.
Völkl, J., Müller G., and Blum, W. (1987). *J. Cryst. Growth* **83**, 383.
Wagner, R., Wöhler, F. D., and Haasen, P. (1971). *Phys. Stat. Sol. (b)* **44**, 381.
Weertman, J. (1955). *J. Appl. Phys.* **26**, 1213.
Weertman, J., and Weertman, J. R. (1987). *Proc. 8th Riso Int. Symp. Metallurgy and Materials Science*, p. 191. Riso National Laboratory, Roskilde.
Weiler, D., and Mehrer, H. (1984). *Phil. Mag. A* **49**, 309.
Weiss, L. (1975). Thesis, University of Cologne.
Wilsdorf, H. G. F., and Ruddle, G. E. (1977). In "Surface Effects in Crystal Plasticity" (R. M. Latanision and J. T. Fourie, eds.), p. 565. Nordhoff, Leyden.
Yonenaga, I., and Sumino, K. (1987). *J. Appl. Phys.* **62**, 1212.
Yonenaga, I., and Sumino, K. (1989a). *J. Appl. Phys.* **65**, 85.
Yonenaga, I., and Sumino, K. (1989b). *J. Mater. Res.* **4**, 355.
Yonenaga, I., Onose, U., and Sumino, K. (1987). *J. Mater. Res.* **2**, 252.
Yu, S., Gösele, U., and Tan, T. Y. (1989). *J. Appl. Phys.* **66**, 2952.

CHAPTER 4

Mechanical Behavior of Compound Semiconductors

Sivaraman Guruswamy

DEPARTMENT OF METALLURGICAL ENGINEERING
UNIVERSITY OF UTAH
SALT LAKE CITY, UTAH

Katherine T. Faber

DEPARTMENT OF MATERIALS SCIENCE AND ENGINEERING
NORTHWESTERN UNIVERSITY
EVANSTON, ILLINOIS

John P. Hirth

DEPARTMENT OF MECHANICAL AND MATERIALS ENGINEERING
WASHINGTON STATE UNIVERSITY
PULLMAN, WASHINGTON

I. INTRODUCTION	190
II. GENERAL DEFORMATION BEHAVIOR OF ELEMENTAL AND COMPOUND SEMICONDUCTORS	190
1. *Nature of Dislocations*.	190
2. *Temperature Dependence of Flow*	192
3. *Solid Solution Strengthening Model*	194
III. HARDNESS OF III–V AND II–VI COMPOUNDS.	196
IV. COMPRESSIVE/TENSILE STRENGTH OF COMPOUND AND ISOVALENT-DOPED COMPOUND SEMICONDUCTORS.	200
4. *InP and GaP*	205
5. *GaAs and In-Doped GaAs*	206
6. *Other Isovalent-Doped GaAs*	214
7. *CdTe and Related II–VI Compounds*.	215
V. NON-ISOVALENT GROUP II AND VI DOPANTS IN III–V COMPOUNDS. . . .	218
8. *Dislocation Velocity*	218
9. *Yield Stress*	219
VI. THE ROLE OF Si DOPING IN GaAs	220
VII. CRYSTAL GROWTH	222
VIII. SUMMARY	226
REFERENCES	226

I. Introduction

The semiconductor industry has long been aware of the need to avoid deformation during bulk single-crystal semiconductor growth or during subsequent device processing. The generation of dislocations during liquid-encapsulated Czochralski (LEC) growth of Si, Ge, GaAs, and other III-V compounds is believed to occur when the thermal stress imposed on the crystal during growth exceeds the critical resolved shear stress (CRSS) and the crystal is deformed (Mil'vidskii and Bochkavev, 1978; Jordan et al., 1980, 1984, and 1986). The dislocation density increase that results from deformation has a deleterious effect on the device yield, performance and reliability (Nanishi et al., 1982; Petroff and Hartman, 1973; Miyazawa and Hyuga, 1986). The reduction of thermal stresses during crystal growth and enhancement of crystal strength by doping, such as with In and Si in GaAs (Jordan and Parsey, 1986, 1988; McGuigan et al., 1986) makes possible growth of large single crystals with reduced defects. The deformation behavior of Si and Ge has been a subject of much research for the past three decades while limited studies have been performed on III-V and II-VI compound semiconductors such as GaAs, InP, and CdTe. The mechanical response of these crystals depends on the crystal structure, nature of atomic bonding, concentration of dopants, temperature, and strain or loading rate. In this paper we review the current understanding of the deformation behavior in these materials. Their expected behavior at low, intermediate, and high temperatures, predicted from the current understanding of dislocation motion in solids, is presented along with compressive, tensile, and hardness data on GaAs, InP, CdTe, and Si. The effects of ternary dopants on the mechanical behavior of these materials are analyzed using ideas of solid solution strengthening and defect chemistry. Finally, implications for defect-free crystal growth and device fabrication are examined.

II. General Deformation Behavior of Elemental and Compound Semiconductors

1. NATURE OF DISLOCATIONS

Si and Ge both have the diamond cubic structure, which consists of two face-centered cubic (fcc) sublattices both occupied by Si or Ge, respectively, and the nature of bonding is purely covalent. The slip system that is operative in this diamond cubic lattice is $a/2 \langle 110 \rangle \{111\}$. The Peierls barrier in these materials with strong directional bonding has deep troughs along $\langle 110 \rangle$ directions. Hence, the dislocations lie primarily along this direction and may be pure screw or 60° dislocations at small strains. At high dislocation densities, curved dislocations are observed in these crystals because of interactions with other dislocations. Two different core structures, the shuffle set and the glide set, are possible for the 60° dislocation, depending on how the extra half-plane is terminated, as shown in Fig. 1 (Hirth and Lothe, 1982). The glide set is formed by cutting out an incomplete plane bounded by the surface 1-5-6-4, and the shuffle set by cutting out an incomplete plane bounded by 1-2-3-4. In the case of the III-V compounds, GaAs and InP, the structure is zincblende, which again consists of two interpenetrating fcc sublattices but with each sublattice occupied by a Group III or Group V element, respectively. The nature of bonding, while predominantly covalent, has some ionic character. The II-VI compounds have either a wurtzite or zincblende structure (Ray, 1969). Wurtzite consists of interpenetrating close-packed hexagonal lattices. While the ionic character of the bonds has increased for these compounds over the Group IV elements, a strong covalent character persists, as evidenced by the tetrahedral arrangement of nearest neighbors in all of these alloys. In view of the differences in valence of the species occupying the two sublattices, a distinction is made for 60° dislocations depending on whether they end on a row of Group III or II atoms (α type) or whether they end on a row of Group V or VI atoms (β type). α dislocations have a positive charge associated with them and act as

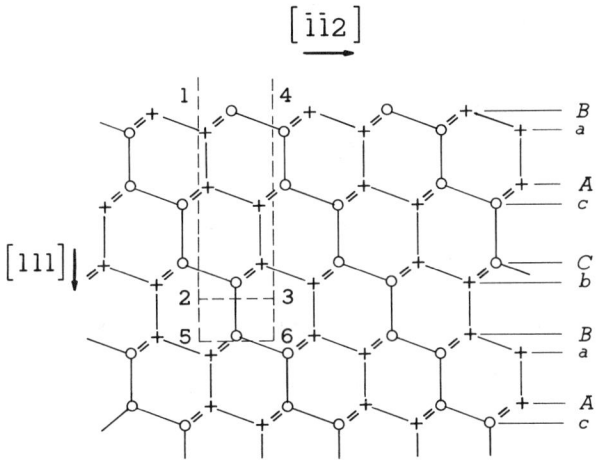

FIG. 1. The diamond cubic lattice projected normal to (110). ○ represents atoms in the plane of the paper, and + represents atoms in the plane below. (111) is perpendicular to the plane of the paper and appears as a horizontal trace.

acceptors, while β dislocations have an associated negative charge and act as donors. Most of these crystals belong to the same isomechanical groups (Frost and Ashby, 1984).

2. Temperature Dependence of Flow

Materials with covalent character have strong temperature dependences for flow at low temperatures, corresponding to a rate-controlling mechanism related to the intrinsic lattice resistance or Peierls barrier. An athermal plateau region appears at intermediate temperatures, corresponding to extrinsic effects such as solid solution hardening, while at high temperatures other creep-type mechanisms would be applicable (Haasen, 1982). At low temperatures, the Peierls periodic potential barrier is so large that movement of a dislocation from one low-energy position to the next requires the application of a large stress. This barrier is overcome by the process of nucleation and lateral propagation of double kinks on the dislocation line (Hirth and Lothe, 1982). At very low temperatures, where even the nucleation and motion of these kinks is limited, the stress for fracture or cleavage along close-packed planes or along planes of minimal charge is exceeded before the yield stress is reached. Plastic yielding could occur at these temperatures if deformation is imposed in the presence of a large hydrostatic stress component such as in an indentation. Above a certain transition temperature (the ductile-to-brittle transition temperature or DBTT), significant plastic flow occurs without fracture. This temperature is about 0.6 T_M in compounds with diamondlike structures (Gridneva et al., 1969; Haasen, 1957). With an increase in temperature, the nucleation and motion of double kinks occur much more easily, and the flow stress drops sharply (Fig. 2). Misfit strain centers promote double-kink nucleation while impeding the subsequent lateral propagation of the two kink segments (Wolfson, 1975). Solute additions at low concentration levels could result in either hardening or softening, depending on the magnitude of these opposing effects. However, the softening effect should be most prominent at the lowest temperatures where the double-kink model is dominant and should be less pronounced as the extrinsic athermal region is approached. Hence, a transition from softening to hardening with increasing temperature below the plateau region would be expected, analogous to observations for body-centered cubic (bcc) metals where the Peierls mechanism is operative at low temperatures (Meshii et al., 1982).

In the athermal region, double-kink nucleation no longer controls dislocation motion. In this region, the resistance to dislocation glide derives mainly from the interaction with solutes and other defects such as intersecting

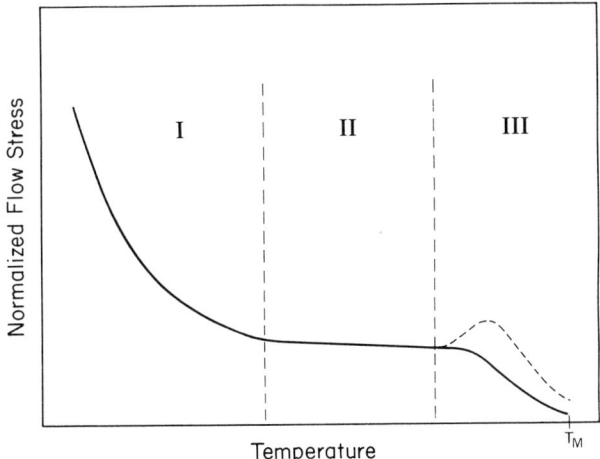

FIG. 2. Schematic of normalized flow stress versus temperature curve for GaAs and other covalent materials at constant strain rate. [Reproduced from Guruswamy et al. (1987).]

dislocations, while the dislocation motion occurs by breakaway from these pins. As in the case of solution-strengthened metals (Haasen, 1982), a nominally temperature-independent flow stress region would be expected, where the thermal component of the flow stress is significantly less than the athermal component. In actuality, the flow stress is not truly temperature-independent in the plateau region, but the slope of the flow stress–temperature plot is significantly less in this region than at lower temperatures (Haasen, 1982). The temperature-independent flow stress level (plateau stress) should increase with increasing solute content. This region is expected to extend to a significant fraction of the melting temperature. Above this temperature, the flow stress should be determined by diffusion-controlled deformation processes, namely, dislocation climb-controlled power-law creep and diffusional creep.

In the region of power-law creep, the dislocation velocity, and hence the flow stress level, should depend on the lattice interdiffusion coefficient or diffusivity. In this regime, the dislocation velocity for a given applied stress is written as (Frost and Ashby, 1984)

$$v \approx D_v \sigma_n \Omega / bkT. \qquad (1)$$

Here D_v is the bulk diffusivity, σ_n is the local normal stress that produces a climb force on the dislocation, Ω is the atomic volume, b is the length of the Burgers vector, and k is Boltzmann's constant. Because σ_n is proportional to the flow stress σ_s and the average dislocation velocity is proportional to the strain rate ε, the latter is given by (Frost and Ashby, 1982)

$$\varepsilon \approx (AD_v\mu b/kT)(\sigma_s/\mu)^3, \qquad (2)$$

where μ is the shear modulus and A is a dimensionless constant. For a binary compound such as GaAs, D_v is replaced by $D_{eff} = D_{Ga}D_{As}/D_{Ga} + D_{As}$ (Frost and Ashby, 1984; Shewmon, 1989). Thus, the stress for a given strain rate would be controlled in the GaAs lattice by the slower of the diffusing species, which is arsenic. The influence of solute elements on the diffusion in the arsenic sublattice would thus be an important factor in determining the deformation behavior. The trapping of arsenic vacancies because of elastic or electronic interactions with solute centers in doped GaAs would be a possible factor that could reduce D_{eff}, and hence provide a hardening effect in the power-law creep regime.

At high temperatures, where solute atoms are mobile, resistance to dislocation motion also comes from the drag of a solute atmosphere by the dislocation. The force–velocity relationship for this Cottrell drag has been treated in detail (Hirth and Lothe, 1982). For a given dislocation velocity, v, the drag stress increment caused by the solute atmosphere varies as the square of the strength of solute–dislocation interaction, dV, and inversely as the effective diffusivity, D_{eff}. The solute drag stress versus temperature curve would exhibit a maximum and, if present, would be superposed on the strength versus temperature plot, as shown by the dotted line in Fig. 2. Because of very low diffusivities in the GaAs lattice, for example, the solute drag effect is expected only at high temperatures.

3. Solid Solution Strengthening Model

As in other materials, the difference in atomic size between solute and solvent atoms and the difference in valence lead to elastic misfit strains and electronic interactions between solute atom and dislocations causing an increase in resistance to dislocation motion. In the case of an isovalent solute in a lattice site, the size misfit would be the primary factor determining the ease of dislocation motion or lack thereof. However, the magnitude of the misfit strain is related to the difference in bond lengths with the nearest neighbors, which are at the corners of the tetrahedron surrounding the lattice site under consideration. EXAFS measurements of bond lengths in (Ga, In)As by Mikkelson and Boyce (1983) showed that the In–As and Ga–As bond lengths varied only by a very small amount (0.004 nm), while the cation–cation distance changed according to Vegard's law across the GaAs–

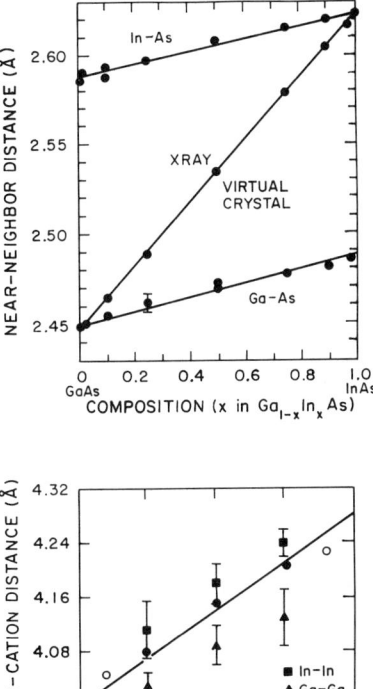

FIG. 3. Variation in bond lengths in GaAs-InAs binary system determined by EXAFS measurements. [From Mikkelson and Boyce (1983).]

InAs binary system (Fig. 3). Ehrenreich and Hirth (1985) suggested that $InAs_4$ tetrahedral clusters act analogously to solute atoms in metals in hardening the system. With the assumption that bond lengths remain constant in other systems, other isovalent solutes may be evaluated for their strengthening of compound semiconductor matrices in terms of the variation in lattice parameters. The volumetric strain associated with $InAs_4$ tetrahedral cluster in GaAs is about 21%, resulting in large solid solution strengthening potential. Photoluminescence studies of In-doped GaAs by Fujiwara et al. (1986), as well as electroreflectance and photocapacitance measurements performed on undoped and In-doped GaAs by Raccah et al. (1986), confirm the presence of large localized strains in the system. Expected local strains associated with various isovalent substitutions in III-V and II-VI com-

TABLE I

Misfit strain by dopants in III–V and II–IV compounds based on the Ehrenreich and Hirth model

System	($\delta a/a$)	Strengthening Effect
CdTe–HgTe	0.003	—
CdTe–ZnTe	0.061	effective
CdTe–BeTe	0.145	—
CdS–Hgs	0.006	—
CdS–ZnS	0.700	effective
CdS–BeS	0.166	—
GaAs–AlAs	0.001	ineffective
GaAs–InAs	0.072	effective
GaAs–BAs	0.155	ineffective
GaP–AlAs	0.003	ineffective
GaP–InAs	0.077	effective
GaP–InAs	0.167	ineffective

pounds are listed in Table I (Hirth and Ehrenreich, 1985b). In view of the strain fields associated with the solute clusters, several possibilities exist for hardening of the crystal: (i) pinning of glide dislocations by the Orowan–Fleischer process; (ii) formation of core or Cottrell atmospheres, and hardening associated with the drag of these atmospheres, and (iii) pinning of climbing dislocations by the core or Cottrell atmospheres with the solute drag associated with dislocation motion. There are other possible effects of isovalent solute additions, including influences of defect concentrations and mobilities, influences on core reconstructions, and weak effects on electronic states at dislocations.

III. Hardness of III–V and II–VI Compounds

General trends in the flow stress variation with temperature can be easily checked by means of hardness measurements. Hardness measurements are frequently used in semiconductors as a measure of strength because of the simplicity of the technique, the small sample sizes required, and the generally brittle behavior of these crystals at low temperatures. Under hardness indentations, flow occurs under a large compressive hydrostatic stress component with minimal or no fracture, while tensile and compressive tests at temperatures below the DBTT are quite difficult to perform, since fracture occurs before the yield point is reached and a large scatter is observed in the fracture stresses measured. For an ideal plastic material, slip line field theory predicts that hardness will be about three times the yield point (McClintok and Argon, 1966). When strain hardening is considered, hardness values correspond to the flow stress of the material at around 7 or 8% strain, the average plastic strain under the indentation (Cahoon et al., 1971). Hardness versus critical resolved shear stress values would, therefore, be sensitive to the work hardening coefficient and the orientation of the crystal surface on which the indentation is made, factors that determine the operative slip systems.

Hardness indentations at low temperatures result in the formation of cracks eminating from the corners of the hardness impression, the size of which can be used to predict the fracture toughness of the material. This technique is widely used for brittle materials (Anstis et al., 1981). However, in single crystals, anisotropy of the material leads to asymmetric cracking patterns, and the use of indentation fracture toughness measurements based on the elastic analysis for polycrystalline materials must be done with caution. Also, the presence of surface residual stresses that result from crystal growth will influence the cracking process. Fracture toughness measurements on GaAs using both the indentation technique and the double cantilever beam test, a standard fracture mechanics technique, have been reported by Chen (1984).

Measured hardness values are also sensitive to the surface orientation and the polarity of the surface (Watts and Willoughby, 1984; Hirsch et al., 1985). In III-V and II-VI compounds, the extension of hardness measurements to high temperatures is difficult because of the high vapor pressure of one or more of the components. Hardness measurements under protective glassy films and liquid boric oxide were attempted by the authors, but were ineffective because of surface pitting or cracking around the indentation that makes indentation size measurements difficult.

Fig. 4 shows the variation of Vickers hardness as a function of temperature from room temperature to 900°C for three semi-insulating GaAs wafers containing 0, 5.5×10^{19} and 2.2×10^{20} In atoms/cm^3. All three alloys also contained boron at levels of 2×10^{18} atoms/cm^3. All hardness data are normalized by the shear modulus to give the residual temperature dependence related to the deformation mechanisms. The Voigt average shear modulus was calculated from the anisotropic elastic constants given as a function of temperature by Jordan (1980) and Burenkov et al. (1973). A significant difference in hardness between undoped and In-doped GaAs is seen only above 300°C. No cracks were observed at indentation corners above 400°C (Guruswamy et al., 1986). This transition from cracking to deformation by slip during indentation has also been observed in GaAs by Ogura et al. (1979) and Hirsch et al. (1985) in the temperature range 300-400°C. However, these earlier hardness studies were confined to temper-

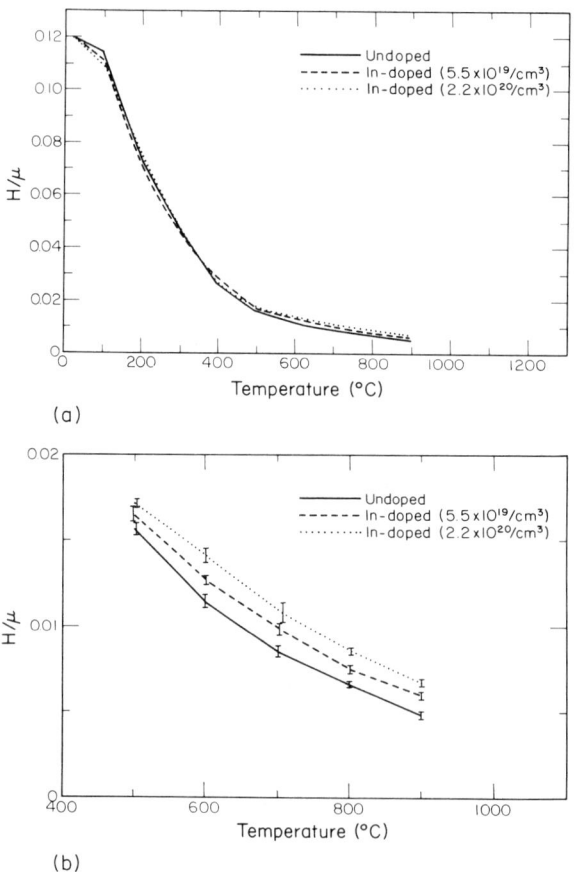

FIG. 4. Normalized hardness versus temperature plot for GaAs, $Ga_{0.9975}In_{0.0025}As$ and $Ga_{0.99}In_{0.01}As$. [From Guruswamy et al. (1987).]

atures less than 500°C. Hardness drops sharply with temperature and flattens out at higher temperatures, indicative of the existence of the plateau stress region. The shape of the curve is very similar to that observed for metals exhibiting plateau behavior (Haasen, 1982). Both the In-associated effects of the softening–hardening transition at low temperatures and the presence of the plateau region are consistent with the model discussed previously of the Peierls barrier control superseded by solid solution hardening. The plateau behavior for undoped GaAs in Fig. 4 would then be associated with pinning caused by dislocation intersections, residual impurities, jogs, or antisite defects. By virtue of their higher hardness, the alloys containing indium are inferred to have a higher plateau stress level. Boron, an isoelectronic substitution for gallium, would produce a larger size mismatch, $\delta V/V = 50\%$, than indium, but an attendant lower solubility (Hirth and Ehrenreich, 1985b). Hence, it should also be a strong hardener if uniformly distributed in solution, an effect verified experimentally (Thomas, 1985; Tada et al., 1982), but could lead to internal stresses and dislocation generation if present in an inhomogeneous distribution or as precipitates. A similar temperature dependence of hardness is also observed in the case of Si and Ge (Gilman, 1969; Nikitenko, 1967), as seen in Fig. 5.

Since GaAs has a structure similar to Si and since the deformation map for GaAs is not available at the present time, the Si deformation map (Frost and Ashby, 1982) is used here as a guideline for possible transitions in deformation mechanisms. Fig. 6 presents data extracted from the Si map for strain rates of 10^{-3} to 10^{-7} s^{-1}. The flow behavior undergoes a transition from flow following the double-kink model to that of a power-law creep model at about $0.5\ T_m$. For GaAs, the intervention of solid solution hardening is expected to move the transition to power-law creep to higher homologous temperatures, but on the basis of all available deformation map data (Frost and Ashby, 1984), still to leave a region of power-law creep below the melting point. The cellular dislocation networks present in as-grown GaAs crystals with dislocation densities of the order of 10^3–10^4 cm/cm^3 provide indirect

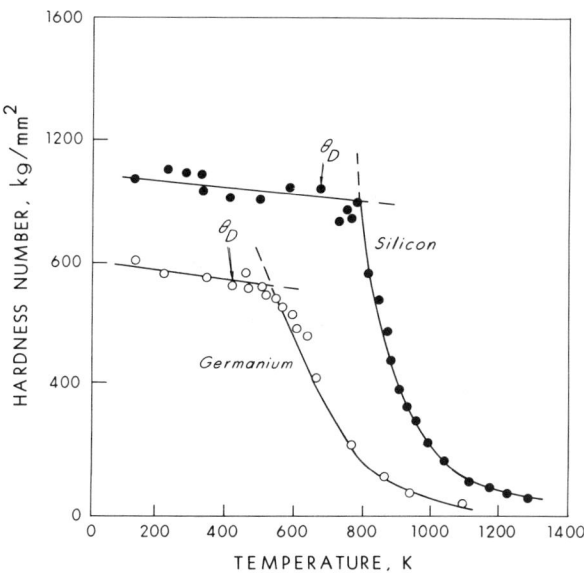

FIG. 5. Effect of temperature on hardness of Ge and Si. [From Nikitenko (1967).]

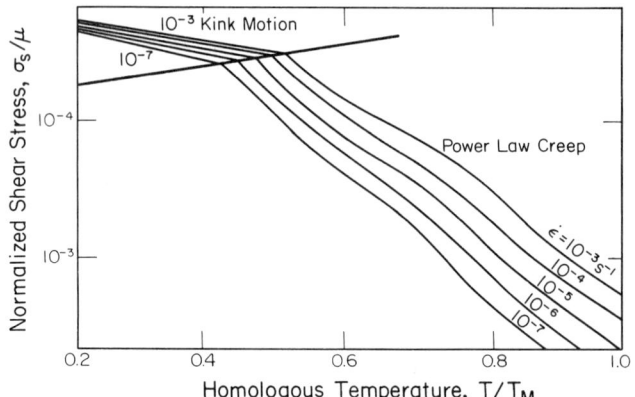

FIG. 6. A section of deformation mechanism map for Si that belongs to the same isomechanical group as the III-V compounds. [Reprinted with permission from H.J. Frost and M.F. Ashby, "Deformation Mechanism Maps," © 1982 Pergamon Press PLC.]

evidence that climb processes are involved in the deformation occurring during crystal growth. As indicated in Fig. 6, above the transition temperature the stress for power-law creep at a given strain rate and temperature is less than that for double-kink flow extrapolated to the same conditions. However, even for the power-law creep region, indium should have a hardening influence.

IV. Compressive/Tensile Strength of Compound and Isovalent-Doped Compound Semiconductors

Problems associated with acquiring bulk deformation data in compression or tension on compound semiconductors are numerous. First, vaporization of one or more of the elements at the deformation temperature is likely. In order to prevent vaporization of one or more constituents, the deformation experiments can be performed under liquid encapsulation, often B_2O_3, and/or with an inert gas overpressure (Guruswamy et al., 1987; Siethoff and Behrensmeier, 1990). Second, alignment of the test fixtures and the specimen is essential. In compression, parallel compression faces in the fixture are required. A typical compression fixture is shown in Fig. 7. Results are also sensitive to the aspect ratio of the test specimen at aspect ratios less than two (Swaminathan and Copley, 1975). Surface finish may also influence the measured yield properties. A typical surface treatment consists of a fine

(0.3 μm alumina) mechanical polish followed by a chemical polish in a 1% bromine–methanol solution.

Most of the deformation studies in semiconductors are done at temperatures above the DBTT (above about 0.5 T_M). The deformation of these crystals in constant strain-rate tensile or compressive tests is characterized by a pronounced yield point drop. Also, the upper and lower yield stress values are sensitive to the initial dislocation density (Fig. 8), surface preparation, strain rate, crystal orientation, and temperature. The theory of the yield point drop in constant strain-rate tests is based upon the need for the instantaneous mobile dislocation density and dislocation velocity to be sufficient to accommodate the imposed strain rate on the crystal. Otherwise, the stress

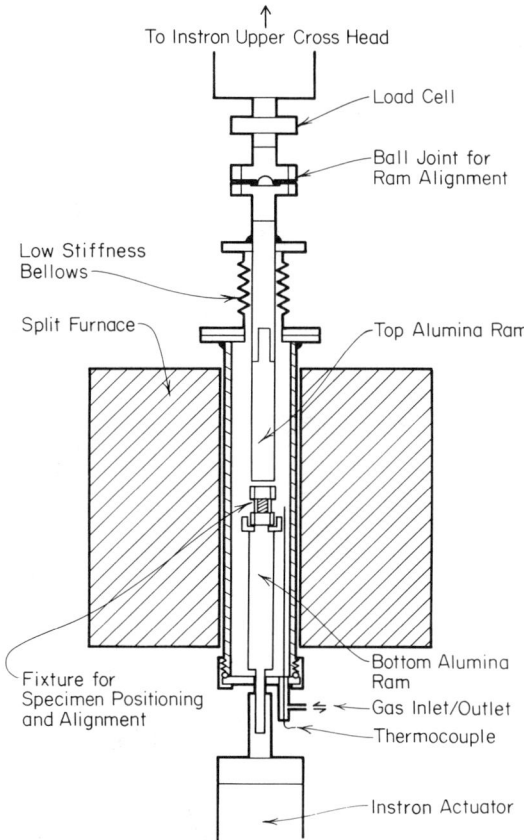

FIG. 7. Schematic of the high-temperature compression testing fixture. [From Guruswamy et al. (1987).]

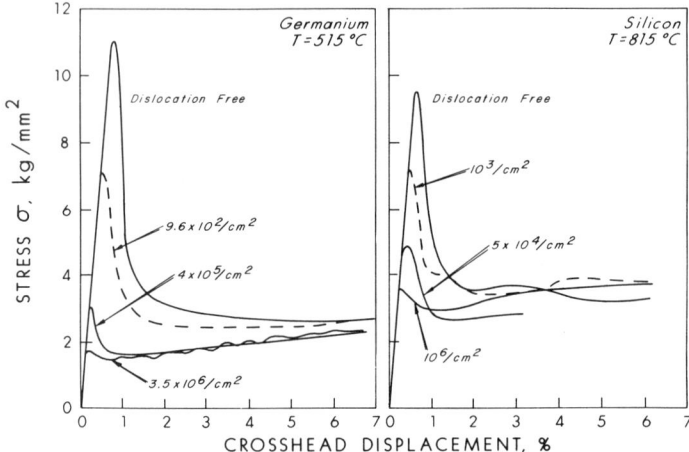

FIG. 8. Influence of initial dislocation density on the stress–strain curves of (a) Ge at 515°C and (b) Si at 815°C. Initial slopes adjusted and vary with dislocation density. Crystal orientation was 14, which is a single-slip orientation and lies between [123] and [011], and approximate strain rate was 2×10^{-3} min^{-1}. [From Patel and Chaudhury (1963).]

will continue to rise, and fracture will result. The microscopic details of this concept are treated in detail by Hirth and Lothe (1982). The general forms of stress–strain curves for single crystals in single-slip and multiple-slip orientations are shown in Fig. 9. The pre-yield region at low temperatures represents primarily elastic deformation, but increased dislocation activity with increasing temperature results in significant plastic deformation or microyielding, prior to macroscopic yielding. In the Stage I or easy slip deformation region, dislocation motion is limited to a single-slip system. Strain hardening in this region comes mainly from the interaction of tangled primary dislocations on this slip plane, although some fine secondary slip is usually present. In stage II, two or more slip systems are in operation simultaneously, and the intersection of dislocations moving in different slip systems leads to rapid work-hardening of the crystal. At high stresses, recovery processes, such as cross-slip at low temperatures or climb at elevated temperatures, become operative, leading to stage III of deformation, shown in Fig. 9. However, in the intermediate temperature region, it is possible for cross-slip and climb processes to occur in succession, leading to stage IV and stage V deformation, as suggested by the work of Schroter and Siethoff (1984) and Siethoff and Haasen (1988) in Si, Ge, and InP, and Guruswamy et al. (1987) in GaAs.

As a prologue to the discussion of the bulk deformation of compound semiconductors, Si and Ge are first examined. Extensive literature is available on the deformation of Si and Ge (Alexander and Haasen, 1968, 1972; Myshlyzaev *et al.*, 1969; Frost and Ashby, 1982; Schroter *et al.*, 1983). Of note are studies by Patel and Chaudhuri (1963), who deformed single crystals of Si and Ge in the temperature range of 0.4–0.6 T_M. In this temperature range, a semilogarithmic plot of flow stress against temperature is linear but, as indicated by other studies, flattens out at higher temperatures. The values of upper and lower yield point as a function of temperature for Si and Ge are shown in Fig. 10. For the Ge crystals, the upper yield point increases significantly with decreasing initial dislocation density. With increasing temperatures and slower strain rates, well-defined yield points are not observed, and the yield point phenomenon becomes less pronounced. Schroter *et al.* (1983) and Myshlyzaev *et al.* (1969) have examined Si and Ge and Si, respectively, at temperatures up to 0.8 T_M. The stress at the lower yield point at a given strain rate is related to the stress for an equivalent static creep rate through the stress exponent and the activation energy for flow (Schroter *et al.*, 1983). Myshlyaev *et al.* (1969) examined the steady-state creep rate at different stress levels and observed that the dislocation density increases much more slowly than in a dynamic tensile test, but structures obtained at

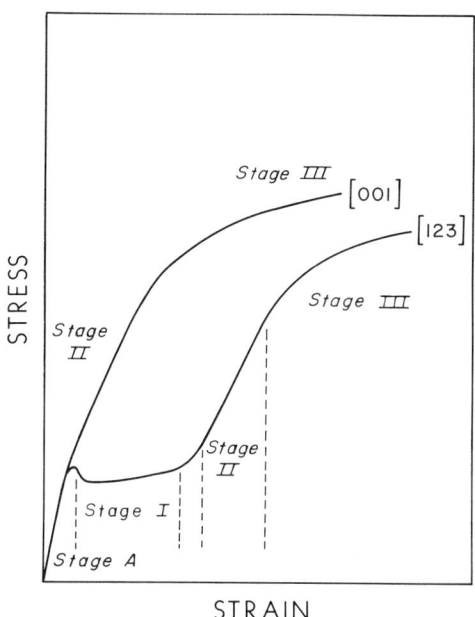

FIG. 9. Schematic diagram of stress–strain curves of a single crystal in single-slip [123] and multiple-slip [001] orientations.

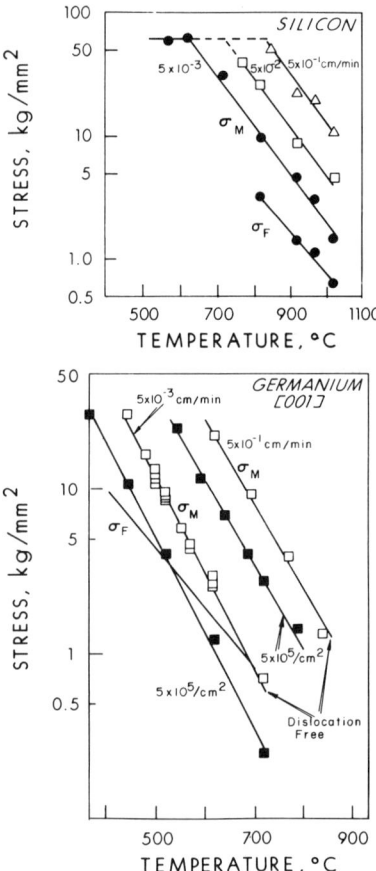

FIG. 10. Temperature dependence of maximum stress and flow stress in (a) Si and (b) Ge. [From Patel and Chaudhuri, (1963).]

the end of steady-state creep correspond to structures at the lower yield point in a dynamic test. They also observed that the activation energy for flow is a strong function of stress, varying between 3.5 and 5.2 eV as compared to the sublimation energy and self-diffusion activation energy values of Si (5.6 and 5.13 eV, respectively). Deformation mechanism maps showing the conditions of temperature and stress level at which different rate-controlling deformation mechanisms operate are available for both Si and Ge (Frost and Ashby, 1984).

4. InP AND GaP

The deformation behavior of InP single crystals has been examined by Gall et al. (1987), Müller et al. (1985), Brown et al. (1980), Brasen and Bonner (1983), and Siethoff et al. (1988). Gall et al. examined InP crystals deformed by compression along the [123] orientation in the temperature range of 300–800°C. The crystals were pre-deformed to a strain of about 2% at 0.77 T_M to make the dislocation density uniform in all the specimens. The lower yield stress values determined as a function of temperature show two distinct regions. In region I, deformation is very sensitive to temperature. The activation energy is large, the activation volume is low, and the deformation is controlled by double-kink nucleation and motion. At higher temperatures, the thermal activation is very nearly zero and the stress level is very low. These two regimes correspond to regions I and II in Fig. 11. The activation

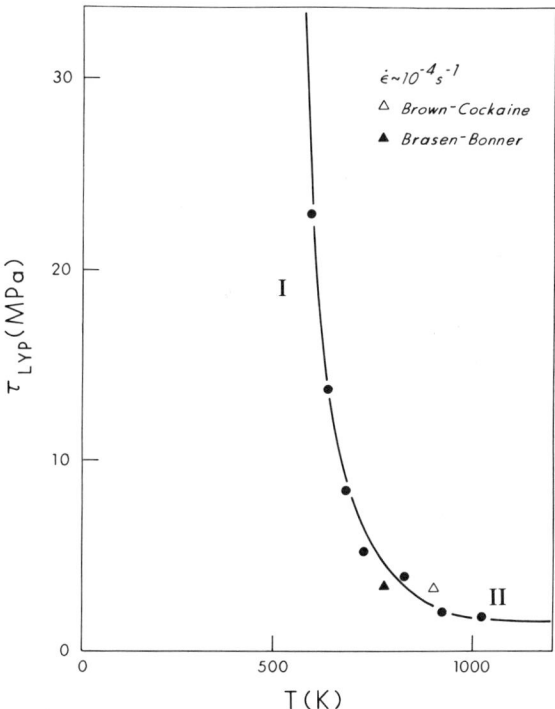

FIG. 11. Variation of lower yield point versus temperature in InP at a strain rate of about $10^{-4}\,s^{-1}$. [From Gall et al. (1987).]

volume increases dramatically on entry to region II, indicating that the resistance to dislocation comes from long-range interactions from other dislocations and solutes with high misfit strain fields. InP crystals exhibit brittle fracture below about 250°C.

The mechanical behavior of GaP and GaAs$_{0.8}$P$_{0.2}$ has been investigated by Yonenaga and Somino (1989a) and Yonenaga et al. (1989), who found their behavior to be similar to Si and Ge. The GaAs$_{0.8}$P$_{0.2}$ had a much higher strength compared to GaP or GaAs at higher temperatures, indicating a large athermal stress component.

5. GaAs AND In-Doped GaAs

GaAs crystals have deformation behavior similar to that of both Si and InP. A summary of deformation studies on undoped and In-doped GaAs in single-slip [123] and in multiple-slip [001] orientations by a number of investigators is presented in Table II. Many of the studies on GaAs have been performed in reducing or inert gas atmospheres in the temperature region below 600°C (Swaminathan and Copley, 1975; McGuigan et al., 1986a; Djemel and Castaing, 1986; Tohno et al., 1986), where the yield point in both single-slip and multiple-slip orientations is sharp. The semilogarithmic plot of flow stress versus temperature in this temperature range shows a strong temperature dependence and is linear, indicative of double-kink nucleation and motion being the rate-controlling process.

Table III summarizes the different materials used in our investigations and the relevant material characteristics such as composition, average dislocation etch pit density, resistivity, and mobility. All the crystals except crystal A were grown by the LEC process. Crystal A was grown by a vertical gradient freeze process.

Table III clearly shows the strong strengthening effects caused by the isovalent dopant indium in GaAs and explains the dramatic reduction in dislocation density produced by In doping in LEC-grown GaAs single crystals. We now consider the general observations on the deformation characteristics of both undoped and isovalent-doped GaAs crystals.

In crystals E and C, tested in the [001] orientation (Fig. 12), no easy-glide, stage I deformation is observed. Only at 700°C for the In-doped case is there an indication of the yield point phenomena characteristic of a similar material (Hobgood et al., 1986) tested at 350 to 590°C. The tendency for a less pronounced yield point with increasing temperature is consistent with results in other studies (Hobgood et al., 1986; Tabache et al., 1986), and the absence of well-defined yield points at high temperatures agrees with the work of Djemel and Castaing (1986). The slope of the stress–strain curve below the

TABLE II

CRITICAL RESOLVED SHEAR STRESS VALUES FOR UNDOPED AND IN-DOPED GaAs[a,b]

Material	T(K)	[123] Orientation σ_Y (MPa)	[001] Orientation σ_Y (MPa)	$\sigma_{.002}$ (MPa)	$\sigma_{.04}$ (MPa)
In-doped	973	4.64	—	9.34	27.84
	1,173	3.33	—	8.80	23.07
	1,273	—	—	8.53	17.22
	1,373	3.27	—	7.84	11.59
Undoped GaAs	973	2.55	—	2.78	21.76
	1,173	1.91	—	2.45	10.04
	1,373	1.83	—	2.45	6.04
In-doped (LEC) GaAs	1,253	1.4[c]			
	1,353	1.05[c]			
	873		8.5[d]		
			5.5[d]		
	800			9.6[f]	
	900			7.6[f]	
	1,053			7.2[f]	
Undoped (LEC GaAs	1,053	1.22[c]			
	1,253	1.0[c]			
	1,353	0.53[c]			
	873		6.9[d]		
	800			2.2[f]	
	900			2.1[f]	
	1,053			2.2[f]	
Undoped non-LEC GaAs	773	3.9[g]			
	823		2.4[e]		

[a]From Guruswamy et al. (1987).
[b]σ_Y is upper yield point, $\sigma_{0.002}$ and $\sigma_{0.04}$ are offset yield strengths at strains of 0.002 and 0.04, respectively.
[c]Tabache et al. (1986).
[d]Hobgood et al. (1986).
[e]Swaminathan and Copely (1975, 1976).
[f]Djemel and Castaing (1986).
[g]Yonenaga et al. (1987).

TABLE III
Characteristics of Isolvalent-Doped GaAs Crystals.[a]

Material	Dopant Concentration (atoms/cm^3)	Dislocation Density (cm/cm^3)	Mobility (cm^2 V^{-1} s^{-1})	Resistivity (Ω cm)
A. Undoped (low B)	B = 1 × 10^{16}	< 3,000	6,500	9.3 × 10^7
B. P-doped	P = 2 × 10^{19}	5 × 10^4	3,600	4.6 × 10^8
C. In-doped (high B)	In = 1–2 × 10^{20} B = 5 × 10^{17}	0–10^2	2,150	3 × 10^7
D. Sb-doped	Sb = 2 × 10^{19}	0–10^5	4,200	1.2 × 10^4
E. Undoped (high B)	B = 5 × 10^{17}	10^4–10^5	4,670	1.4 × 10^8
F. (In, Si) codoped	In = 1 × 10^{20} Si = 1 × 10^{18}	1.1 × 10^4		conducting ~ 1.0

[a]From Guruswamy et al. (1989).

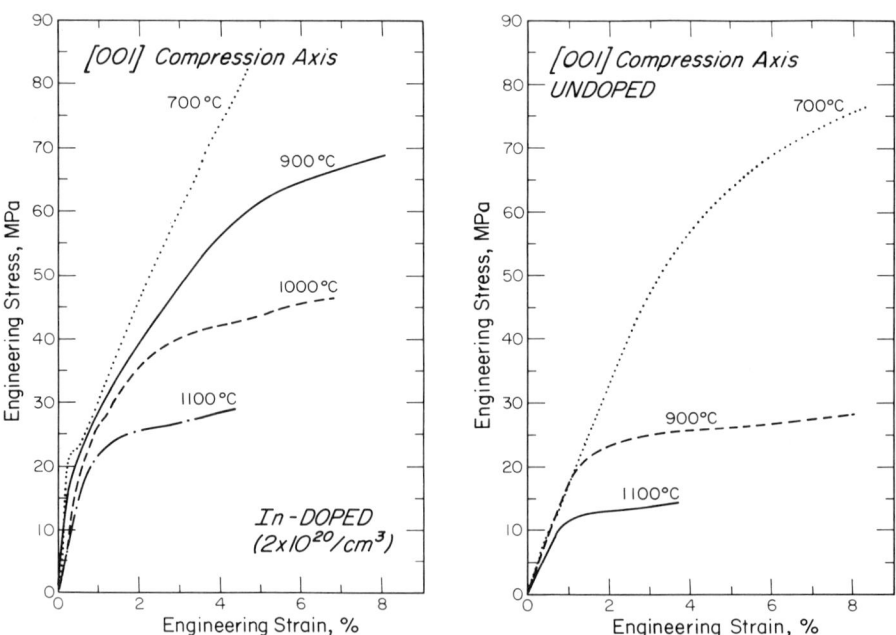

FIG. 12. Engineering stress–strain curves at different temperatures for (a) In-doped GaAs and (b) undoped GaAs specimens tested in the [001] orientation. [From Guruswamy et al. (1987).]

yield point at 700°C for the In-doped case is less than the reported elastic modulus, so that microyield and plastic flow are already occurring. This characteristic has been observed in other measurements (Hobgood et al., 1986; Tabache et al., 1986; Djemel and Castaing, 1986), including those at lower temperatures. Indeed, *in-situ* x-ray topographic measurements made during tensile testing of In-doped and undoped GaAs at 450-700°C demonstrate directly that dislocation formation and propagation occur at stresses well below the upper yield point (Tohno et al., 1986). Such phenomena are well known for metals where microyield also occurs well below the macroscopic yield point (Nabarro et al., 1964) with sufficient mobility and multiplication rates for dislocations available at lower temperatures than for the covalent crystal case.

The microyield regime A (Guruswamy et al., 1987) merges smoothly into stage II deformation behavior. The stress-strain curves, with the exception of In-doped GaAs at 700°C, are characterized by a stage II linear work hardening regime of flow, followed by stage III, characterized by a decrease in work hardening rate, and the beginning of the stage IV linear hardening regime characteristic of diamond cubic materials (Siethoff and Schroter, 1984; Schroter and Siethoff, 1984; Brion and Haasen, 1985). The results indicate hardening effects by indium in all stages of deformation. The stage A slopes increases with decreasing temperature and are larger at all temperatures for the In-doped crystals. The onset of stage III occurs at larger stresses in the In-doped case, and the slope in stage IV is greater for the In-doped case.

In contrast to tests in [001] orientation, compression tests on [123] orientation for crystals E and C revealed six-stage behavior (A, I, II, III, IV, and V) at 900°C and four-stage behavior above and below 900°C, with the stage III mechanism switching above and below that temperature (Guruswamy et al., 1987; Rai et al., 1989; Fig. 13). Transmission electron microscopy, together with analogous work on Si, Ge, InSb, and InP (reviewed by Brion and Haasen, 1985), indicated that throughout microyield stage A, the yield stress at the start of stage I and the stage I dislocation motion was glide controlled. Stage II was controlled by dislocation intersection and jog drag. Stage III at temperatures below 900°C represents recovery associated with thermal assistance of jog drag by vacancy motion, and at temperatures above 900°C by the climb of extended jogs in hexagonal dislocation networks that form as recovered substructures (stage V at 900°C). Stage IV is associated with changes in mesh size of network dislocations as extrinsic dislocations are absorbed into the boundaries. For both stages II and IV, the role of In is deduced to provide a friction stress, via the size effect discussed earlier, resisting the motion of dislocation segments bowing out between jogs.

Crystals E and C tested in [123] orientation showed sharp yield points.

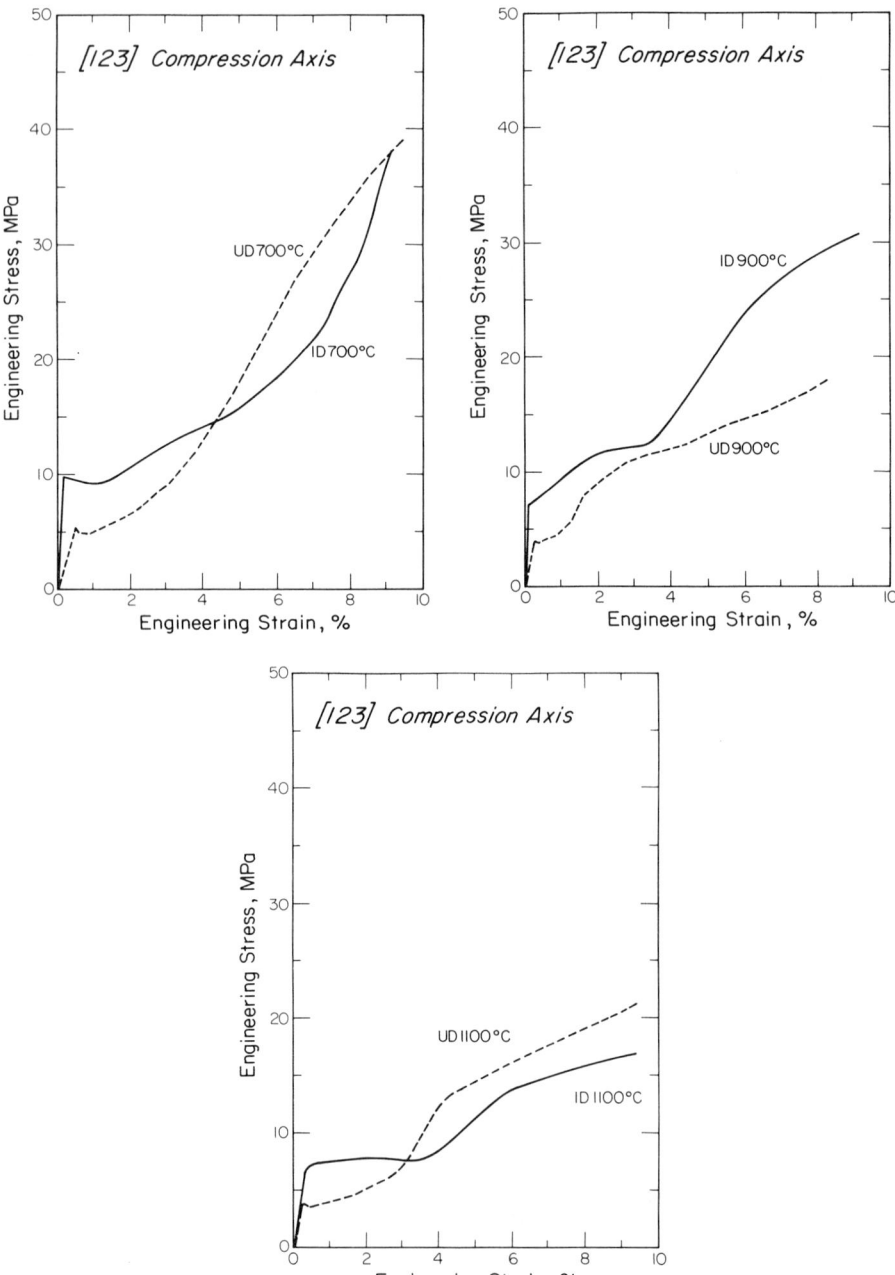

FIG. 13. Engineering stress strain curves for undoped (UD) and In-doped (ID) GaAs in the [123] orientation. [From Guruswamy et al. (1987).]

The single-slip orientation crystals showed extensive stage I deformation with a small yield drop at 700°C (Fig. 12), but not at 900 or 1,100°C. The pre-yield slopes in the microyield region A varied just as in the [001] case, with smaller slopes for the undoped case at each temperature and with a decrease in slope with increasing temperature. In comparison, the suppression of the yield point phenomena at temperatures above 700°C for [001] crystals (Guruswamy et al., 1987; Djemel and Castaing, 1986) indicates a strong impediment to dislocation intersection in these crystals. This impediment, favored by the relatively low stacking fault energy of ≈ 45 mJ/m^2 in these materials (Jiminex et al., 1986; Rai et al., 1989), suppresses easy glide and contributes to the large stage II slope for the [001] crystal. Nevertheless, the dislocation mobility in stage II is sufficient to prevent macroscopic yield point behavior. The [001] yield point behavior below 700°C is associated with the reduction in dislocation multiplication/mobility by the presence of the Peierls barrier and the attendant mechanism of double-kink dislocation motion.

The overall form of the stress–strain curves with stages II to IV for [001] crystals, and stage I as well for [123] crystals, is consistent with the behavior at elevated temperatures of the much more extensively studied Si and Ge crystals (Siethoff and Schroter, 1984; Schroter and Siethoff, 1984; Brion and Haasen, 1985). The analogy suggests that glide processes are important in determining the flow stress in stages I and II, while recovery processes are important in addition in stages III, IV, and V. The relative temperature independence of the critical resolved shear stress above 700°C (Fig. 14) implies plateau-type behavior (Haasen, 1982), characteristic of athermal solid solution hardening. Hence, the hardening indicated at yield and in stage I is consistent with the suggestion (Ehrenreich and Hirth, 1985) that indium should provide strong solid solution hardening in $Ga_{1-x}In_xAs$.

The larger stress for the onset of stage III and the larger slope in stage IV for In-doped crystals imply a decrease in recovery rate for the doped crystals. This could be caused, in principle, by suppression of cross-slip or by retardation of climb. Several factors indicate that the latter climb effect is predominant. First, transmission electron microscopy studies (Jiminex et al., 1986; Rai et al., 1989) indicate that In-doping, at the level studied here, does not affect the stacking fault energy, implying that there should be little influence on the cross-slip probability. Second, transmission electron microscopy studies of Rai et al. (1987, 1989) of [001] specimens strained into stage IV show well-developed dislocation networks similar to structures observed for stage IV deformation of Si (Brion and Haasen, 1985), a case where climb is known to be the dominant recovery mechanism. Third, stress relaxation measurements (Fig. 15) show consistently lower relaxation rates for the In-doped case at a given stress over the entire range of relaxation, implying a lower climb rate.

For observations of stage A and stage I, the behavior in the low-

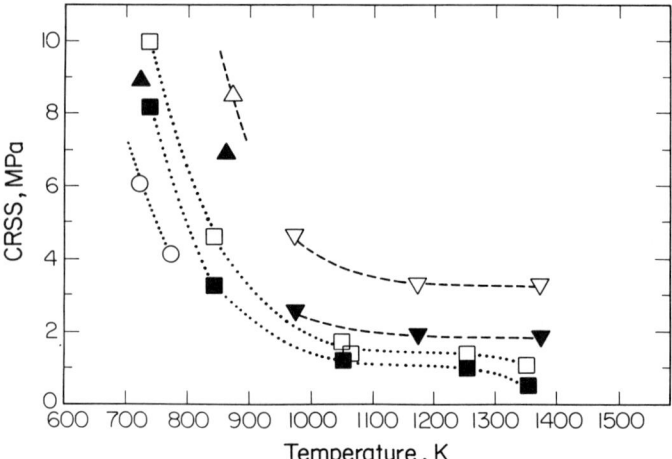

FIG. 14. Comparison of critical resolved shear stress values for undoped and In-doped GaAs as a function of temperature. Dashed line with inverted open triangle represents In-doped GaAs (Guruswamy et al., 1987); dashed line with inverted solid triangle represents undoped GaAs (Guruswamy et al., 1987); dashed line with open triangle represents In-doped GaAs (Hobgood et al., 1986); solid triangles represent undoped GaAs (Hobgood et al., 1986); dotted line with open squares represents In-doped GaAs (Tabache et al., 1986); dotted line with solid squares represents undoped GaAs (Tabache et al., 1986); dotted line with open circles represents undoped GaAs (Swaminathan and Copley, 1975). [From Guruswamy et al. (1987).]

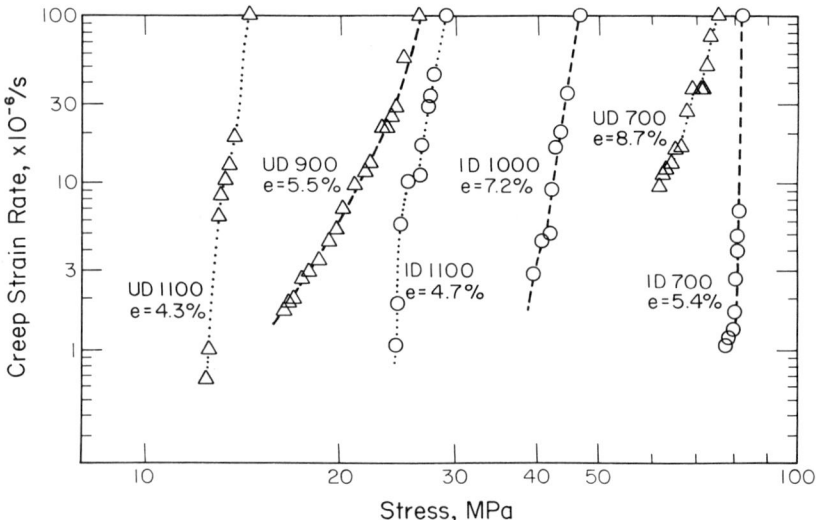

FIG. 15. Creep rate during relaxation after unloading during compressive tests.

temperature region 500 to 700°C is consistent with the dislocation motion by double-kink motion (Hirth and Lothe, 1982; Haasen, 1982), with solute hardening retarding the motion of kinks along the dislocation line (Brion and Haasen, 1985; Guruswamy et al., 1986a, 1987; Djemel et al., 1988; Rai et al., 1989). The strong hardening effect of In and other isovalent solutes in this regime could also be related to the elastic field of an $InAs_4$ unit (Ehrenreich and Hirth, 1985), in this case effective through an impediment to the propagation of kinks along a dislocation line. Because of the apprent change of mechanism below about 700°C, results obtained below this temperature, e.g., on dislocation velocity (Yonenaga et al., 1986; Matsui and Yokoyama, 1985) may not be applicable to behavior above 700°C. Above 700°C, the solute hardening directly provides a friction stress resisting dislocation glide motion. To further test this interpretation, the yield stress was determined as function of strain rate for undoped crystal A. As shown in Fig. 16, there was a strong temperature and strain-rate dependence of the yield stress at 500–700°C, consistent with a small activation area and the double-kink model, but a weak temperature dependence at higher temperatures, in accordance with a larger activation energy and the model for bowout between pinning points.

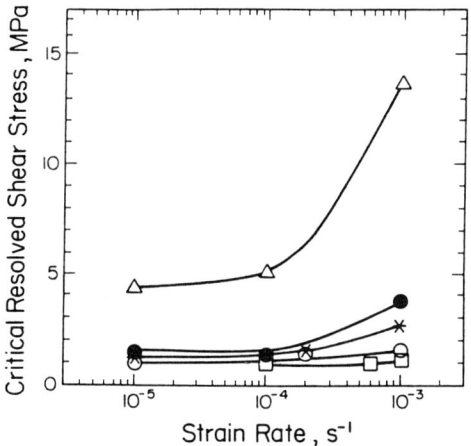

FIG. 16. Variation of critical resolved shear stress with strain rate in the case of undoped GaAs crystal A at different temperatures. Open triangles: 500°C; filled circles: 700°C; asterisks: 900°C; open circles: 1,100°C; open squares: 1,150°C. [From Guruswamy et al. (1987).]

6. OTHER ISOVALENT-DOPED GaAs

Influence of other isovalent dopants P, B, and Sb on the high-temperature yeild strength of GaAs is shown in Table IV. The solute hardening analysis indicates that yield strength should scale with the size effect of the appropriate tetrahedral cluster, provided that the dopants remain in solution. The relative volume change associated with the hypothetical insertion of a solute tetrahedron into the matrix can be calculated using the Ehrenreich and Hirth model. In Table V, the changes in flow stress $\delta\sigma$ produced by the change in concentration by δc for different dopants are given, along with relative volume change associated with doping. The quantitative relationship between $\delta\sigma/\delta c$ and $\delta v/v$ is not usually linear, and therefore data obtained at a single concentration value are insufficient to describe the behavior. But very

TABLE IV

YIELD STRESS (CRITICAL RESOLVED SHEAR STRESS) VALUES AT A STRAIN RATE OF 10^{-4} s^{-1} AT DIFFERENT TEMPERATURES FOR MATERIALS DESCRIBED IN TABLE III[a]

Material	CRSS (MPa)				
	500°C	700°C	900°C	1,100°C	1,150°C
A. Undoped (low B)	5.10	1.33	1.45	1.00	0.96
B. P-doped	6.60	2.39	1.26	0.91	
C. In-doped (high B)	9.90	4.64	3.33	3.27	
D. Sb-doped	10.25	2.45	1.49	1.25	
E. Undoped (high B)	7.30	2.55	1.91	1.83	
F. (In, Si) codoped	9.37	2.04	1.26	1.28	

[a]From Guruswamy et al. (1989).

TABLE V

THE CHANGE IN YIELD STRESS (CRITICAL RESOLVED SHEAR STRESS) PRODUCED BY DOPING AND THE SIZE EFFECT $\delta v/v^a$

Specimen	$\delta a/a$	$\delta v/v$	$\delta\sigma/\delta c$ (MPa/at.%)		A [MPa/(at.%)$^{1/2}$]	
			500°C	1,100°C	500°C	1,100°C
A. Undoped	0	0	0	0	0	0
B. P-doped	3.6	12	641	—	136	—
C. In-doped	7.2	23	796	502	535	337
D. Sb-doped	7.8	25.4	8710	553	1,852	118
E. B-doped	15.5	40	44,200	36,700	2,102	371
F. (In, Si) codoped			676	62	—	—

[a]From Guruswamy et al. (1989).

dilute solutions often produce strengthening of the form

$$\delta\sigma = Ac^{1/2}. \tag{3}$$

For this relationship, the value of A would vary linearly with $\delta v/v$. Consistent with the trends in the table, dislocation densities are reduced for In, (In, B), and Sb (Jacob et al., 1983; Mil'vidskii et al., 1981) additions at levels equivalent to those listed in Table V. On the other hand, P, which is expected to be a weak strengthener as shown in the table, is less effective at reducing dislocation densities even at levels of 10^{19}–10^{20} cm^{-3}.

From the observed broadening of the stoichiometric concentration region of GaAs, by the addition of dopants, Blom and Woodal (1988) suggest that the addition of In or Sb to GaAs is equivalent to adding more As to the melt. For Al or P the stoichiometric range should shrink in contrast to the case for In and Sb. A shift towards higher As potential results in increased As$_{Ga}$ defects (EL$_2$ centers) described in detail by Bourgoin et al. (1988). These antisite defects would also contribute to hardening.

7. CdTe AND RELATED II–VI COMPOUNDS

Both CdTe and HgTe have a borderline Philips' ionicity parameter, $F = 0.72$ and 0.65, close to the borderline ionicity value of 0.785 that separates tetrahedrally coordinated (predominantly covalent) compounds from octahedrally coordinated (predominantly ionic) compounds. The II–VI compounds are softer than III–V compounds, and many deform plastically at room temperature. Edge or mixed dislocations end in a row of like sign-ions Cd^q or Te^{-q}. The jogs on the dislocation lines carry an effective charge $\pm q/2$, and q is nearly equal to 1. Cationic jogs serve as donors and anionic jogs serve as acceptors. Transitions from p- to n-type behavior after plastic deformation and electroplastic effects observed in these crystals have been explained on the basis of charged dislocations and jogs by Hirth and Ehrenreich (1985a). However, because of difficulties in growth of large quality single crystals, only limited mechanical property studies have been performed. Petrenko and Whitworth (1980) performed deformation of a number of II–VI compounds such as ZnO, ZnS, ZnSe, ZnTe, CdS, and CdTe. They observed that the critical resolved flow stress varied linearly with the charge on the dislocation and also observed n-type to p-type transitions on plastic deformation. Hirth and Ehrenreich explained this transition as due to creation of charged jogs on the dislocations. The flow stress is sensitive to applied voltage (the electroplastic effect) as well as to light (the photoplastic effect). (Pellegrino and

Galligan, 1986). The critical resolved shear value for CdTe in darkness and with no applied voltage was about 7 MPa at room temperature.

Cole et al. (1982) studied the deformation behavior of the $Cd_xHg_{1-x}Te$ system at room temperature using four point bend tests. The strength of HgTe increases with Cd content (Fig. 17). At $x = 0.2$, the yield stress at room temperature was 20–30 MPa. In contrast, at these stress levels InSb deforms at 285°C, GaAs at 450°C, and Si at 1,015°C. The activation energy for plastic flow is estimated to be about 0.16 eV. Hardness studies by Schenk and Fissel (1988) on (Hg, Cd)Te over the range $x = 0$ to $x = 1$ are summarized in Fig. 18. They observed that the addition of HgTe to CdTe produces a marked increase in hardness. In similar studies, the mechanical behavior of $Cd_xHg_{1-x}Te$ single crystals has been investigated by Borbot et al. (1988, 1990) using hardness measurements and compression testing to temperatures of 350°C. Hardness values show a maximum at around $x = 0.7$. At low temperatures, deformation is controlled by the Peierls barrier, while at high temperatures the mechanism has yet to be defined. While the mechanical strength depends upon composition, dissociation width and stacking fault

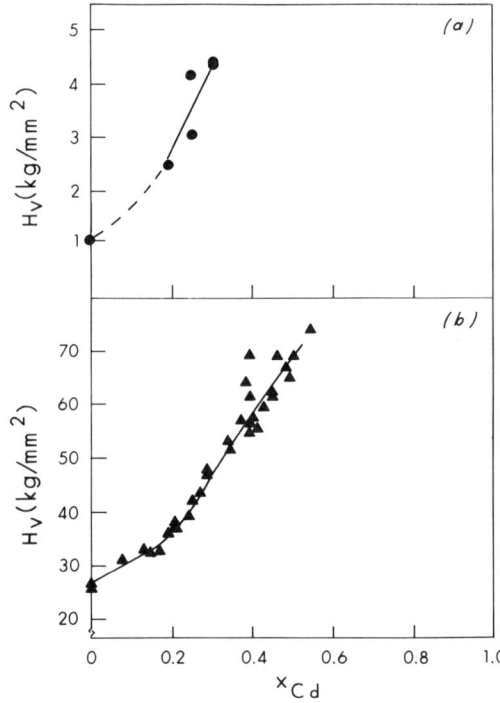

FIG. 17. (a) Yield stress and (b) Vickers hardness as a function of mole fraction of Cd in $Cd_xHg_{1-x}Te$ single crystals. [From Cole et al. (1982a).]

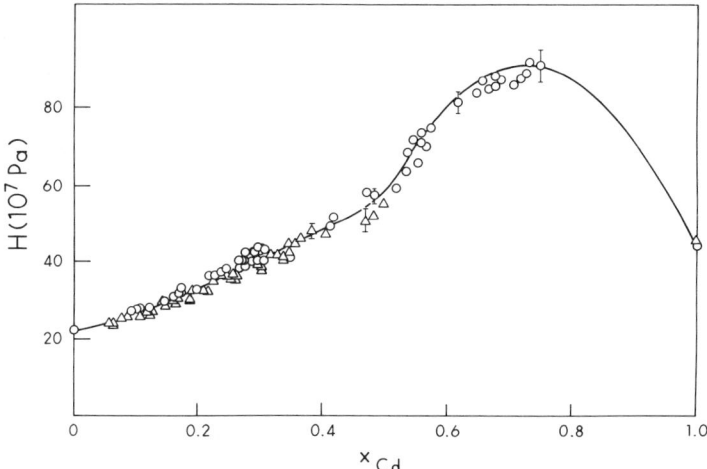

FIG. 18. Vickers hardness as a function of composition x of $Cd_xHg_{1-x}Te$. Data fit to a second-order curve. [From Schenk and Fissel (1988).]

energy determined from transmission electron microscopy remain constant at high temperatures.

ZnTe and CdSe additions to CdTe have a similar influence, but the addition of ZnTe to HgTe has a much larger hardening effect compared to the addition of CdTe to HgTe, as expected from the solid solution strengthening model (Table I). This may explain the low dislocation densities and subgrains in Zn-doped CdTe (Yoshikawa et al., 1986; James and Stroller, 1984; Bell and Sen, 1985; Tranchart et al., 1985). Guergouri et al. (1988) investigated the $Cd_xZn_{1-x}Te$ and $Cd_xMn_{1-x}Te$ systems using compressive deformation of parallelopiped samples as well as hardness tests. The yield stress at 0.2% strain was used to estimate the critical resolved shear stress. Zn additions at levels of 4% increased the CRSS value from 6.4 MPa to 27.3 MPa, while Mn had no strengthening effect (Table VI), a result

TABLE VI

COMPARISON OF PLASTIC AND STRUCTURAL PROPERTIES OF CdTe, $Zn_{0.04}Cd_{0.96}Te$, AND $Mn_{0.1}Cd_{0.9}Te$ CRYSTALS

	CRSS ([123]Compression) (MPa)	CRSS (Microhardness) (MPa)	EPD (cm^{-2})	Cellular Structure (μm)
CdTe	6.4	84	5×10^5	125
$Zn_{0.04}Cd_{0.96}Te$	27.3	159	5×10^4	200
$Mn_{0.1}Cd_{0.9}Te$	5.2	90	5×10^5	100

[a]From Guergouri et al. (1988).

consistent with the hardness studies of Triboulet *et al.* (1990) and in agreement with the solid solution hardening model of Ehrenreich and Hirth (1985).

V. Non-Isovalent Group II and VI Dopants in III-V Compounds

8. DISLOCATION VELOCITY

Dislocations in semiconductors and ionic solids carry a line charge, and we could therefore expect that the dislocation velocities would be influenced by non-isovalent doping. Different factors that need to be considered are elastic and electrical interactions between dislocations and impurities, the structure of the dislocations, and the influence of impurities on the concentration of charged vacancies and other point defects. Patel and Chaudhuri (1963) showed that in Ge at 500°C, As doping increased dislocation velocities while Ga doping decreased velocities. Similar effects on dislocation velocity with *n*- or *p*-type doping have since been observed in Si and III-V compounds. Unlike Si or Ge, in III-V or II-VI compounds, 60° dislocations may consist of either a positive α or negative β dislocation. Therefore, interactions with each of them separately need to be considered. The mobility of dislocations in undoped GaAs has been investigated by Steinhardt and Haasen (1978), Osvenskii *et al.* (1973), and Osvenskii and Kholodnyi (1973). It is observed that α dislocations have much higher mobility than screw or β dislocations, with activation energies of 0.93, 1.11, and 1.57 eV for α, screw, and β dislocations, respectively, in the temperature range of 200 to 450°C.

The influence of doping on dislocations arises from electrostatic interactions, but the exact nature of interactions is not well understood. The dopants can influence dislocation motion by changing the energy for kink nucleation (Steinhardt and Haasen, 1978). They can influence motion as well, by changing the vacancy concentrations (Longini and Green, 1956; Reiss, 1953; Valenta and Ramasastry, 1957), presumably through electrostatic interactions with dislocations.

Osvenskii *et al.* (1973) investigated effects of Zn and Te doping on the dislocation velocities in GaAs in the temperature range of 200–500°C and at stress levels of 2.5 to 50 MPa. Te at a level of $10^{19}/cm^3$ drastically reduced the dislocation velocities, while Zn at a level of $10^{18}/cm^3$ increased velocities slightly. Other investigators found similar effects (Steinhardt and Haasen, 1978; Choi *et al.*, 1977). Steinhardt and Haasen observed that $v_\alpha > v_\beta > v_{screw}$ at 400°C and 10 MPa in undoped GaAs, while additions of Zn decreased the velocity of α and increased the velocity of β and screw dislocations. The addition of the *n*-type dopant Te decreased the velocity of all three types of dislocations by orders of magnitude, with a concomitant reduction in the creep rate. With the exception of the work of Yonenaga et al., (1987) and Yonenaga and Somino (1989), all reported dislocation velocity studies have been confined to low temperatures where the double-kink nucleation and growth model describes the deformation behavior. In the former, the influence of isovalent In-doping at levels of $10^{20}/cm^3$ in GaAs at intermediate temperatures has been investigated. α dislocations were found to be immobile below a stress level of 10 MPa in the temperature range 350–700°C. Such immobility was not found for β dislocations in In-doped and α and β dislocations in undoped GaAs.

Deep level transient spectroscopy (Dobrilla and Blakemore, 1986), electron paramagnetic resonance (Suezawa and Sumino, 1986) and positron lifetime measurements (Dannefaer and Kerr, 1986) have been performed to analyze the nature and distribution of point defects in GaAs. Complex formation occurs between defects of opposite charge or opposite dilatation. Vacancy–impurity complexes such as $Zn_{Ga}V_{As}$ and $Si_{Ga}V_{As}$ are dominant at high dopant concentrations, while divacancies are dominant at low dopant concentrations (Dannefaer and Kerr, 1986). The stability of vacancy–impurity complexes decreases at high temperatures. $Zn_{Ga}V_{As}$ defects are not observed above 600°C, whereas the $Si_{Ga}V_{As}$ complex is observed up to 800°C (Swaminathan and Copley, 1976). These complexes are, therefore, unlikely to play a major role in the strengthening process at high temperatures. The hardening due to the divacancy or these complexes can result from an asymmetric strain distribution, as in the Ehrenreich and Hirth model, as well as from charge-related effects. Deep level antisite defects of the EL_2 type are also formed athermally during plastic deformation (Suezawa and Sumino, 1986; Weber et al., 1982).

9. YIELD STRESS

Sazhin et al. (1966) studied the yield behavior of Zn- and Te-doped GaAs crystals at doping levels of 6×10^{16} to $10^{19}/cm^3$ by uniaxial compression in a $\langle 111 \rangle$ direction and found that while Te increased the strength, Zn had little effect on the yield stress. More recent work by Boivin et al. (1990a, 1990b) on undoped, n-type, and p-type GaAs in the temperature range 10 to 600°C demonstrates results in general agreement with earlier studies.

Doping effects of Zn doping (p-type) and S doping (n-type) on InP have been studied by Muller et al. (1985) in the temperature range 600–800°C at a strain rate of $10^{-4} s^{-1}$. The yield stress, σ, was fitted to an equation of the form $\sigma = A \exp(-B/kT)$. The values of A and B for Zn-doped ($2.2 \times 10^{18}/cm^3$) InP were 0.006 MPa and 0.548 eV, respectively, compared to 0.031 MPa and 0.33 eV for an undoped InP. The S-doped ($5 \times 10^{18}/cm^3$)

InP crystal had values of 0.162 MPa and 0.22 eV. Extrapolated values at the melting point were 0.7, 1.15, and 0.6 MPa, respectively (Fig. 18). Three different types of behavior with change in strain rate were observed. At low strain rates, they observed microcreep prior to the macroscopic yield point, and the impurity cloud drag was rate-controlling with large activation energies for deformation. At high strain rates, dislocations break away from their impurity clouds, and the activation energy for deformation then corresponds to the double-kink nucleation and growth model. At intermediate strain rates, repeated pinning and breakaway of dislocations were observed. From etch pit studies, the onset of dislocation generation at a strain rate of $1 \times 10^{-4} \, s^{-1}$ seemed to occur at stresses about 20% below the yield stress. Seki *et al.* (1978) studied the influence of additions of Zn, S, and Te in InP and Zn, S, Te, Al, and N in GaAs on the dislocation density. Except Zn, all the dopants (all of them either *n*-type or isovalent) were effective in reducing the dislocation density. Seki *et al.* (1978) relate the reduction in dislocation density to the bond strength between the dopant and any one of the atoms of the host crystal. However, only limited experimental information is available at this time on the influence of different dopants on defect types, concentrations, and interactions at elevated temperatures — insufficient to allow any generalizations to be made.

VI. The Role of Si Doping in GaAs

Si occupies both Ga and As sites and is an amphoteric dopant. The relative occupancy or compensation ratio is sensitive to melt stoichiometry, temperature (Von Neida *et al.*, 1987; Hurle, 1988; Laithwaite and Newman, 1976; Chen *et al.*, 1980) and stress (Otsuki *et al.*, 1986). Work of Otsuki *et al.* (1986) shows that the compensation ratio Si_{Ga}/Si_{As} decreases with a change from compressive to tensile stress in the crystal (Fig. 19). At no applied stress, this ratio is much greater than 1, resulting in *n*-type behavior for the crystal. This *n*-type doping may result in the strong strengthening observed in Si-doped crystals. Bourett *et al.* (1987) studied the effects of a Si dopant at a level of $5 \times 10^{18}/cm^3$ in the temperature range 350–1080°C and found that the Si had a strong strengthening effect at temperatures below 900°C, while at higher temperatures it reduces the strength below the undoped crystal level. In (In, Si) co-doped GaAs crystals, similar behavior is observed (Guruswamy *et al.*, 1989). At low temperatures, the (In, Si) doped crystal (crystal F in Table II) has a strength comparable to an (In, B)-doped crystal (crystal C). At high temperatures, crystal F has a much lower strength than crystals C or E but is still stronger than the undoped crystal A. The addition of Si at levels greater than $5 \times 10^{18}/cm^3$ results in precipitation of elemental Si and $B_{13}As_3$

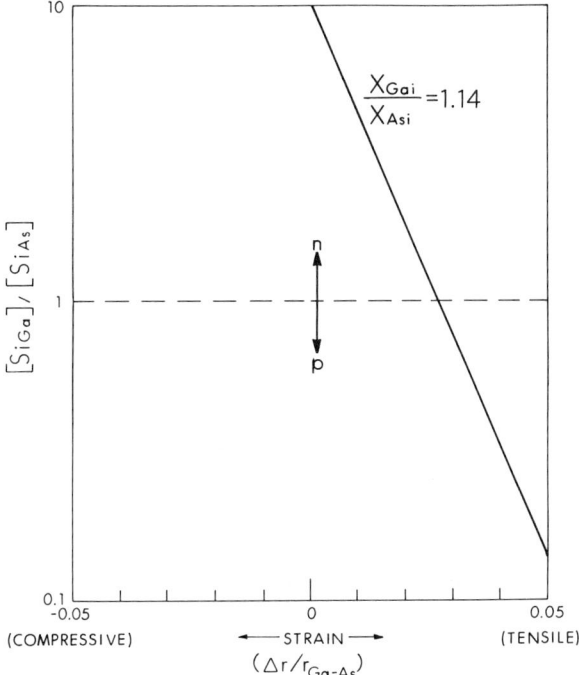

FIG. 19. Calculated compensation ratio versus strain in Si-doped GaAs. [From Otsuki et al. (1986).]

precipitates (Cockayne et al., 1988). Interesting recent work by Frigerio et al. (1989) shows that the addition of Si increases the B concentration in LEC grown crystals. Boron incorporation likely occurs through reactions such as (Cockayne et al., 1988; Thomas et al., 1984)

$$3Si_{GaAs} + 2B_2O_{3(L)} = 4B_{GaAs)} + 3SiO_{2(B2O3)}. \tag{4}$$

This B incorporation may lead to some strengthening as suggested by Ehrenreich and Hirth (1985). The effect of changes in point defect concentration with doping has been considered by Blom and Woodall (1988). As-rich melts are characterized by higher grown-in dislocation densities. As-rich precipitates are also observed. Si doping in Ga-rich and stoichiometric melts results in low dislocation densities. Swaminathan and Copely (1975) also observed that Si and Cr doping increased yield strength in the temperature range 250–500°C. The strengthening was attributed to $Si_{Ga}V_{As}$ pairs. Si diffusion coefficients in GaAs are influenced by the presence of other donor elements. A decrease in Si diffusivity is observed in GaAs containing column VI donors compared to column IV donor doped GaAs (Deppe et al., 1988). This is attributed to greater binding energy of column VI donor–gallium vacancy complexes. In an analogous system, Brown et al. (1983) showed that Ge doping of InP at low temperatures weakens InP, but above 600°C strengthens the crystal.

Apart from the charge-related strengthening effects, the addition of Si also results in lattice dilatation. A large negative dilation is observed in Ga-rich GaAs, while a small positive dilation is observed in As-rich crystals (Brice, 1985). The strengthening of the lattice is also expected because of this size misfit.

VII. Crystal Growth

The formation of dislocations in LEC-grown crystals is related to thermally induced stressess during growth and to stoichiometry. Thermal stress-induced formation of dislocations in Si, Ge, GaAs, GaP, and InP has been known for the past three decades (Jordan et al., 1980). The growth of III–V compounds free from dislocations is much more difficult compared to Si or Ge, because these compounds are inherently softer and have lower thermal conductivities. The addition of dopants to strengthen the crystal is used both to attempt to decrease thermal gradients during crystal growth and to eliminate or reduce dislocation densities. Thermoelastic analyses of stresses during LEC growth of Si, Ge, and III–V compounds have been performed extensively by Jordan and coworkers. Recent calculations by Jordan et al. (1986) indicate that dislocations can be completely removed in 75-mm diameter GaAs crystals grown under low thermal gradient conditions if the CRSS for plastic flow initiation is a factor of four larger than a GaAs base level value of 0.6 MPa, determined by extrapolation to the melting point. The basis for the extrapolation is the data for the yield point of undoped GaAs (Swaminathan and Copely, 1975). The work of Guruswamy et al. (1987) (Fig. 14) suggests a value of 3.3 MPa for the yield point extrapolated to the melting point at an In level of $1-2 \times 10^{20}$ cm^{-3}. Hence, such a doping level of indium should prevent the profuse dislocation generation characterstic of stage I deformation according to the theory of Jordan et al. (1986a).

However, the high-temperature deformation studies (Guruswamy et al., 1987; Hobgood et al., 1986; Tabache et al., 1986; Djemel and Castaing, 1986) show that microyield in GaAs occurs well below the yield point in compression tests above 700°C and particularly at 1,100°C. These results, together with the *in situ* x-ray topographic studies in tension (Tohno et al., 1986) at temperatures less than 700°C, indicate that dislocations, once formed, are highly mobile at high temperatures. This suggests that if GaAs crystals are grown in a dislocation-free condition, a major impediment to the presence of dislocations must be the nucleation process. In all of the compression tests (Swaminathan and Copely, 1975; Hobgood et al., 1986; Tabache et al., 1986; Djemel and Castaing, 1986; Guruswamy et al., 1987), the crystal end in contact with the loading platens is an easy site for dislocation nucleation because of stress intensification at contact asperities and because of the weak singularity in the stress field at the edge of the contact surface of the specimen. Hence, such data, while indicative of the yield point, may not correspond in microyield behavior to the crystal growth case, where nucleation of dislocations would be more difficult. There is some indication of such a possibility in the low-temperature tensile test observations of Tohne et al. (1985, 1986), a case where such end effects are absent and where dislocation nucleation does not occur at stress levels of at least 20% of the yield stress. All of the preceding discussion has been in terms of quasi–steady-state flow corresponding to deformation at and subsequent to the lower yield point. An additional hardening-type influence of isovalent hardeners could be manifested if the deformation during crystal growth occurred mainly in the transition region reflected by the microyield stress, the upper yield point, and deformation between the upper and the lower yield points. A retarding effect of a solid solution hardener on dislocation motion during multiplication would tend both to increase the upper yield point and to prolong the strain range between the upper and lower yield points (Muller et al., 1985; Tohno et al., 1985). The microyield stress, at which dislocation motion begins, would also be increased by such a retarding effect.

Increasing additions of the hardener indium in GaAs results in an earlier onset of constitutional supercooling and an increased tendency to fracture. Low thermal gradients make diametral control difficult, and high thermal gradients result in cracking of the crystal. Thus, there is a trade-off between low thermal gradients and high In doping (McGuigan et al., 1986b). Poor diametral control may make nucleation of dislocations easier. Constitutional supercooling ahead of the liquid–solid interface will result in the formation of multiple crystal nuclei, and therefore the loss of single crystal growth. Large dopant additions may result in second phase precipitation, which helps nucleate the dislocations due to elastic incompatability stresses. It appears, therefore, that isovalent dopants with an optimal combination of solubility and size misfit in combination with donor dopants would be the most effective in reducing dislocation density. The use of multiple solute additions would provide strengthening while tending to obviate the precipitation and constitutional supercooling problems.

A deviation from stoichiometry in undoped or lightly doped materials and a shift towards *p*-type behavior lead to an increase in dislocation density, as shown by Lagowski et al. (1984, 1986). These dislocations result from the condensation of vacancies at temperatures well below the melting point. Fig. 20 shows that an increase in acceptor concentration increases the dislocation density, while donor doping decreases the dislocation density. In Fig. 21, the variation of the Fermi energy level with carrier concentration is shown. These dislocations result from the condensation of excess vacancies to form dislocation loops, and dislocation densities as high as $10^5/cm^3$ may result from nonstoichiometry. The addition of $10^{17}/cm^3$ of an *n*-type dopant can eliminate stoichiometry-related dislocations completely. Once the nonstoichiometry-related dislocations are eliminated, thermal stresses are the only source of dislocations. Hardening of the lattice by impurities can then result in complete elimination of dislocations in the crystal. Stoichiometry-related effects are important at dopant levels of $10^{17}/cm^3$, beyond which impurity-

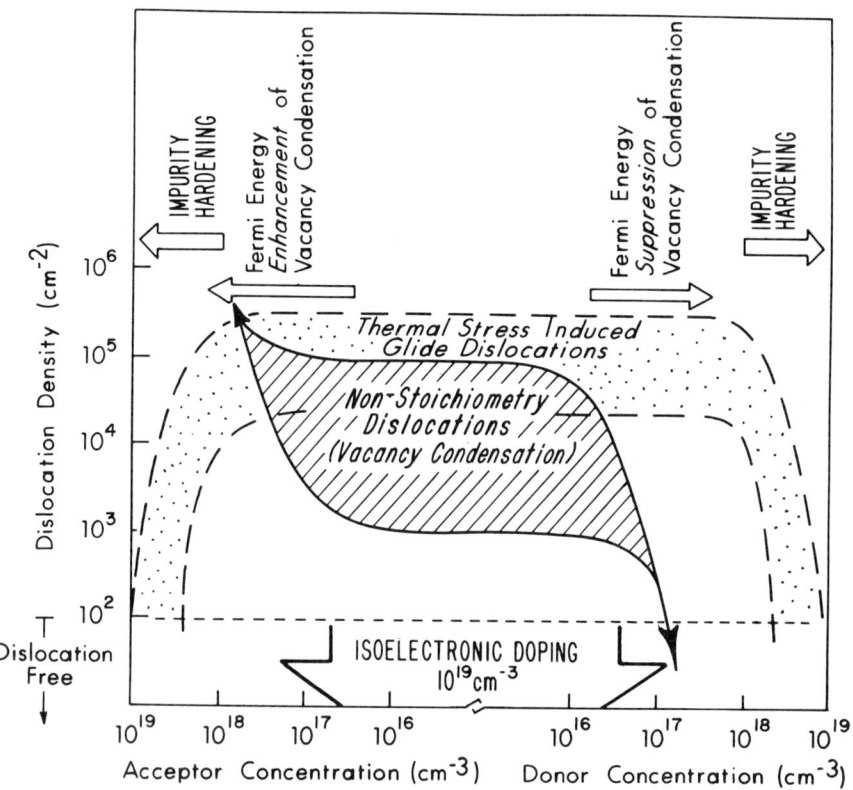

FIG. 20. Schematic diagram of dislocation density versus carrier concentration and conduction type in GaAs. [From Lagowskii et al. (1984).]

4. MECHANICAL BEHAVIOR OF COMPOUND SEMICONDUCTORS

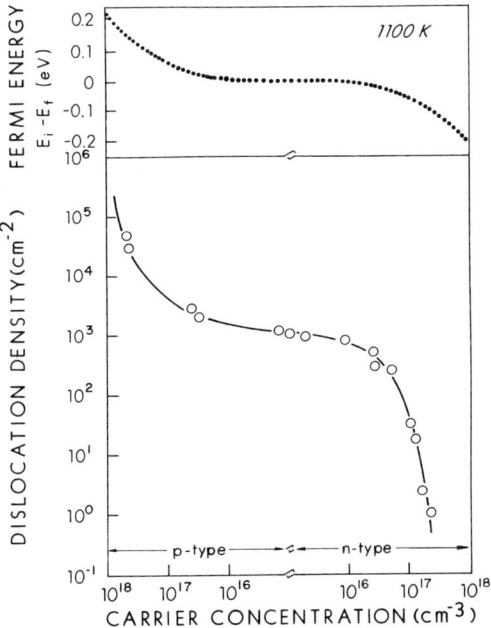

FIG. 21. Variation of Fermi energy level with carrier concentration in GaAs at 300 K. [From Lagowskii *et al.* (1984).]

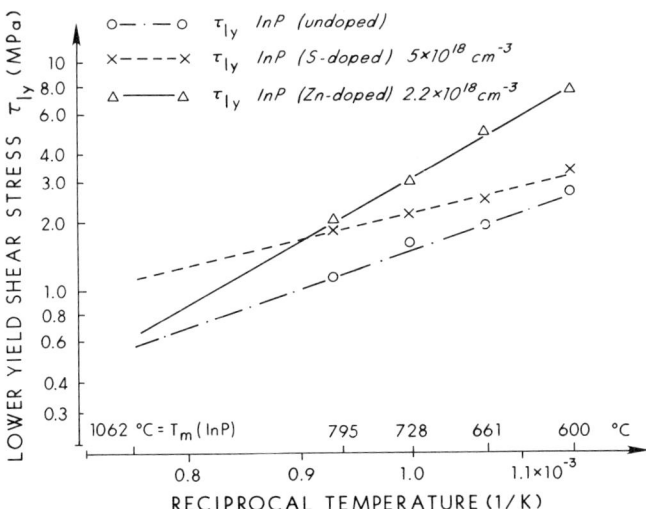

FIG. 22. Lower yield shear stress versus reciprocal temperature for critical S and Zn concentrations (minimum concentrations for growth of dislocation-free crystals) and for an undoped InP at a strain rate of 10^{-4} s^{-1}. [From Müller *et al.* (1985).]

hardening effects dominate. In the transition region between the n-type and p-type regions, the addition of isovalent dopants that do not influence the Fermi level, at levels of $10^{19}/cm^3$, can reduce stress-induced dislocations. In general, an increase (decrease) in donor concentration will suppress (enhance) any defect with an associated negative charge. Strengthening of InP crystals by Zr, a p-type dopant and S, an n-type dopant, at critical concentration levels for the growth of dislocation-free crystals is shown in Fig. 22. As discussed earlier, the dramatic strengthening effect and the dislocation density reduction effect of isovalent undoped GaAs can be explained based on the Ehrenreich and Hirth (1985) model.

VIII. Summary

General deformation behavior in elemental, III–V, and II–VI compounds is similar. Effects of solute additions in the case of isovalent dopants can be understood based on the ideas of the Ehrenreich and Hirth solid solution strengthening model. Antisite defects, whether present at equilibrium or produced athermally, would produce analogous hardening. Amphoteric additions would produce more complicated effects including a size effect, electrostatic hardening effects, and indirect effects associated with vacancy complexes at low homologous temperatures. However, the electrical interactions with the dislocations or other point defects are not clearly understood at elevated temperatures. Only limited data are available on point defect type, density, and interactions at elevated temperatures approaching the melting point. More attention needs to be paid to small strain deformations at high temperatures and the determination of threshold stress for dislocation generation under crystal growth conditions. The deformation of dislocated crystals at very low strain rates is poorly understood.

References

Alexander, H., and Haasen, P. (1968). *Solid State Phys.* **22**, 27.
Alexander, H., and Haasen, P. (1972). *In* "Annual Review of Materials Science" (Huggins, R. A., Bube, R. H., and Roberts, R. W., eds.), Vol. 2, p. 291. Annual Reviews, Inc., Palo Alto, California.
Anstis, G. R., Chantikul, P., Lawn, B. R., and Marshall, D. B., (1981). *J. Am. Ceram. Soc.* **64**, 533.
Barbot, J. F., Rivaud, G., Garem, H., Blanchard, C., Desoyer, J. C., LeScoul, D., Dessus, J. L., and Durand, A. (1990). *J. Mater. Sci.* **25**, 1877.
Barbot, J. F., Rivaud, G., and Desoyer, J. C. (1988). *J. Mater. Sci.* **23**, 1659.

Bell, S. L., and Sen, S. (1985). *J. Vac. Sci. Technol.* **A3**, 112.
Blom, G. M., and Woodall, J. M. (1988). *J. Elec. Mater.* **17**, 391.
Boivin, P., Rabier, J., and Garem, H. (1990a) *Phil. Mag.* **A61**, 619.
Boivin, P., Rabier, J., and Garem, H. (1990b). *Phil. Mag.* **A61**, 647.
Bourett, E. D., Tabache, M. G., and Elliot, A. G. (1987). *Appl. Phys. Lett.* **50**, 1373.
Bourgoin, J. C., von Bardeleben, H. J., and Stiévenard, D. (1988). *J. Appl. Phys.*, **64**, R65.
Brasen, D., and Bonner, W. A. (1983). *J. Mater. Sci. Eng.* **61**, 167.
Brice, J. C., (1985). *In* "Properties of Gallium Arsenide" (Blackmore, J. C., and Price, J. C., eds.). INSPEC, Institution of Electrical Engineers, London, EMIS Data Review RN2005.
Brion, H. G., and Haasen, P. (1985). *Phil. Mag.* **A51**, 879.
Brion, H. G., Haasen, P., and Siethoff, H., (1971). *Acta Metall.* **19**, 283.
Brown, G. T., Cockayne, B. and MacEwan, W. R. (1980). *J. Mater. Sci.* **15**, 1469.
Brown, G. T., Cockayne, B., and MacEwan, W. R. (1981). *J. Crystal Growth* **51**, 369.
Brown, G. T., Cockayne, B., MacEwan, W. R., and Ashen, D. J. (1983). *J. Mater. Sci. Lett.* **2**, 667.
Burenkov, Yu, A., Burdukov, Yu. M., Davydov, S. Yu., and Nikanarov, S. P. (1973). *Sov. Phys. Solid State* **15**, 1175.
Cahoon, J. P., Broughton, W. H., and Kutzak, A. R. (1971). *Met. Trans.* **2**, 1979.
Chen, C. P. (1984). *JPL Publication 84–81*.
Chen, R. T., Rana, V., and Spitzer, W. G. (1980). *J. Appl. Phys.* **51**, 1532.
Choi, S. K., Mihara, M., and Ninomiya, T. (1977). *Jpn. J. Appl. Phys.* **16**, 737.
Cockayne, B., MacEwan, W. R., Pope, D. A.O., Harris, I. R., and Smith, N. A. (1988). *J. Crystal Growth* **87**, 6.
Cole, S., Willoughby, A. F.W., and Brown, M., (1982a). *J. Crystal Growth* **59**, 370.
Cole, S., Brown, M., and Willoughby, A. F.W., (1982b). *J. Mater. Sci.* **17**, 2061.
Dannefaer, S., and Kerr, D. (1986). *J. Appl. Phys.* **60**, 591.
Deppe, D. G., Holonyak, N., Jr., and Baker, J. E. (1988). *Appl. Phys. Lett.* **52**, 129.
Djemel, A., and Castaing, J. (1986). *Europhys. Letts.* **2**, 611.
Djemel, A., Castaing, J., and Duseaux, M. (1988). *Phil. Mag.* **A57**, 198.
Dobrilla, P., and Blakemore, J. S. (1986). *J. Appl. Phys.* **60**, 169.
Ehrenreich, H., and Hirth, J. P. (1985). *Appl. Phys. Lett.* **46**, 668.
Frigerio, G., Muchino, C., Weyher, J. L., Zanotti, L., and Paorici, C. (1989). Paper presented at the International Conference on Crystal Growth IX, Sendai, Japan, August, 1989.
Frost, H. J. and Ashby, M. F. (1982). "Deformation Mechanism Maps" Pergamon Press, Elsmford, New York.
Fujiwara, Y., Kita, Y., Tonami, T., and Hamakawa, Y. (1986). *Appl. Phys. Lett.* **49**, 161.
Gall, P., Peyrade, J. P., Coquille, R., Reynaud, F., Gabillet, S., and Albacete, A. (1987) *Acta Metall.* **45**, 143.
Gilman, J. J. (1969). "Micromechanics of Flow in Solids" McGraw-Hill, New York.
Gridneva, I. V., Milman, Yu, V., and Trefilov, V. I. (1969). *Phys. Stat. Sol.* **36**, 59.
Guergouri, K., Triboulet, R., Tromson-Carli, A., and Marfaing, Y. (1988). *J. Crystal Growth* **86**, 61.
Guruswamy, S., Hirth, J. P., and Faber, K. T. (1986a). *J. Appl. Phys.* **60**, 4136.
Guruswamy, S., Rai, R. S., Faber, K. T., and Hirth, J. P. (1986b). Unpublished.
Guruswamy, S., Rai, R. S., Faber, K. T., and Hirth, J. P. (1986c). *Mater. Res. Soc. Symp. Proc.* Vol. **53**, 329.
Guruswamy, S., Rai, R. S., Faber, K. T. and Hirth, J. P. (1987). *J. Appl. Phys.* **62**, 4130.
Guruswamy, S., Rai, R. S., Faber, K. T., Hirth, J. P., Clemans, J. E., McGuigan, S., Thomas, R. N., and Mitchel, W. (1989). *J. Appl. Phys.* **65**, 2508.
Haasen, P. (1982). *In* "Dislocations in Solids." (F. R.N. Nabarro, ed.), Vol. 4, p. 147. North-Holland, Amsterdam.

Haasen, P. (1957). *Acta Metall.* **5**, 598.
Hirsch, P. B., Pirouz, P., Roberts, S. G., and Warren, P. D. (1985). *Phil. Mag.* **B52**, 759.
Hirth, J. P., and Ehrenreich, H. (1985a). *J. Vac. Sci. Technol.* **A3**, 367.
Hirth, J. P., and Ehrenreich, H. (1985b). Private communication.
Hirth, J. P., and Lothe, J. (1982). "Theory of Dislocations," 2nd edition. Wiley, New York.
Hobgood, H. M., McGuigan, S., Spitznagel, J. A., and Thomas, R. N. (1986). *Appl. Phys. Lett.* **48**, 1654.
Hurle, D. T. J. (1988). *In* "Semi-insulating III-V Materials, Proc. 5th Conf. on Semi-insulating III-V materials, Malmo, Sweden," p. 11. Adam-Hilger, Bristol and Philadelphia.
Jacob, G., Duseaux, M., Farges, J. P., Van Den Boom, M. M.B., and Raksoner, P. J. (1983). *J. Crystal Growth* **61**, 417.
James, R. W., and Stroller, R. E. (1984). *Appl. Phys. Lett.* **44**, 56.
Jimenex-Melendo, M., Djemel, A., Riviere, J. P., and Castaing, J. (1986). *In* "Defects in Semiconductors," (H. J. Von Bardeleben, ed.) Materials Science Forum, Vol. 10-12, p. 791. Trans. Tech. Publications, Switzerland.
Jordan, A. S. (1980). *J. Crystal Growth* **49**, 631.
Jordan, A. S., and Parsey, J. M., Jr, (1986). *J. Crystal Growth* **79**, 280.
Jordan, A. S., and Parsey, J. M., Jr, (1988). *Mater. Res. Soc. Bull.* **13**, 36.
Jordan, A. S., Von Neida, A. R., and Caruso, R. (1986). *J. Crystal Growth* **76**, 243.
Jordan, A. S., Caruso, R., and Von Neida, A. R. (1980). *Bell System Technical Journal*, **59**, 593.
Jordan, A. S., Von Neida, A. R., and Caruso, R. (1984). *J. Crystal Growth* **70**, 555.
Lagowski, J., Gatos, H. C., Aoyama, T., and Lin, D. G. (1984). *In* "Semi-insulating III-V Materials" (Cook, D. C., and Blakemore, J. C., eds.), p. 60. Shiva Publishing Ltd., Cheshire, United Kingdom.
Lagowski, J., Gatos, H. C., Aoyama, T., and Lin, D. G. (1986). *Appl. Phys. Lett.* **45**, 680.
Laister, D., and Jenkins, G. M. (1973). *J. Mater. Sci.* **8**, 1218.
Laithwaite, K., and Newman, R. C. (1976). *J. Phys.* **C9**, 4503.
Longini, H. L., and Green, R. F. (1956). *Phys. Rev.* **102**, 992.
Matsui, M. and Yokoyama, T. (1985). "Int. Symposium on GaAs and Related Compounds, Japan Int. Prop. Conference Series No. 79," Chap. 1, p. 13.
McClintok, F. A., and Argon, A. S. (1966). "Mechanical Behavior of Materials," p. 43. Addison-Wesley, Reading, Massachusetts.
McGuigan, S., Thomas, R. N., Barret, D. L., Hobgood, H. M., and Swanson, B. W., (1986a). *Appl. Phys. Lett.* **48**, 1377.
McGuigan, S., Thomas, R. N., Hobgood, H. M., Eldridge, G. W., and Swanson, B. W. (1986b). *In* "Semi-insulating III-V Materials, 1986" (Kukimoto, H., and Miyazawa, S., eds.) p. 29. OHM, North Holland, Amsterdam and Ohmsha, Tokyo.
Meshii, M., Nagakawa, J., and Bang, G. W. (1982). *In* "Hardening in B.C.C. Metals" (Meshii, M., ed.). p. 95. Metallurgical Soc. of AIME, Warendale, Pennsylvania.
Mikkelson, J. C., Jr., and Boyce, J. B. (1983). *Phys. Rev.* **B28**, 7130.
Mil'vidskii, M. G., and Bochkarev, E. P. (1978). *J. Crystal Growth* **44**, 61.
Mil'vidskii, M. G., Osvensky, V. B., and Shifrin, S. S. (1981). *J. Crystal Growth* **52**, 396.
Miyazawa, S., and Hyuga, F., (1986). *IEEE Trans. Electron Devices* **ED-33**, 227.
Müller, G., Rupp, R., Volkl, J., Wolf, M., and Blum, W. (1985). *J. Crystal Growth* **71**, 771.
Myshlyaev, M. M., Nikitenko, V. I., and Nestenenko, V. I. (1969). *Phys. Stat. Sol.* **36**, 89.
Nabarro, F. R.N., Basinski, Z. S., and Holt, D. B. (1964). *Adv. Phys.* **13**, 193.
Nanishi, Y., Ishida, S., Honda, T., Yamazaki, H., and Miyazawa, S. (1982). *Jpn. J. Appl. Phys.* **21**, L335.
Nikitenko, B. I. (1967). *In* "Dislocations and Physical Properties of Semiconductors," p. 58. Acad. Sci., U. S.S. R., Moscow.

Ogura, M., Adachi, Y., and Ikoma, T. (1979). *J. Appl. Phys.* **50**, 6745.
Osvenskii, V. B., and Kholodnyi, L. P. (1973). *Soviet Physics — Solid State* **14**, 2822.
Osvenskii, V. B., Kolodnyi, L. P., and Mil'vidskii, M. G. (1973). *Soviet Physics — Solid State* **15**, 661.
Otsuki, T., Shimano, A., Aoki, H., Takagi, H., Kano, G., and Teramoto, I. (1986). *In* "Semi-insulating III-V Materials" (Kukimoto, H., and Miyazawa, A., eds.), p. 243. OHM * North-Holland, Amsterdam.
Patel, J. R., and Chaudhuri, A. R. (1963). *J. Appl. Phys.* **34**, 2788.
Pellegrino, J., and Galligan, J. M. (1986). *Appl. Phys. Lett.* **48**, 1127.
Petrenko, V. F., and Whitworth, R. W. (1980). *Phil. Mag.* **A41**, 681.
Petroff, P., and Hartman, R. L. (1973). *Appl. Phys. Lett.* **23**, 469.
Raccah, P. M., Garland, J. M., Zhang, Z., Mioc, S., De, Y., Chu, A. H.M., McGuigan, S., and Thomas, R. N. (1986). "Society of Photo-optical Instrumentation Engineers, Vol. 623: Advanced Processing and Characterization of Semiconductors III," SPIE, Bellingham, WA., p. 40.
Rai, R. S., Faber, K. T., Guruswamy, S., and Hirth, J. P. (1987). "Proc. 45th Annual Meeting of Electron Microscopy Society of America"(Bailey, G. W., ed.), p. 320. San Francisco Press, San Francisco.
Rai, R. S., Guruswamy, S., Faber, K. T., and Hirth, J. P. (1989). *Phil. Mag.* **A60**, 339.
Ray, Brian (1969). "II-IV Compounds" Pergamon Press, New York.
Reiss, H. (1953). *Chem. Phys.* **21**, 1209.
Ruda, H. E., Lagowskii, J., Gatos, H. C., and Walukiewicz, W. (1984). "Semi-insulating III-V Materials," p. 263. Shiva Publishing, Cheshire, United Kingdom.
Sazhin, N. P., Milvidskii, M. G., Osvenskii, V. B., and Stolyarov, O. G. (1966). *Soviet Phys. — Solid State* **8**, 1223.
Schenk, M. and Fissel, A. (1988). *J. Crystal Growth* **86**, 502.
Schroter, W., and Siethoff, H. (1984). *Z. Metall.* **75**, 482.
Schroter, W., Brion, H. G., and Siethoff, H. (1983). *J. Appl. Phys.* **54**, 1816.
Seki, Y., Watanabe, H., and Matsui, J. (1978). *J. Appl. Phys.* **49**, 822.
Shewmon, Paul (1989). "Diffusion in Solids," 2nd edition TMS, AIME, Reading, Pennsylvania.
Siethoff, H. and Behrensmeier, R. (1990). *J. Appl. Phys.* **67**, 3673.
Siethoff, H., and Schroter, W. (1984). *Z. Metall.* **75**, 475.
Siethoff, H., Ahlborn, K., Brion, H. G., and Volkl, J. (1988). *Phil. Mag.* **A57**, 235.
Steinhardt, H., and Haasen, P. (1978). *Phys. Stat. Sol. (a)* **49**, 93.
Suezawa, M., and Sumino, K. (1986). *Jpn. J. Appl. Phys.* **25**, 533.
Swaminathan, V., and Copley, S. M. (1975). *J. Amer. Ceram. Soc.* **58**, 482.
Swaminathan, V., and Copley, S. M. (1976). *J. Appl. Phys.* **47**, 4405.
Tabache, M. G., Bourett, E. D., and Elliot, A. G. (1986). *Appl. Phys. Lett.* **49**, 289.
Tada, K., Kawasaki, A., Kotani, T., Nakai, R., Takebe, T., and Akai, S. (1982). *In* "Semi-insulating III-V Materials," (Makram-Ebied, S., and Tuck, B., eds.), p. 36. Shiva Publishing Ltd, Cheshire, United Kingdom.
Thomas, R. N. (1985). Private communication.
Thomas, R. N., Hobgood, H. M., Eldridge, G. W., Barret, D. L., Braggins, T., Ta, L. B., and Wang, S. K. (1984). *In* "Semi-insulating GaAs," Vol. 20 in "Semiconductors and Semimetals" (Willardson, R. K., and Beer, A. C., eds), p. 1. Academic Press, New York.
Tohno, S., Shinoyama, S., Katsui, A., and Takaoka, H. (1985). *J. Crystal Growth* **73**, 190.
Tohno, S., Shinoyama, S., Katsui, A., and Takaoka, H. (1986). *Appl. Phys. Lett.* **49**, 1204.
Tranchart, J. C., Latorre, B., Foucher, C. and Le Gouge, Y. (1985). *J. Crystal Growth* **72**, 468.
Triboulet, R., Heurtel, A., and Rioux, J. (1990). *J. Crystal Growth* **101**, 131.
Valenta, W. M., and Ramasastry, C. (1957). *Phys. Rev.* **106**, 73.

Von Neida, A. R., Pearton, S. J., Stavola, M., and Caruso, R. (1987). *In* "GaAs and Related Compounds 1986" (Lindley, W. T., ed.), Inst. of Phys. Conf. Serivces No. 83, p. 57, Institute of Physics, Bristol.
Watts, D. Y., and Willoughby, A. F.W. (1984). *J. Appl. Phys.* **56**, 1869.
Weber, E. R., Ennen, H., Kaufmann, U., Windscheif, J., Schneider, J., and Wosinski, T. (1982). *J. Appl. Phys.* **53**, 6141.
Wolfson, R. G. (1975). *In* "Treatise on Materials Science and Technology" (Arsenault, R. J., ed.), Vol. 6, p. 333. Academic Press, New York.
Yonenaga, I., Somino, K., Izawa, G., Watanabe, H., and Matsui, J., (1989). *J. Mater. Res.* **4**, 361.
Yonenaga, I. and Sumino, K. (1989a). *J. Mater. Res.* **4**, 355.
Yonenaga, I. and Sumino, K. (1989b). *J. Appl. Phys.* **65**, 85.
Yonenaga, I., Sumino, K. and Yamada, K. (1986). *Appl. Phys. Lett.* **48**, 326.
Yonenaga, I., Onose, U., and Sumino, K., (1987). *J. Mater. Res.* **2**, 252.
Yoshikawa, M., Maruyama, K., Saito, T., Maekawa, T., and Takigawa, H. (1987). "1986 U. S. Workshop on Physics and Chemistry of CdHgTe, Dallas, Texas" *J. Vac. Sci. Technol.* **A5**, 3052.

CHAPTER 5

Deformation Behavior of Compound Semiconductors

S. Mahajan

DEPARTMENT OF MATERIALS SCIENCE
CARNEGIE MELLON UNIVERSITY
PITTSBURGH, PENNSYLVANIA

I.	INTRODUCTION	231
II.	DISLOCATIONS AND DEFORMATION-INDUCED STACKING FAULTS	233
III	DEFORMATION CHARACTERISTICS OF BINARY COMPOUND SEMICONDUCTORS	237
	1. *Deformation Behavior at Low Temperatures and High Stresses*	237
	2. *Deformation Behavior at Intermediate Temperatures*	245
	3. *Deformation Behavior at High Temperatures*	250
IV.	PHOTOPLASTIC EFFECTS IN COMPOUND SEMICONDUCTORS	252
V.	ATOMIC ORDERING AND SURFACE PHASE SEPARATION IN TERNARY AND QUATERNARY III–V SEMICONDUCTORS AND THEIR RAMIFICATIONS IN DEGRADATION RESISTANCE OF LIGHT-EMITTING DEVICES	257
VI.	SUMMARY	264
	ACKNOWLEDGMENTS	265
	REFERENCES	265

I. Introduction

That dislocations have deleterious effects on the performance and reliability of minority carrier devices is well documented. Petroff and Hartman (1973), Hutchinson and Dobson (1975) and Ishida and Kamejima (1979) have shown that dark line defects (DLDs) observed in the degraded regions of GaAs/GaAlAs double heterostructure laser diodes originate from existing dislocations. A consensus has emerged that DLDs oriented along the $\langle 110 \rangle$ directions in the (001) plane evolve by the glide of threading dislocations, whereas the $\langle 100 \rangle$ oriented DLDs could form by glide and climb. The work of Mahajan et al. (1979b) on optically degraded InGaAsP epitaxial layers has demonstrated that nonluminescent regions, as revealed by spatially resolved photoluminescence, consist of dislocation networks. It has also been observed that dislocations and stacking faults generated in GaAs/GaAlAs structures as a result of poor preparation of the GaAs substrate surface have a marked effect on the performance of light-emitting devices (Dutt et al., 1981).

Meier et al. (1985) have carried out a systematic study on the influence of dislocations on the minority carrier diffusion length in web-silicon that contains numerous growth twins whose habit planes are parallel to the growth surface. In addition, they have correlated the minority carrier diffusion length with the efficiency of solar cells fabricated from the companion portions of the silicon ribbons. They have observed that the increase in the density of dislocations lowers the minority carrier diffusion lengths. These decrements are in turn reflected in the reduced efficiency of the solar cells. They have also shown that the decorated dislocations could be responsible for lowering the minority carrier diffusion lengths.

The recent work of Beam et al. (1990) has shown that during homoepitaxy, dislocations present in an underlying substrate and whose Burgers vectors are either inclined or parallel to the substrate surface are replicated into an epitaxial layer. It is therefore reckoned that substrates having low dislocation densities are essential for the growth of epitaxial wafers required for the fabrication of a variety of minority carrier devices, i.e., as-grown crystals should have a low dislocation density.

The introduction of dislocations during crystal growth could result from three different sources. First, dislocations present in seed crystals could propagate into a growing crystal. Second, dislocations in peripheral regions of the growing crystal could, under the influence of thermal-gradient-induced stresses, propagate into the crystal interior. Third, supersaturation of point defects could lead to clustering to form dislocation loops during cooldown following growth. Since it is possible to grow from dislocated seeds crystals with large areas that are macroscopically dislocation-free, it appears that the quality of seeds does not have an important influence on the perfection of as-grown crystals (Mahajan et al., 1981).

Figure 1 shows an etch-pitted (001) surface of a highly dislocated S-doped InP crystal (Mahajan, 1989). It is apparent that the majority of the etch pits are aligned along the [110] and [1$\bar{1}$0] directions, which define the lines of intersection of the {111} slip planes in the zincblende lattice with the (001) plane. Following Penning (1958) and Jordan et al. (1980), the observed discloation distribution can be rationalized in terms of slip induced by thermal-gradient-induced stresses.

It is apparent from the preceding discussion that we require substrates with low dislocation density for the growth of epitaxial wafers needed for the fabrication of reliable minority carrier devices. To achieve this objective we must understand how dislocations are introduced into as-grown crystals. The accumulated evidence indicates that thermal-gradient-induced stresses are a

5. DEFORMATION BEHAVIOR OF COMPOUND SEMICONDUCTORS

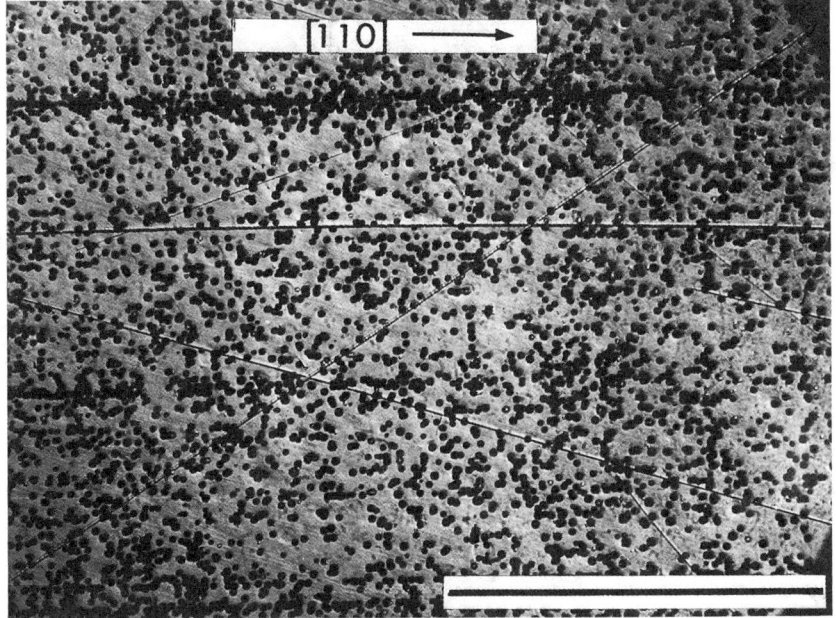

FIG. 1. Typical dislocations etch pit distribution observed on a (001) slice of highly dislocated S-doped InP wafer (Mahajan, 1989). Marker represents 0.1 mm.

major source of dislocations. Therefore, to devise ways to reduce the dislocation density in as-grown crystals, we must understand how bulk undoped semiconductor crystals deform and how doping affects their strength. In this chapter, we have highlighted these two topics for the compound semiconductors. We have also presented recent results on the occurrence of phase separation and atomic ordering in ternary and quaternary epitaxial layers of III-V compound semiconductors. The influence of the two microstructural features on the multiplication of dislocations by glide and climb is discussed.

II. Dislocations and Deformation-Induced Stacking Faults

Before discussing the deformation behavior of compound semiconductors, it is pertinent to elaborate on the types of dislocations and stacking faults that can form in these materials. This discussion should enhance the appeal of this review to a wider audience.

The compound semiconductors crystallize in the zincblende structure, which consists of two interpenetrating fcc unit cells. The two units are displaced from each other by $(a/4)\langle 111 \rangle$, where a is the lattice parameter of the semiconductor. One of the units is occupied by group III or II atoms, whereas group V or VI atoms reside on the second unit. The stacking arrangement of the $\{111\}$ planes is $A(\text{III})\ a(\text{V})\ B(\text{III})\ b(\text{V})\ C(\text{III})\ c(\text{V})\ A(\text{III})\ a(\text{V})\ldots$. Since the zincblende structure is noncentrosymmetric, the $\{111\}$ and $\{\bar{1}\bar{1}\bar{1}\}$ planes are not equivalent. For example, in GaAs, if the $\{111\}$ plane is bounded by the Ga atoms, then the $\{\bar{1}\bar{1}\bar{1}\}$ plane consists of As atoms. In the current accepted terminology, the $\{111\}$ group III atom plane is denoted as $\{111\}_A$, whereas the $\{\bar{1}\bar{1}\bar{1}\}$ group V atom plane is labeled as $\{\bar{1}\bar{1}\bar{1}\}_B$. Furthermore, these materials are covalently and ionically bonded solids, and the ionic component of the bonding increases in going from III-Vs to II-VIs.

The slip systems in this structure are $(a/2)\langle 1\bar{1}0 \rangle\ \{111\}$, as in the case of fcc crystals. Hornstra (1958) and Hirth and Lothe (1982) have shown that two types of $(a/2)\langle 110 \rangle$ perfect dislocations can form in the diamond-cubic and zincblende structures. Referring to Fig. 2, a 60° dislocation can be formed by removing the material bounded by the surfaces 15, 56, and 64, and subsequently welding the surfaces 15 and 64. The extra half-plane of the resulting

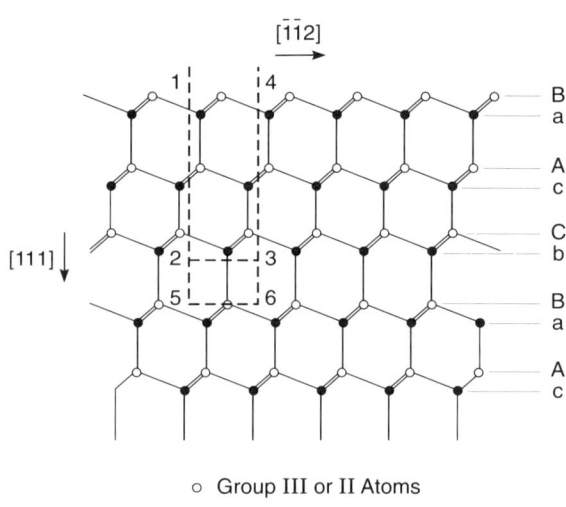

○ Group III or II Atoms

● Group V or VI Atoms

FIG. 2. Zincblende lattice projected onto the $(1\bar{1}0)$ plane. The (111) planes are perpendicular to the plane of the paper and appear as horizontal lines. A glide dislocation can be formed by removing the material along the surfaces 15, 56, and 64, followed by welding of the 15 and 64 surfaces, whereas a shuffle set dislocation can be formed by removing the material along the surfaces 12, 32, and 34 and subsequently welding together the 12 and 34 surfaces.

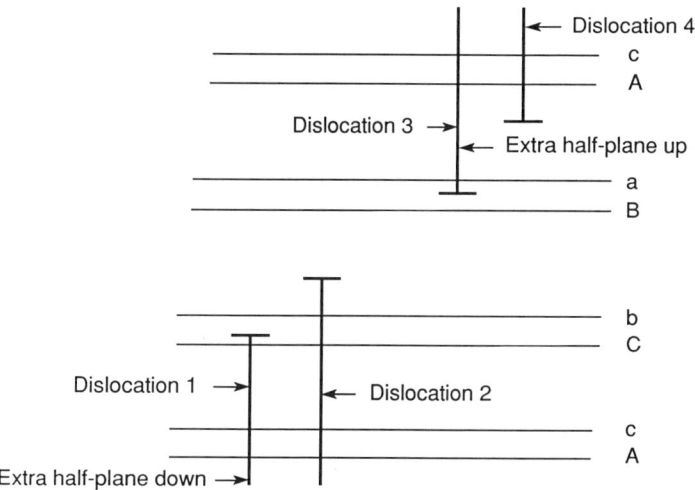

FIG. 3. Schematic depicting core structures of perfect dislocations in compound semiconductors. Dislocation 1 belongs to the shuffle set, and its extra half-plane terminates on the $\{111\}_A$ plane. According to the Hünfeld convention, dislocation 1 is $\beta(s)$, where s stands for the shuffle set. Dislocations 2, 3, and 4 are, respectively, $\alpha(g)$, $\alpha(s)$, and $\beta(g)$, where g implies the glide set.

perfect dislocation terminates between the narrowly spaced $\{111\}$ planes and belongs to the "glide set" dislocation. On the other hand, when the material bounded by the 12, 23, and 34 surfaces is removed and the crystal is subsequently welded along the 12 and 34 surfaces, the resulting imperfection is called a "shuffle set" dislocation, and its extra half-plane terminates between the widely separated $\{111\}$ planes.

Since the zincblende structure consists of two types of atoms, a complication occurs in the description of dislocation cores (Haasen, 1957). This situation is schematically illustrated in Fig. 3. The extra half-plane of dislocation 1 terminates on the $\{111\}_A$ plane and belongs to the shuffle set. Following the convention agreed upon at the Hünfeld Conference (1979), this dislocation can be termed as $\beta(s)$, where s stands for the shuffle set.

As in the case of fcc crystals, a perfect dislocation in the glide set can dissociate into two Shockley partials bounding a stacking fault. For example, an $(a/2)[10\bar{1}]$ dislocation gliding on the (111) plane could dissociate into Shockley partials according to the following reaction:

$$\frac{a}{2}[10\bar{1}]_{(111)} \to \frac{a}{6}[2\bar{1}\bar{1}]_{(111)} + \frac{a}{6}[11\bar{2}]_{(111)}.$$

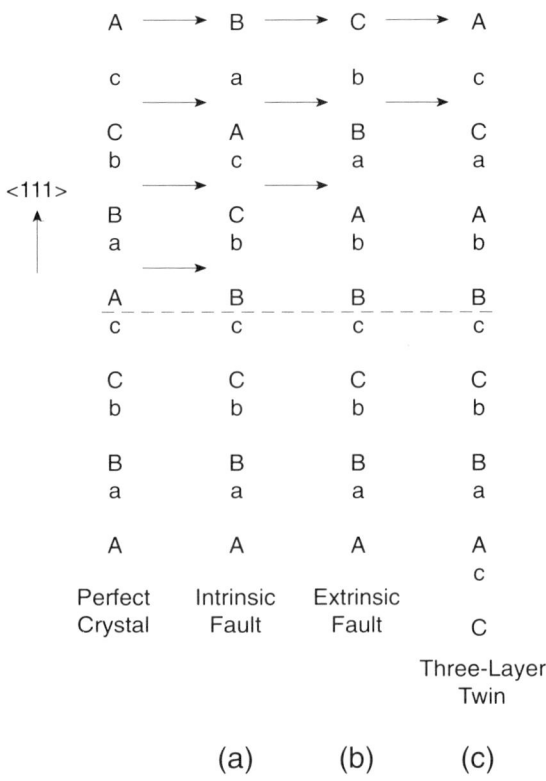

FIG. 4. Schematic illustrating the formation of (a) intrinsic and (b) extrinsic stacking faults, and (c) a three-layer twin in the zincblende structure by shearing of the {111} planes.

Depending upon the orientation of the perfect dislocations, the resulting partials could be 30°, 90°, etc. Furthermore, the structure of a dissociated dislocation in the shuffle set is much more complex (Hirth and Lothe, 1982).

Two types of stacking faults (SFs), intrinsic and extrinsic, can be produced by shearing of the zincblende structure. This situation is illustrated schematically in Fig. 4. As shown in Fig. 4, an intrinsic SF can be formed by shearing some of the atoms in the A-a layers into the B-b position. This is achieved by the passage of a Shockley partial between the A and c {111} planes. Referring to Fig. 4, the imposition of the shear on the A-a layers also causes the displacement of the layers above these layers. The resulting layer arrangement constitutes an intrinsic SF and is shown in Fig. 4a. To produce an extrinsic fault, another Shockley partial must be passed between the C-b layers above the first two shear planes; the resulting layer arrangement for an extrinsic SF is shown in Fig. 4b. If we continue this shearing process, a three-layer twin can be produced, Fig. 4c. It is emphasized that, unlike in the case of fcc crystals, the "true" mirror symmetry does not exist across the habit planes of SFs and twins in the zincblende structure.

III. Deformation Characteristics of Binary Compound Semiconductors

With a brief introduction to the types of dislocations and stacking faults that can form in the zincblende structure, we are now in a position to discuss the deformation behavior of compound semiconductors. For a cogent and comprehensive discussion, this section has been divided as follows: (1) deformation behavior at low temperatures and high stresses that encompasses the effects of doping, isoelectronic substitutions, and deformation-induced microstructures; (2) deformation behavior at intermediate temperatures; (3) deformation behavior at high temperatures; and (4) photoplastic effects. Most of the current information pertains to GaAs. However, wherever applicable an attempt will be made to include other materials such as InP and CdTe to establish the generic nature of the results.

Recently, George and Rabier (1987) and Rabier and George (1987) have reviewed some of the preceding topics with reference to GaAs, and the reader is encouraged to refer to those reviews.

1. Deformation Behavior at Low Temperatures and High Stresses

To study the deformation behavior of naturally occurring minerals, Griggs and Kennedy (1956) have devised a technique to conduct deformation studies under hydrostatic pressure. The principal effect of the imposed pressure is to prevent the propogation of microcracks in brittle materials. Rabier et al. (1985), Rabier and Garem (1985), Lefebvre et al. (1985), Androussi et al. (1989), and Boivin et al. (1990a) have utilized the Griggs-Kennedy approach to evaluate the mechanical behavior of GaAs at low temperatures.

Figures 5a and 5b show the stress–strain curves under confining pressure for intrinsic GaAs and n- (Se: 2.2×10^{18} cm^{-3}) and p-type (Zn: 2×10^{18} cm^{-3}), respectively (Boivin et al., 1990a). Two interesting observations emerge from Fig. 5. First, as expected, the stress–strain behavior of the intrinsic material is strongly temperature-sensitive. Second, at the same deformation temperature and pressure, the p-type crystals are considerably

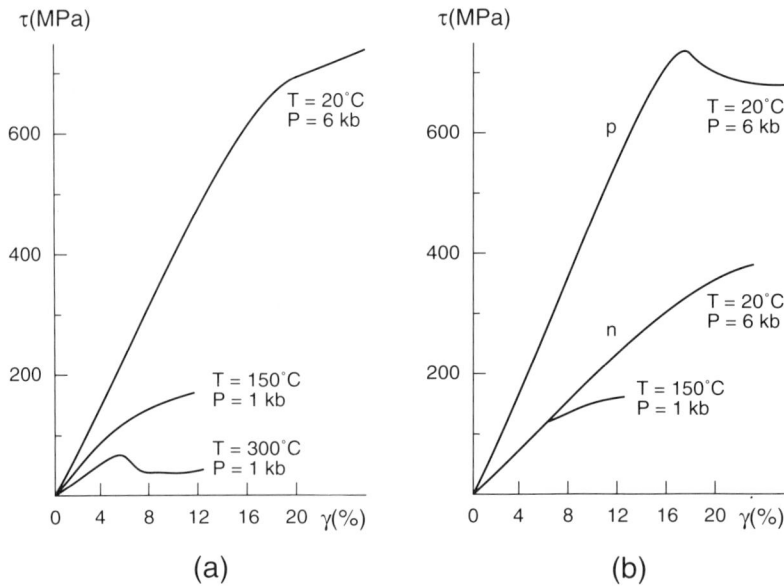

FIG. 5. Stress-strain curves under confining pressure for (a) intrinsic and (b) n-type and p-type GaAs ($\varepsilon = 2 \times 10^{-5}$ s^{-1}) (Boivin et al., 1990a).

harder than the n-type. Boivin et al. (1990a) have shown by comparing the deformation behaviors of the prestrained and the confined samples that it is not an artifact of the confined deformation. Similar results have been obtained by Androussi et al. (1989), who in addition have examined the effect of isoelectronic substitutions such as In on the mechanical behavior. They find that the In-substituted crystals are softer than the intrinsic material, which is in turn softer than the p-type crystals.

The slip patterns observed on the $(11\bar{1})$ and $(1\bar{5}4)$ faces of crystals, deformed under constraint along the $[\bar{3}1\bar{2}]$ direction, are shown in Fig. 6 (Androussi et al., 1989). For this orientation, at small strains single slip should occur, and slip bands should not be observed on the (111) cross-slip plane because the activated slip vector lies in that plane. This assessment essentially holds in the case of undoped GaAs crystals, deformed in compression to a strain of 2.3% (Fig. 6a), whereas this is not true for the Zn-doped crystals, Fig. 6b. Even at very small strains, the blotchy features are observed on the $(11\bar{1})$ surface, implying deformation activity on the secondary system. In addition, slip lines are very fine and are more or less evenly distributed across the $(1\bar{5}4)$ surface. On the other hand, slip patterns are coarser in the undoped crystal, Fig. 6a, and slip lines are inhomogeneously distributed. If one compares the results presented in Fig. 6 with those of Boivin et al. (1990a)

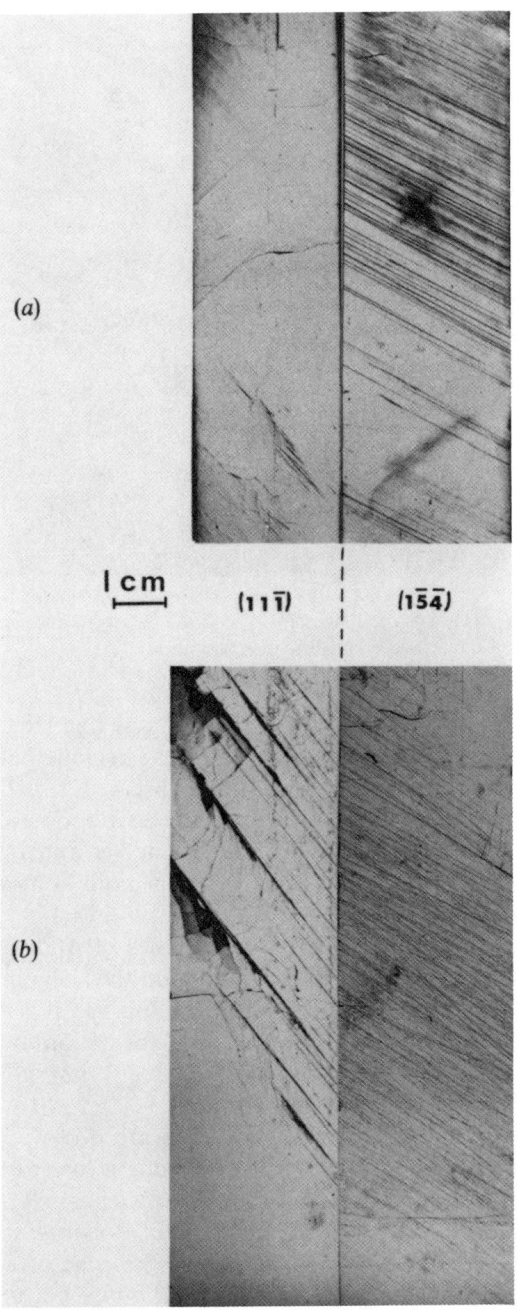

FIG. 6. Optical micrographs of the $(11\bar{1})$ and $(1\bar{5}\bar{4})$ surfaces of GaAs deformed samples: (a) undoped GaAs, $\varepsilon = 2.3\%$; (b) Zn-doped GaAs, $\varepsilon = 0.3\%$ (Androussi et al., 1989).

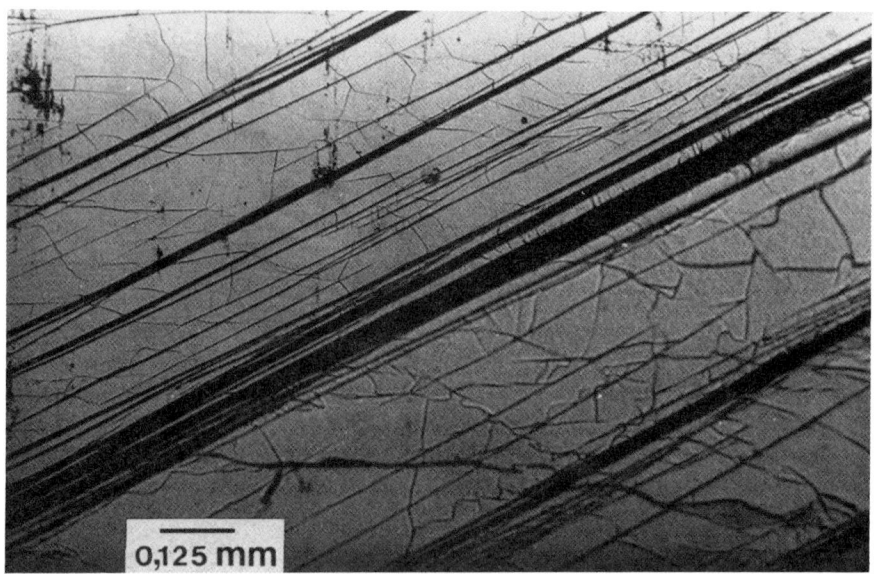

FIG. 7. Slip lines on (1$\bar{5}$4) lateral face of n-type GaAs deformed at room temperature. Note the inhomogeneity of the deformation (Boivin et al., 1990a).

on n-type (Se: 2.2×10^{18} cm^{-3}) GaAs, Fig. 7, one finds that the slip lines in this case are clustered into broader bands, the regions between the bands contain very few slip lines, and cross-slip is observed. It is therefore inferred that n- and p-type doping in GaAs changes the distribution of slip observed in the undoped crystals. For n-doping, slip is coarser and is distributed into broad bands, whereas p-doping leads to finer, more homogeneously distributed slip lines.

The transmission electron microscopy studies of Androussi et al. (1989) and Boivin et al. (1990a) show that twinning on the (111) primary slip plane coexists with slip in the Zn-doped crystals (see Fig. 8a). It is therefore inferred that the blotchy features in Fig. 6b are deformation twins. In fact, by measuring the surface tilts, Androussi et al. (1989) have concluded that the twinning vector is $(a/6)[\bar{2}11]$.

Deformation twins and screw dislocations are observed in the intrinsic material, Fig. 8b, but the propensity for twinning is lower (compare Figs. 8a and 8b). On the other hand, deformation twins are absent in the In-substituted material. Furthermore, the work of Boivin et al. (1990a) shows the presence of perfect screw dislocations at the edge of a glide band and triangular extended stacking faults in deformed n-type GaAs (see Figs 9a and 9b). They have rationalized the formation of faults due to stress-induced

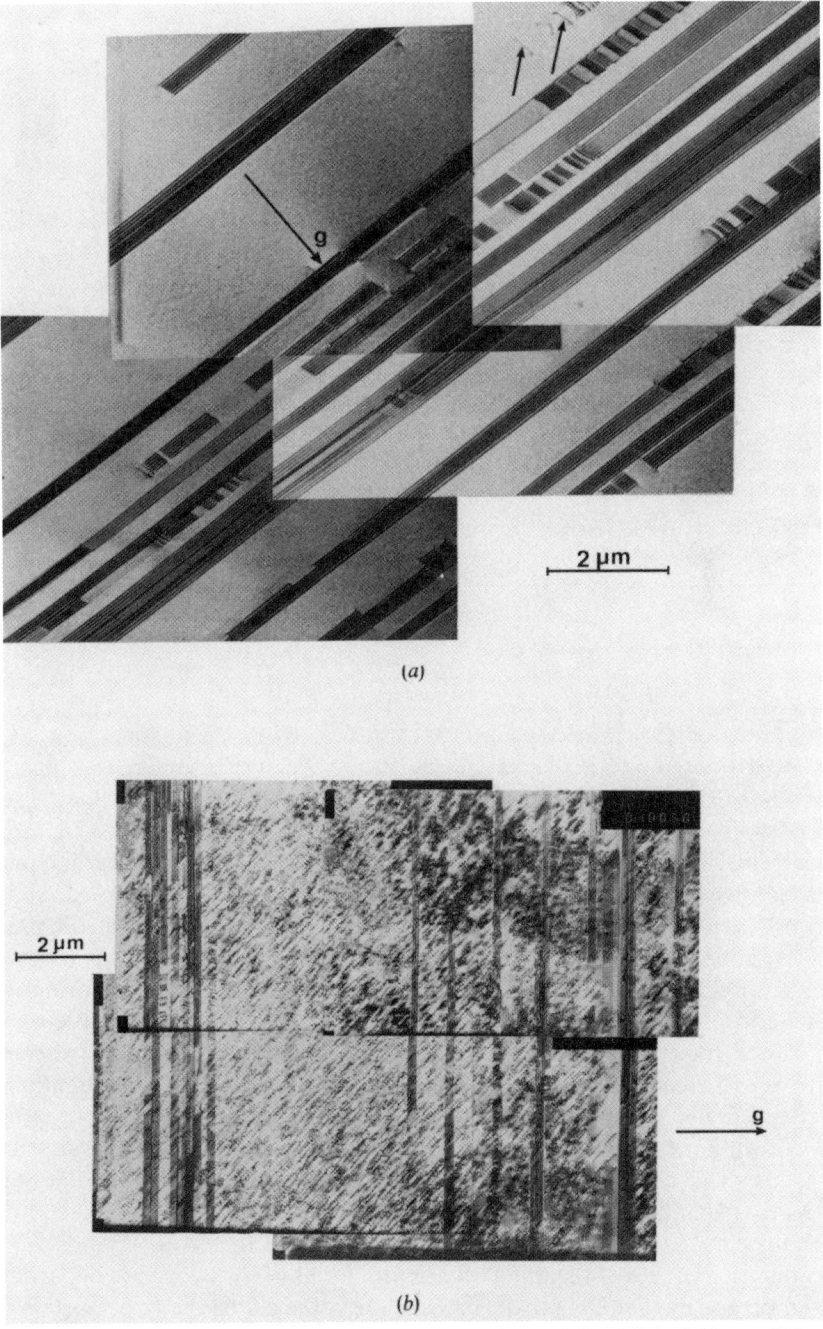

FIG. 8. Electron micrographs of areas characteristic of (100) foils. (bright field; $g = 022$): (a) Zn-doped GaAs showing microtwins, (b) undoped GaAs showing microtwins and screw dislocations; (c) In-doped GaAs showing screw dislocations. [After Androussi *et al.* (1989).]

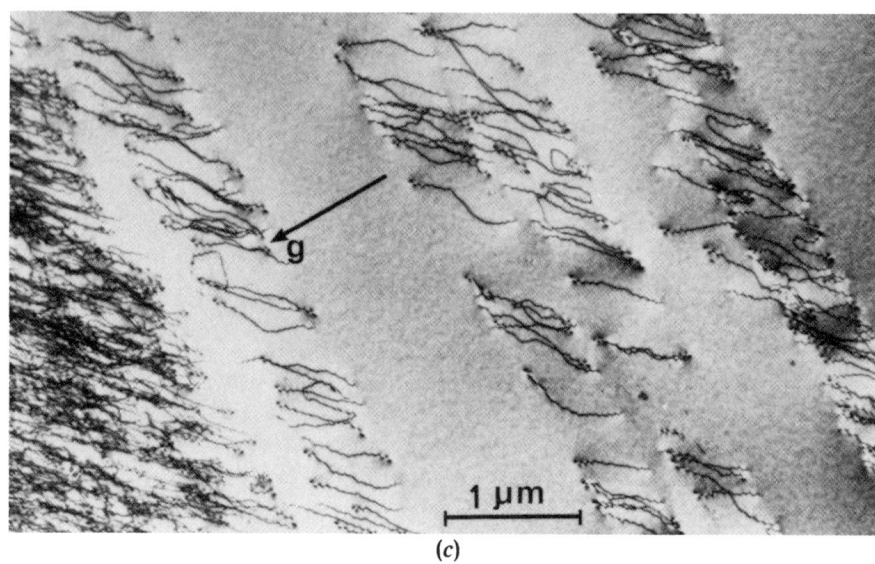

(c)

FIG. 8. (continued)

decoupling of the Shockley partials resulting from the dissociation of a perfect dislocation. This behavior in turn reflects the differences in the mobilities of the partials bounding a shear loop (Boivin et al., 1990a).

Let us now summarize the significant results of this section. First, at room temperature the Zn-doped GaAs crystals are harder than the intrinsic material, which is in turn stronger than the n-type and In-substituted GaAs. Second, even at very small strains slip lines are fine in the p-material, are more or less uniformly distributed across the $(1\bar{5}4)$ surface, and cross-slip is not evident. Furthermore, deformation twins constitute a major deformation mode and are manifested as blotches on the cross-slip plane. On the other hand, deformation twins and screw dislocations coexist in the intrinsic material, whereas coarse slip bands, regions where cross-slip is evident, and extended faults are observed in the n-type GaAs.

Following the work of Rabier and Boivin (1990), the preceding results may be rationalized as follows. Figure 10 shows a dissociated shear loop in n-type GaAs that is under high stress. The separations between the $30\beta/30\alpha$ and the $30\alpha/30\beta$ partials, resulting from the dissociation of the screw segments, are asymmetric because of the difference in mobilities of the two types of partials. Based on Escaig's model for the cross-slip of dissociated dislocations (1968), it can be argued that the $30\beta/30\alpha$ partials will have a strong tendency for cross-slip even though the resolved component of the applied stress on the cross-slip plane is zero. This is inferred because the occurrence of cross-slip would restore the equilibrium separation between the two partials. On the

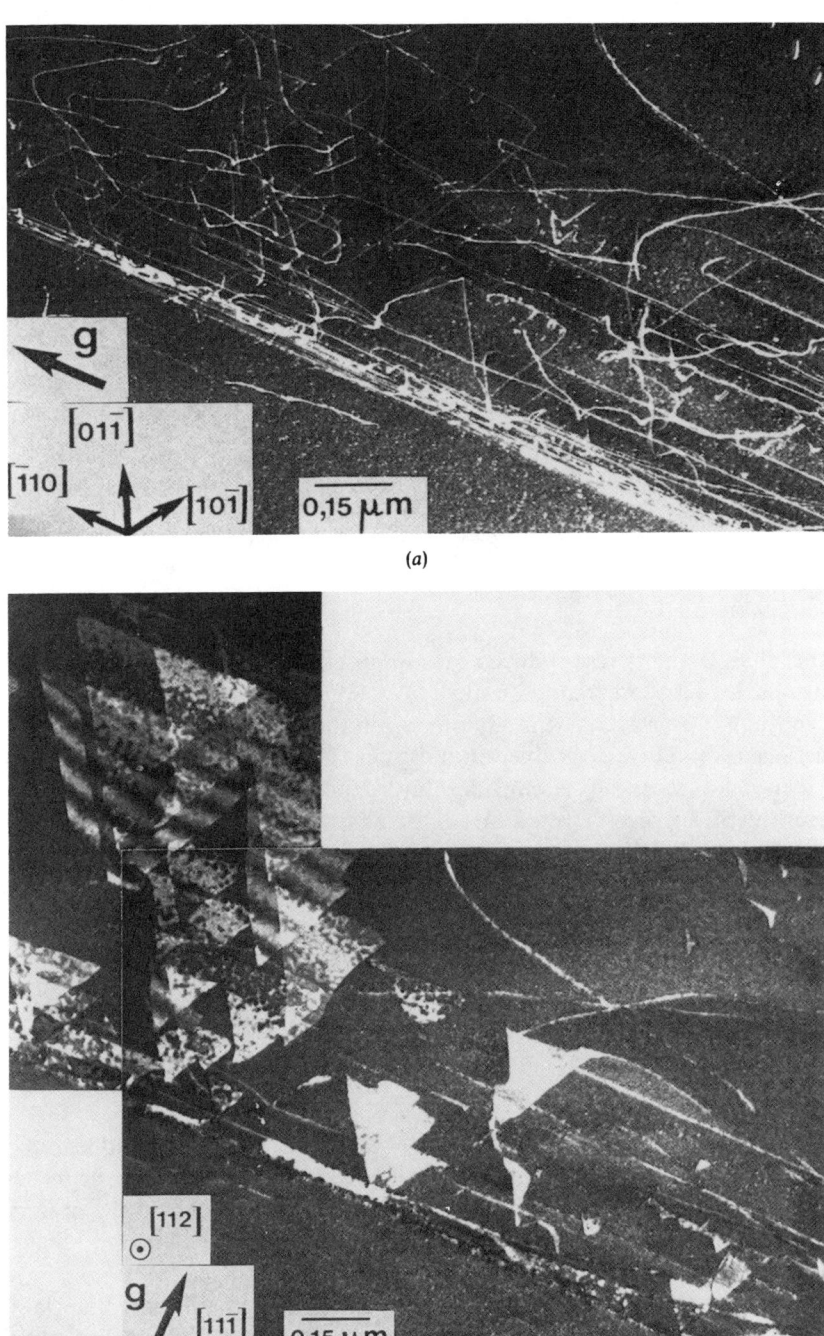

FIG. 9. n-Type GaAs deformed at room temperature in the preplastic range. Observation plane parallel to the glide plane (111). (a) Perfect screw dislocations observed at the edge of a glide band. (b) Triangular extended stacking faults (Boivin et al., 1990a).

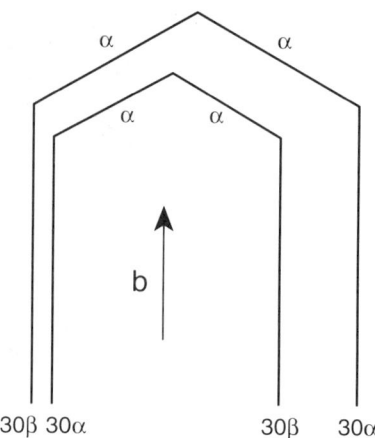

FIG. 10. Schematic of a dissociated shear loop in stressed n-type GaAs.

other hand, the stress-induced decoupling would occur at the $30\alpha/30\beta$ partials, leading to extended faults.

In the p-type material, the asymmetry in the mobilities of the 30β and 30α partials is much less (Rabier and Bovin, 1990), with the result that the tendency for cross-slip is considerably reduced, an assessment that is consistent with the results. In addition, the absence of cross-slip leads to planar slip (Gerold and Karnthaler, 1989). It is envisaged that under these conditions, secondary coplanar slip is activated, a common feature of the fcc metals and alloys deformed in the easy glide region (Pande and Hazzledine, 1971). The primary and coplanar slip dislocations can interact with each other according to the following dislocation reaction:

$$\frac{a}{2}[\bar{1}10]_{(111)} + \frac{a}{2}[\bar{1}01]_{(111)} \to 3 \times \frac{a}{6}[\bar{2}11]_{(111)}.$$

The reaction is energetically favorable (Mahajan, 1975) and could lead to the formation of fault-pairs that are assumed to be precursors of three-layer twins in fcc crystals (Mahajan and Chin, 1973b). It is visualized that the preceding reaction also governs the formation of deformation twins in GaAs, a suggestion consistent with the results of Androussi et al. (1989).

Since twinning occurs early in the deformation of the p-type material, the crystal is broken up into domains by these twins. As a consequence, the slip path of the dislocations is curtailed and their motion is blocked by twins (Sleeswyk and Verbraak, 1961; Mahajan et al., 1970; Mahajan and Chin 1973a). This leads to rapid work hardening in the microstrain region, resulting in a higher yield strength of the p-type material. Since twins do not form in the *n*-type material, their strengthening effect is absent, resulting in a softer crystal. The behavior of the intrinsic material should be intermediate between the two because slip and twinning coexist, but the propensity for twinning is lower than that in the *p*-type crystals.

Rabier and Boivin (1990) have suggested that the *n*-type crystals are softer than the *p*-type because screw dislocations can annihilate each other by cross-slip in the *n*-type material. This *cannot be so*, as cross-slip is confined to screw dislocations of the same sign, i.e., the $30\beta/30\alpha$ combination in Fig. 10.

Pirouz (1987) and Pirouz and Hazzledine (1991) have proposed a model for the formation of deformation twins in diamond-cubic and zincblende III–V compound semiconductors. This mechanism requires double cross-slip of primary dislocations. If this model were valid, we would have expected twinning in the *n*-type material and not in the *p*-type crystals, an assessment that is in contradiction with the experimental findings.

2. Deformation Behavior at Intermediate Temperatures

Boivin *et al.* (1990b) have investigated in detail the deformation behavior of intrinsic *n*- and *p*-type GaAs crystals, oriented for single slip at different temperatures. The *n*- and *p*-type crystals were doped with Se and Zn, respectively, to 2×10^{18} cm^{-3}, and the temperature range covered was 150 to 650°C. The stress–strain curves at different temperatures for the intrinsic, *n*-, and *p*-type crystals, respectively, are shown in Figs. 11a, b, and c. It is evident that at the same deformation temperature, the *n*-type crystals are harder than the intrinsic material, which is in turn stronger than the *p*-type crystals—compare these results with those obtained at room temperature. Also, the three types of crystals exhibit yield drops (YD), and the magnitude of the drop decreases with increasing deformation temperature. In addition, their yield strengths depend strongly on the deformation temperature.

The observed temperature dependence of the yield stress for the three types of crystals is shown in Fig. 12. Two observations can be made. First, the intrinsic and the *p*-type materials exhibit the same temperature dependence, and the behavior of *n*-type crystals is similar. Second, for the intrinsic and *p*-type crystals, the athermal regime starts around 450°C, whereas it is around 600°C for the *n*-type material.

Boivin *et al.* (1990b) have analyzed thermally activated glide in the three types of crystals. They find the following values for the Peierls barrier at 0 K:

1. 2.5 eV for intrinsic GaAs,
2. 2.8 eV for *n*-type GaAs,
3. 2.5 eV for *p*-type GaAs.

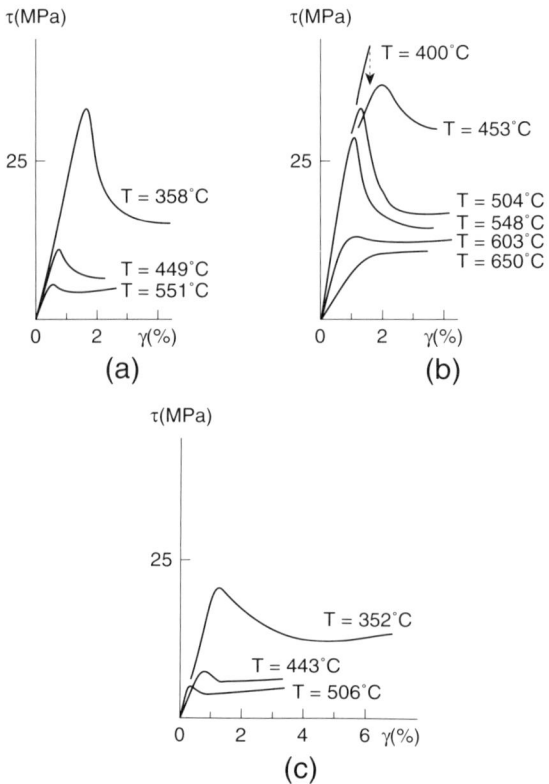

FIG. 11. Stress-strain curves obtained for (a) intrinsic, (b) n-type, and (c) p-type GaAs crystals deformed at different temperatures (Boivin et al., 1990b).

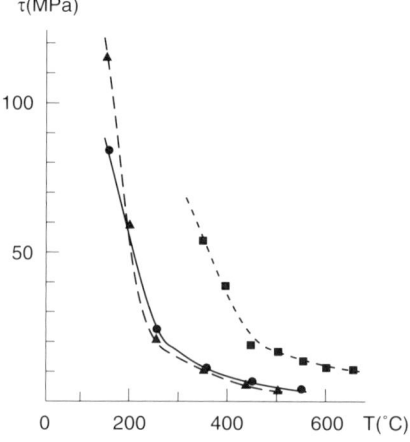

FIG. 12. Temperature dependence of yield stress for intrinsic (■), p- (▲), and n-type (●) GaAs (Boivin et al., 1990b).

Figure 13 shows the dislocation structures observed in the intrinsic, n-, and p-type GaAs crystals deformed to a plastic strain of 0.02 at 350°C (Boivin et al., 1990b). In the intrinsic material, Fig. 13a, edge dislocation dipoles and multipoles, and elongated dislocation loops resulting from the breakup of dipoles, can be seen. Secondary slip dislocations are observed in regions marked A and B.

The dislocation substructure in the n-type crystals deformed at 350°C is characterized by the presence of long screw dislocations on which super kinks α or β are found, Fig. 13b. Dipoles and long mixed dislocations are also observed.

The microstructure of the deformed p-type crystals is similar to that observed in the elemental semiconductors: areas containing a high density of dislocation dipoles and multipoles coexist with areas of low dislocation density that contain long dislocations, Fig. 13c. Dipoles are not in edge orientation. As expected, the loops that result from the breakup of the dipoles are also not of edge character.

For Zn to behave as an acceptor impurity in GaAs, it has to replace some of the Ga atoms on the group III sublattice, while Se atoms, which impart n-type conductivity, must substitute on the group V sublattice. The consequences of the atomic substitutions are twofold. First, ZnAs(4) and Ga(4)Se tetrahedral units are created within the GaAs lattice, and their concentration is linearly related to the dopant concentration. Second, referring to Table I, which shows tetrahedral covalent radii of various atoms according to Pauling (1960), the Zn–As and the Ga–Se bond lengths will be 2.49 (1.31 + 1.18) and 2.40 (1.26 + 1.14) Å, respectively, whereas the Ga–As bond length is 2.44 Å. The percentage changes in the bond lengths in the respective cases are 2.05 and -1.64. According to the solid solution hardening model proposed by Ehrenreich and Hirth (1985), the Zn-doped crystals should be harder than the Se-doped crystals, an assessment that is in contradiction with the experimental results. Even though the results are consistent with the

TABLE I

TETRAHEDRAL COVALENT RADII ACCORDING TO PAULING

	Be	B	C	N	O	F
	1.06	0.88	0.77	0.70	0.66	0.64
	Mg	Al	Si	P	S	Cl
	1.40	1.26	1.17	1.10	1.04	0.99
Cu	Zn	Ga	Ge	As	Se	Br
1.35	1.31	1.26	1.22	1.18	1.14	1.11
Ag	Cd	In	Sn	Sb	Te	I
1.52	1.48	1.44	1.40	1.36	1.32	1.28
	Hg					
	1.48					

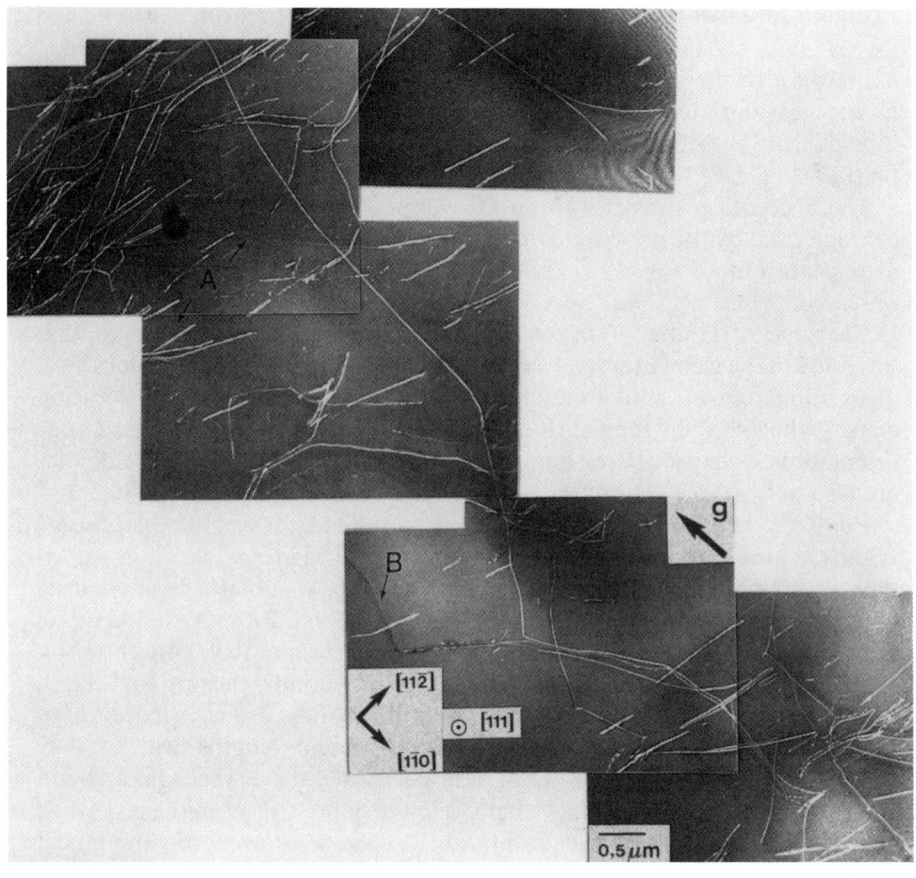

(a)

FIG. 13. Dislocation structures observed in deformed GaAs. (a) Intrinsic GaAs deformed at 350°C ($\gamma_p = 0.02$) following a pre-deformation at 550°C. Observation plane parallel to the primary glide plane (111): dipoles and multipoles parallel to [11$\bar{2}$], rare screw dislocations. (b) n-type GaAs deformed at 350°C ($\gamma_p = 0.02$) following a pre-deformation at 550°C. Observation plane parallel to the primary glide plane (111): numerous screw dislocations. (c) p-type GaAs deformed at 350°C ($\gamma_p = 0.02$). Observation plane parallel to the primary glide plane (111): numerous dipolar loops, rare screw dislocations (Boivin *et al.*, 1990b).

suggestion of Sher *et al.* (1985) that shorter bond lengths would lead to stronger solids, the fit is fortuitous. It will be shown later that this model does not have a general applicability. It is reckoned that the electronic effects dominate in affecting the mobility of dislocations, which in turn affects the mechanical properties and deformation-induced microstructures.

5. DEFORMATION BEHAVIOR OF COMPOUND SEMICONDUCTORS

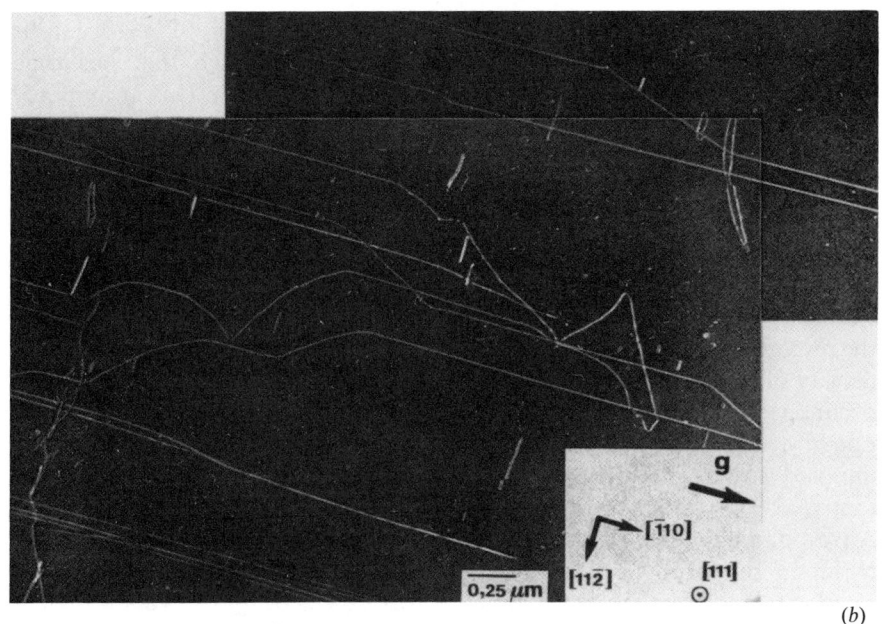

(b)

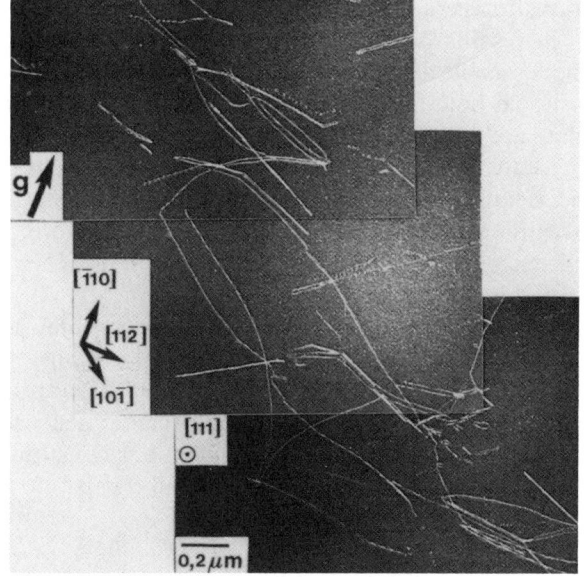

(c)

The decrease in the magnitude of YDs with temperature, Fig. 12, can be rationalized in terms of the Johnston–Gilman approach that has been quantitatively modeled by Haasen (1964). It is well known that an imposed strain rate (ε^*) is related to the number of moving dislocations (n) and their velocity (v) by the following relation:

$$\varepsilon^* = nbv,$$

where b is the Burgers vector. Furthermore, the relationship $v \propto (\tau)^m$, where τ is the applied shear stress and m is the stress exponent, is assumed to hold in the present case. Generally, the as-grown GaAs crystals contain a dislocation density of 5×10^3 cm^{-2}. Assuming that dislocation multiplication does not occur in the microstrain region, the density of mobile dislocations at the upper yield point (UYP) is relatively small. In order to accommodate the imposed strain rate, dislocations multiply rapidly at the UYP. Concomitantly, at constant ε^* the average velocity of the mobile dislocations should correspondingly decrease, thus requiring a much lower shear stress to maintain the imposed strain rate. After the lower yield point (LYP), the elastic interactions between the newly formed dislocations become significant, and this requires increased stress to continue the deformation.

Mahajan et al. (1979a) have investigated in detail the deformation behavior of macroscopically dislocation-free Czochralski and float-zone silicon crystals at various temperatures. They also find that the magnitude of YD decreases with increasing deformation temperature, and that slip becomes more homogeneous as the deformation temperature is increased. We feel that similar trends should hold in the case of GaAs, even though the supporting experimental data are not available. The increased slip homogeneity has been attributed to the activation of dislocation sources with different potency prior to macroscopic yielding. As a result, the mobile dislocation density at the UYP goes up with the increase in deformation temperature. This in turn reduces the difference between the mobile dislocation densities at the upper and LYP, resulting in smaller yield drops.

The evolution of the dislocation substructure in the deformed crystals appears to be controlled by the motion of screw dislocations. It is relatively easy to show that the presence of superjogs on screw dislocations, which in turn could result from double cross-slip, could create dislocation dipoles. Depending upon the spacing between the dislocations constituting a dipole, dipoles could break up into loops by self-glide and climb.

3. Deformation Behavior at High Temperatures

Two recent studies belong to this category. One pertains to the deformation behavior of In-substituted GaAs in the temperature range of

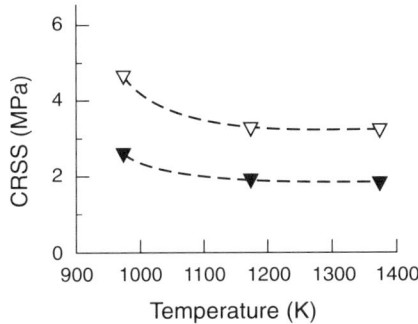

FIG. 14. Comparison of critical resolved shear stress (CRSS) values of undoped and In-doped GaAs specimens tested in [3$\bar{1}$2] orientation: ▽, In-doped; ▼, undoped (Rai *et al.*, 1989).

700–1,100°C (Rai *et al.*, 1989), while the second one deals with the mechanical properties of Zn-substituted CdTe crystals (Rai *et al.*, 1991). The two substitutions are isoelectronic in nature in the respective host lattices and are used to lower dislocation densities in as-grown crystals.

The critical resolved shear stresses of the In-substituted and undoped GaAs crystals are compared at different temperatures in Fig. 14. It is clear that the strengthening in In-substituted crystals is due to the increase in the athermal component of the yield stress. Similar results have recently been obtained on the Zn-substituted CdTe crystals, and they are shown in Fig. 15 (Rai *et al.*, 1991).

Following the work of Sher *et al.* (1985), it can be argued that the In substitution should weaken the GaAs crystals, whereas the replacement of Cd atoms with Zn should strengthen the CdTe crystals. This is inferred because the tetrahedral radius of In atoms is considerably larger than that of the Ga atoms (see Table I), while the tetrahedral radius of the Zn atoms is smaller than that of the Cd atoms. Therefore, the approach of Sher *et al.* (1985) cannot be used in a consistent manner to rationalize the observed strengthening effects.

Recently, Rai *et al.* (1991) have interpreted the increase in the athermal component of the yield stress of the Zn-substituted CdTe crystals in terms of the Ehrenreich–Hirth model (1985). They have computed the magnitude of the athermal component by assuming that ZnTe(4) tetrahedral units act as weak obstacles. The fit between the observed and the computed values is quite good. It is therefore concluded *that the Ehrenreich–Hirth model (1985) can explain adequately the increase in yield strength due to isoelectronic substitutions in compound semiconductors.*

Seki *et al.* (1976, 1978) and Cockayne *et al.* (1983) have grown macroscopically dislocation-free, liquid-encapsulated Czochralski InP crystals by heavily doping them with Zn, S, and Ge, respectively. The Zn and Ge atoms

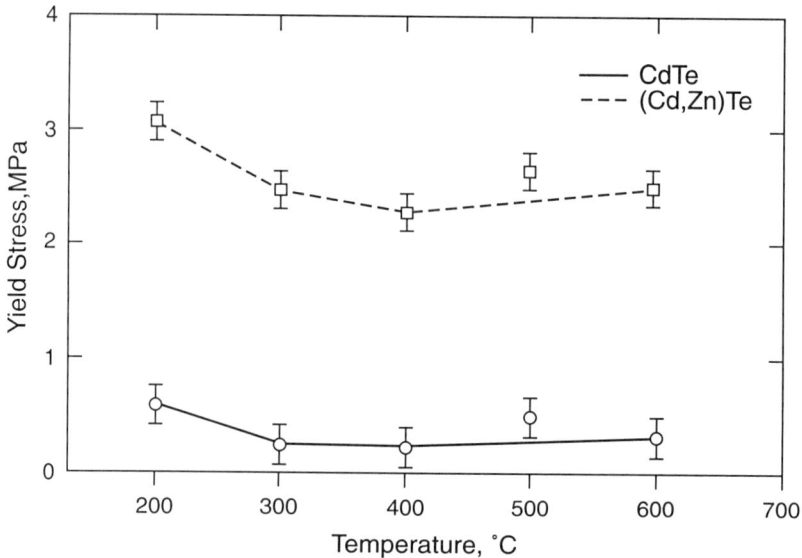

FIG. 15. Temperature dependence of critical resolved shear stress observed in CdTe and (Cd, Zn)Te crystals oriented for single slip (Rai *et al.*, 1991).

should replace In atoms on the group III sublattice, whereas the S atoms should displace P atoms on the group V sublattice. The Zn–P, Ge–P, and In–S bond lengths within the ZnP(4), GeP(4), and In(4)S tetrahedral units would be, respectively, 2.41, 2.32, and 2.48 Å, whereas the In–P bond length is 2.54 Å. The Ehrenreich–Hirth model (1985) correctly predicts that the three types of doped crystals should be stronger than the undoped crystals, but it incorrectly predicts that the Zn-doped crystals should be stronger than the S-doped crystals (Mahajan and Chin, 1981). Again, the complication may be due to different electronic effects of the dopants.

IV. Photoplastic Effects in Compound Semiconductors

The majority of the compound semiconductors are direct bandgap semiconductors, i. e., the minimum in the conduction band and the maximum in the valence band occur at the same value of k, where k is the wave vector. Thus, the electron–hole recombination can occur in these materials without the mediation of a defect level and results in the emission of a photon whose energy is equal to the bandgap.

Imagine a situation where the electron–hole recombination occurs in the vicinity of a dislocation and the released energy somehow gets coupled to the dislocation. It is conceivable that under these conditions, the Arrhenius equation governing the motion of dislocations is favorably modified to facilitate glide. We call this process nonradiative recombination-enhanced glide. As a result, glide may occur at lower stresses, and this phenomenon is referred to as photoplasticity. Both positive and negative photoplastic effects have been observed in compound semiconductors; II–VI materials show positive photoplastic effect, i. e., hardening on irradiation with light (Takeuchi et al., 1983).

Recently, Depraetére et al. (1990) have attempted to quantify the photoplastic effect in intrinsic, n-, and p-type GaAs crystals using a novel approach. Their approach entails micro-indenting of the crystals while they are being optically pumped with a radiation whose energy is greater than the bandgap of GaAs.

Figure 16 shows the photo developed micro-indentations observed in n-type, semi-insulating and p-type materials (Depraetere et al., 1990). In the semi-insulating and n-type materials, the rosettes are equally developed along the two $\langle 110 \rangle$ directions that lie in the (001) plane. On the otherhand, in p-type materials, the rosette along one of the $\langle 110 \rangle$ directions is more developed than along the second $\langle 110 \rangle$ direction. Depraetére et al. (1990) have also examined the microstructural features associated with the rosettes in n- and p-type crystals, and these results are shown in Fig. 17. Elongated dislocations loops, Fig. 17a, are observed along one of the $\langle 110 \rangle$ directions in the n-type material, while overlapping stacking faults or twinlike features are seen along the second direction, Fig. 17b.

The approach used to rationalize the preceding observations is shown schematically in Fig. 18. It is emphasized that our approach is different from that of Depraetere et al. (1990), but is internally consistent with the structural observations on n- and p-type GaAs deformed at room temperature (Androussi et al., 1989, and Boivin et al., 1990a). When an indentation is made on the (001) surface at ambient temperature, glide loops are generated on the $\{111\}$ planes. In our assignment of indices the $(1\bar{1}1)_B$ and $(\bar{1}11)_B$ planes intersect the (001) surface along the [110] direction, whereas the line of intersection of the $(111)_A$ and $(\bar{1}\bar{1}1)_A$ is the $[1\bar{1}0]$ direction. As discussed earlier, in the n-type crystals, $30\beta/30\alpha$ partials resulting from the dissociation of screw segments will be closer together on one side, whereas they may undergo stress-induced decoupling at the opposite segment (Boivin et al., 1990a). Thus, dislocations continue to multiply by double cross-slip, and the rosette branches along the [110] and the $[1\bar{1}0]$ directions are developed. On the other hand, the asymmetry between the $30\beta/30\alpha$ partials in the p-type material is much less. As a result, the total dislocations can react with each

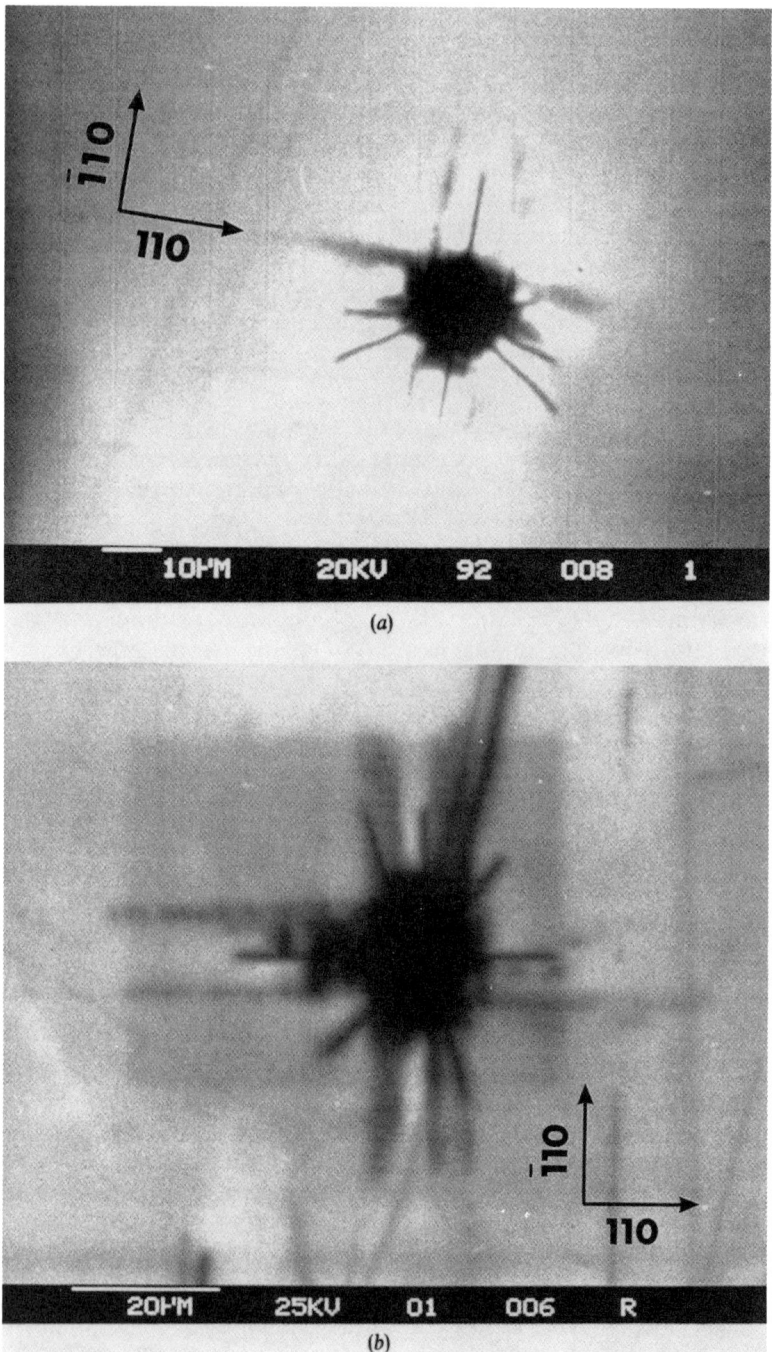

FIG. 16. Cathodoluminescence image of a photodeveloped micro-indentation in (a) *n*-type, (b) semi-insulating, and (c) *p*-type materials. In (a) and (b) rosette arms are observed along the [110] and [1̄10] directions, whereas in (c) the rosette develops along one of the directions (Depraetere *et al.*, 1990).

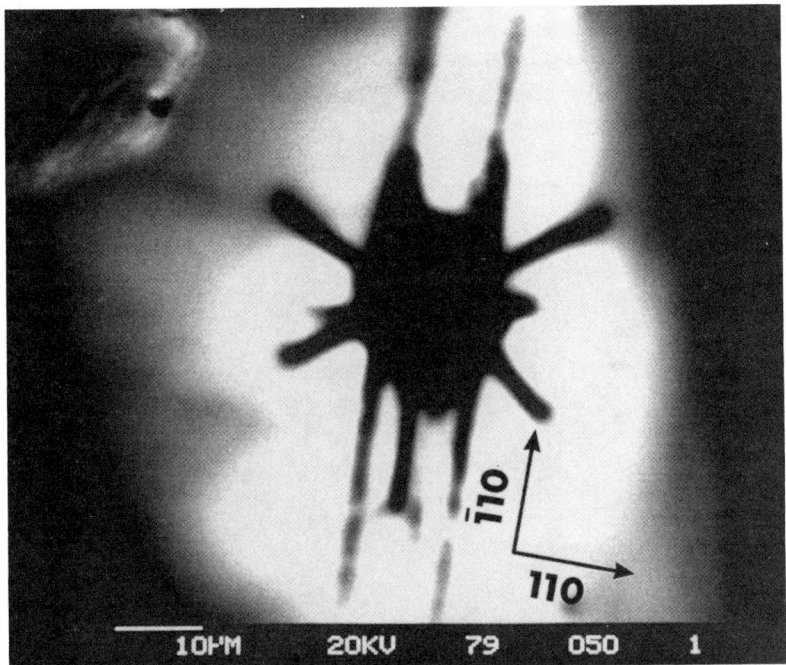

Fig. 16. (c)

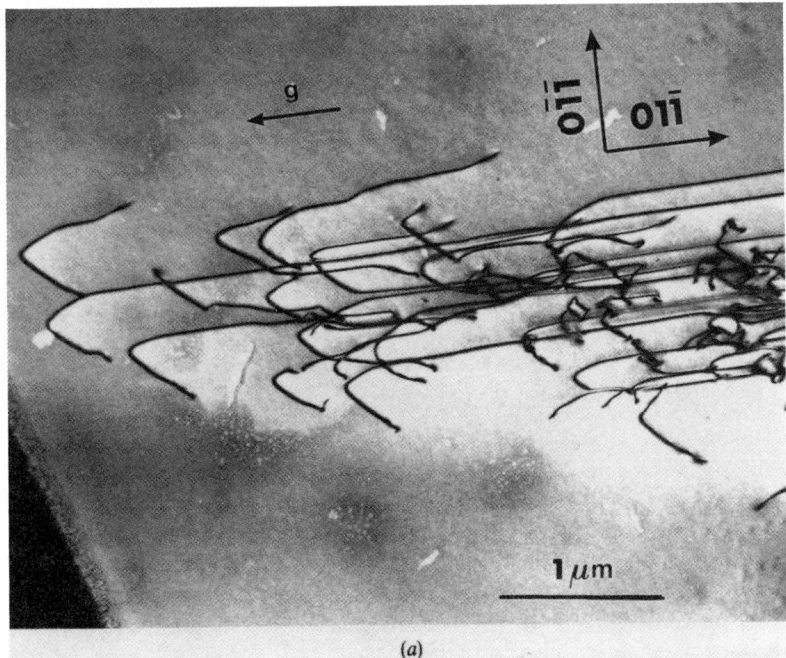

Fig. 17. TEM observations of the microhardness arms developed under a light flux in *n*-type materials. Figures (a) and (b) show that the dislocation substructure is not equivalent in the two orthogonal ⟨110⟩ directions. (Depraetere et al., 1990).

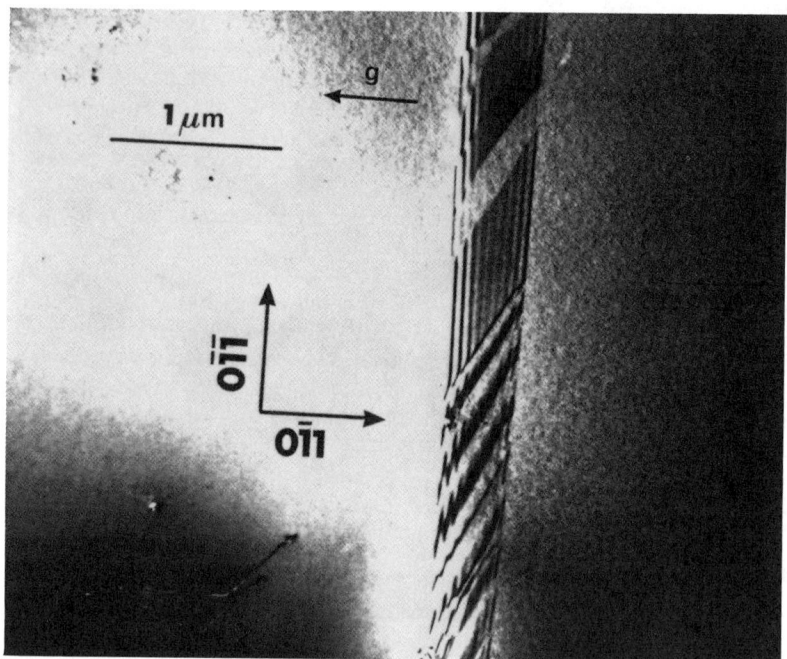

FIG. 17. (continued)   (b)

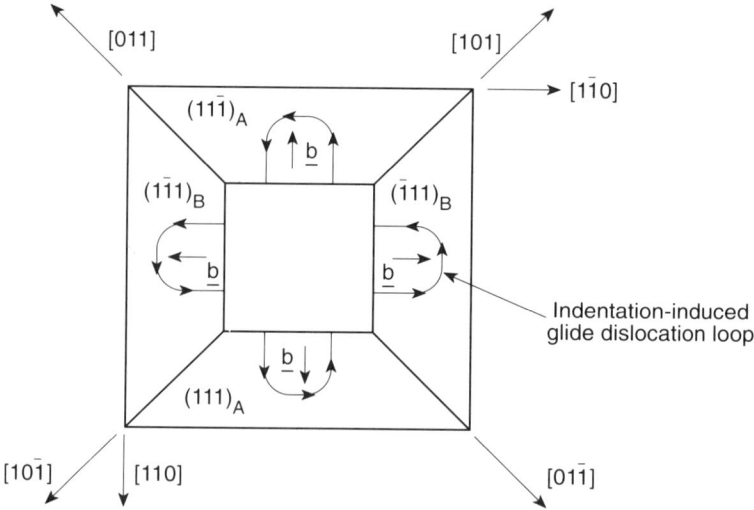

FIG. 18. Schematic showing the generation of glide loops on the four {111} planes due to photodeveloped micro-indentation.

other to form microtwins, as discussed earlier. If we invoke that the [110] and the [1$\bar{1}$0] directions are not equivalent at the surface, slip on the (1$\bar{1}$1)$_B$ and ($\bar{1}$11)$_B$ planes may precede that on the (111)$_A$ and ($\bar{1}\bar{1}$1)$_A$ planes. As a consequence, microtwins may form on the former pair of planes that act as obstacles to slip on the latter pair. Its ramification is that the dark lines would develop extensively along the [110] direction, whereas they will be short along the [1$\bar{1}$0] direction.

Depraetere et al. (1990) have computed the temperature rise during optical pumping and have concluded that it is not significant. They have argued that nonradiative recombinations affect the double kink nucleation and propagation, which in turn control the motion of dislocations in compound semiconductors deformed at low temperatures. They attribute the observed differences along the [110] and [1$\bar{1}$0] directions to these factors.

V. Atomic Ordering and Surface Phase Separation in Ternary and Quaternary III–V Semiconductors and Their Ramifications in Degadation Resistance of Light-Emitting Devices

One of the remarkable developments of the last decade is that of lightwave communication. The production of extremely low-loss fused silica fibers, which are used as a transmitting medium, has been achieved routinely around the world. Furthermore, the replacement of light-emitting devices based on the GaAs/GaAlAs system with the InP/InGaAsP devices, which emit at either 1.33 or 1.55 μm, has further reduced transmission losses. In addition, the devices containing quaternary active layers are considerably more degradation resistant, and the question is "Why?"

The ternary and quaternary compositions can be derived from the binaries by appropriate atomic substitutions on the respective sublattices. An interesting question is whether or not the atoms within the resulting compositions are distributed at random. The answer is an emphatic no. Two types of deviations from randomness are manifested: (1) atomic ordering, and (2) phase separation; the reader is referred to the recent articles by Mahajan et al. (1989) and McDevitt et al. (1990) for a detailed list of bibliography on the two topics.

Figure 19a shows the (110) cross-section of a double heterostructure consisting of InP/InGaAsP/InP layers that was grown by vapor levitation epitaxy (Shahid et al., 1987); the quaternary layer emits at 1.33 μm. The corresponding selected area diffraction pattern is reproduced in Fig. 19b. The superlattice reflections that lie halfway between the 0,0,0 and $\langle 111 \rangle$ zincblende reflections are observed. However, they were not observed in the (1$\bar{1}$0)

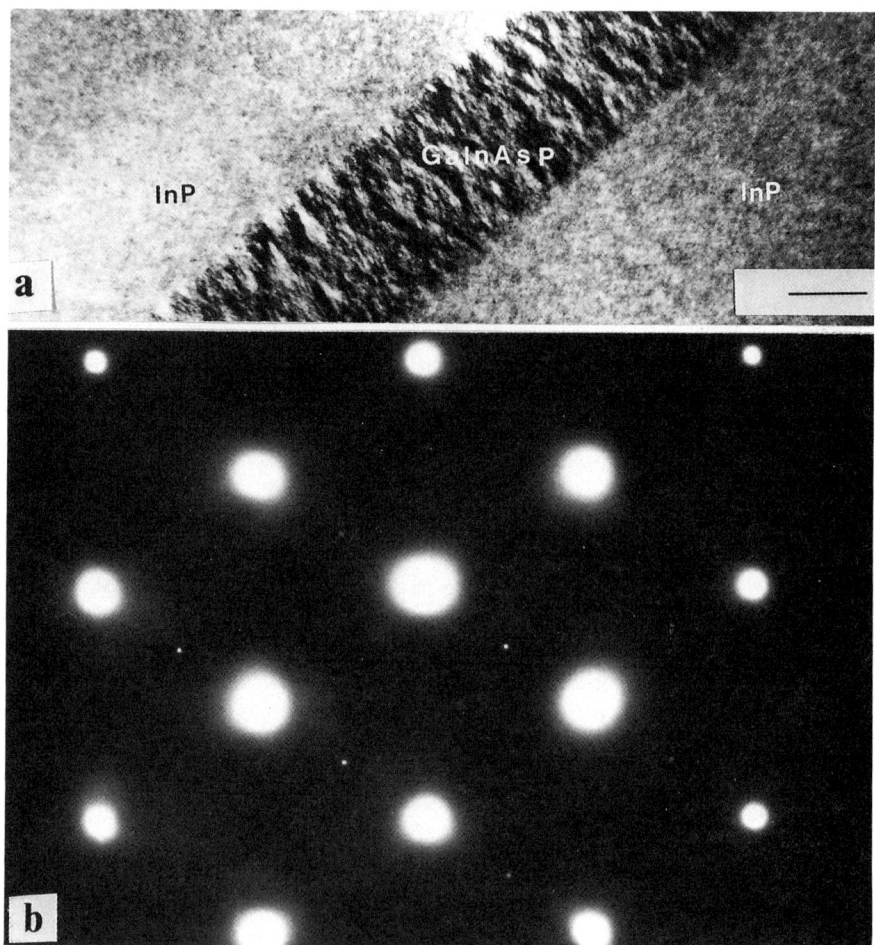

FIG. 19. (a) Electron micrograph showing a (110) cross-section of a heterosturcture consisting of InP/InGaAsP/InP layers grown by vapor levitation epitaxy on a (001) InP substrate. (b) The [110] electron diffraction pattern obtained from the InGaAsP layer shown in (a). Note the presence of superlattice reflections. Marker in (a) represents 0.1 μm.

pattern. The preceding observations indicate that atomic ordering is occurring on two of the possible four $\{111\}$ planes, and that as a result of ordering the real space periodicity along the $\langle 111 \rangle$ direction is doubled. The observed ordering is referred to as CuPt-type in the metallurgical literature. Furthermore, Augarde et al. (1989) have shown experimentally that the atomic ordering is occurring on the $\{\bar{1}\bar{1}1\}_B$ planes.

For example, the doubling of periodicity in GaInAs$_2$ can be rationalized by referring to the schematic shown in Fig. 20. In the disordered material, the layered arrangement of the {111} planes is A(III) a(V) B(III) b(V) C(III) c(V) A(III) a(V). . ., whereas it changes to A(Ga) a(As) B(In) b(As) C(Ga) c(As) A(In) a(As) B(Ga) b(As) C(In) c(As) A(Ga) a(As). . . on ordering. It should not be construed from the preceding discussion that for CuPt-type ordering to occur the composition should be ABC$_2$ or A$_2$BC. Kondow et al. (1989) have shown recently that the CuPt-type ordering can occur even when the composition deviates from ABC$_2$ or A$_2$BC.

The In(3)Ga(1)As and In(1)Ga(3)As tetrahedral units exist in the structure shown schematically in Fig. 20. Within the respective units, three bond lengths, i. e., the In–As and Ga–As bonds, are equal, and the three atoms bonded to As are coplanar and lie in the ordering plane. As suggested by Shahid and Mahajan (1988), the two types of tetrahedral units can be stacked in a coherent manner in the observed ordered structure.

Two models have been proposed to rationalize the formation of the preceeding ordered structures (Suzuki et l., 1988, and Augarde et al., 1989). Both of them suffer from the drawback that they cannot explain an experimental fact that the pair of $\{\overline{1}\overline{1}1\}_B$ planes on which ordering takes place is independent of the sublattice on which atomic substitutions occur to derive

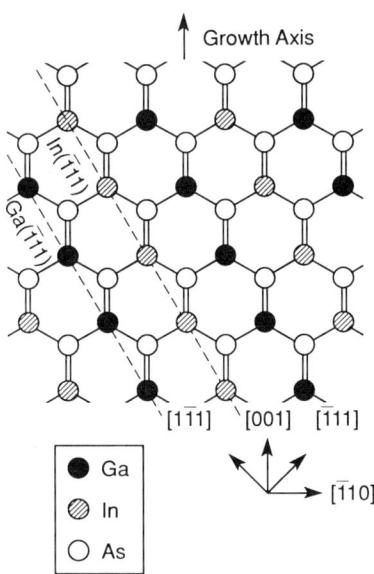

FIG. 20. Schematic illustrating the stacking arrangement of the (111) planes in a GaInAs layer that can be used to explain the origin of superlattice spots in Fig. 19b.

a ternary composition from a binary (Murgatroyd et al., 1990, and Chen et al., 1990). In order to obviate this difficulty, Mahajan (1991) and Philips et al. (1991) have extended the surface-reconstruction–induced ordering model, proposed by Legoues et al. (1990), for the Si–Ge material. Its salient features are shown in Fig. 21, and the model is applied to the growth of a GaInP$_2$ layer on the (001) GaAs surface. Figure 21a shows the [110] projection of the GaAs crystal; the As-terminated (001) surface is unconstructed. To lower the energy, the surface reconstructs, and this is shown in Fig. 21b. As a result of the reconstruction, alternating tensile and compressive regions in the subsurface layer develop along the [1$\bar{1}$0] direction. The plan view of the reconstructed surface is shown in Fig. 21c and shows a row missing As dimers. The

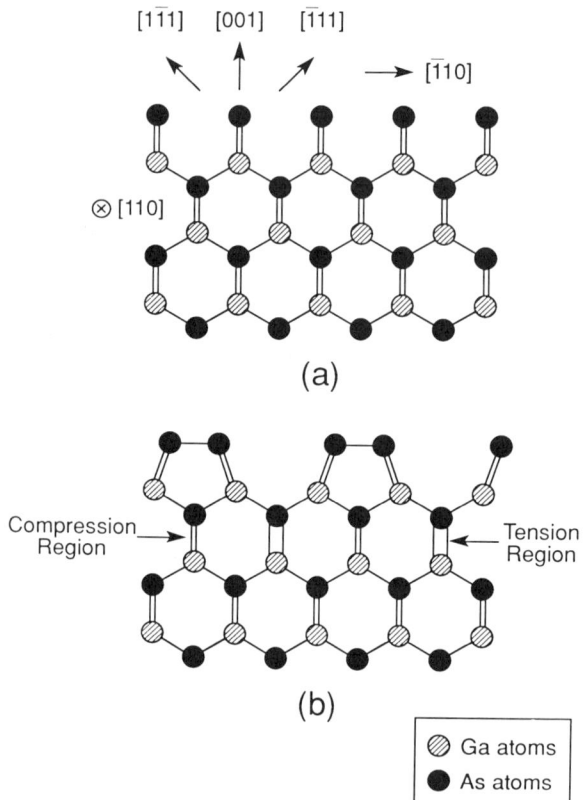

FIG. 21. (a) The (110) perspective of the GaAs crystal; the (001) surface is unconstructed. (b) The (110) perspective showing the formation of As dimers on the As-terminated (001) surface. (c) Plan view of the reconstructed surface.

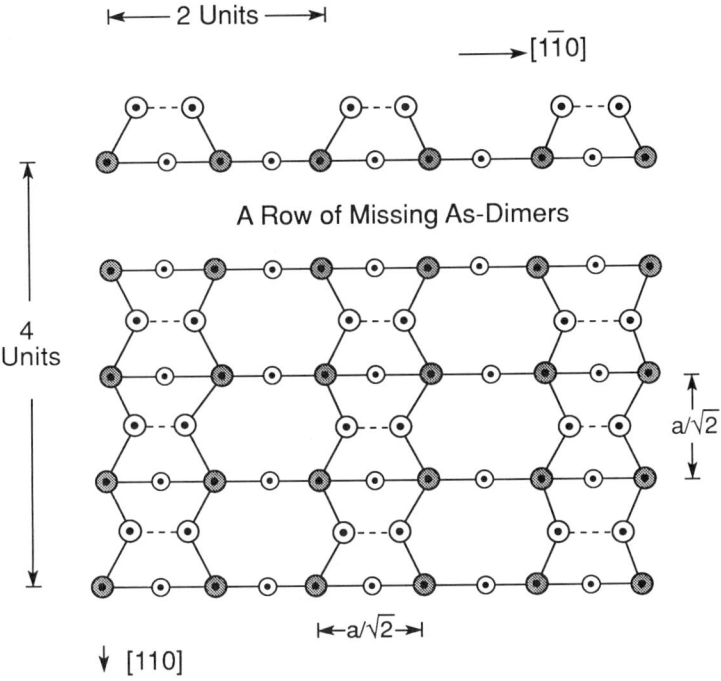

FIG. 21(c).

(2 × 4) reconstruction depicted in Fig. 21c has been observed experimentally by Pashley *et al.* (1988).

When the reconstructed surface is subjected to a flux of Ga, In, and P atoms, the constituents required to grow the layer, the In and Ga atoms, will go to tension and compression regions, respectively. This is inferred because the tetrahedral radius of the In atom is larger than that of the Ga atom — see Table I. As a result, the alternating [110] rows are preferentially occupied by the In and Ga atoms along the [1$\bar{1}$0] direction—the arrangement required to produce CuPt-type ordering. It is relatively easy to show that by propagating the (2 × 4) reconstruction along the growth directions, macroscopic ordered regions can be produced.

Srivastava *et al.* (1985) have shown that in the bulk CuPt-type ordered structures have higher energy than that of CuAu-I type and chalcopyrite. However, the CuPt-type structure has the lowest energy at the surface and is inherited in the bulk because the diffusion is extremely slow in ternary and quaternary materials. Thus, the CuPt-type structure could represent an unstable situation in the bulk.

The size of the ordered domains is very sensitive to the growth temperature and the growth rate (Mahajan et al., 1989). Furthermore, the domains appear to nucleate from steps present on vicinal surfaces (Augarde et al., 1989).

It is evident from Fig. 19a that the contrast of the quaternary layer is considerably different from that of the binary. This difference results from the occurrence of phase separation (Shahid and Mahajan, 1988). McDevitt et al. (1992) have demonstrated unequivocally that phase separation occurs at the surface while the layer is growing and is two-dimensional in nature. It occurs along those directions in the growth plane along which the energy for transformation is a minimum. With the exception of the layers grown by liquid-phase epitaxy that show only phase separation, phase separation coexists with atomic ordering in a variety of ternary and quaternary materials grown by vapor-phase epitaxy, organometallic vapor-phase epitaxy, and molecular-beam epitaxy. Its ramification is that we will not be able to achieve the long-range order parameter of 1 in these materials because the atomic arrangements required for the two microstructural features are not compatible with each other. Furthermore, like the ordered structure, the phase-separated microstructure is unstable in the bulk (Mahajan, 1991).

As indicated in the introduction, GaAs/GaAlAs double heterostructure degrade by the formation of DLDs in the active layer. An example reproduced from the work of Ishida and Kamejima (1979) is shown in Fig. 22. DLDs in Fig. 22 are resolved into irregularly shaped dislocation dipoles by

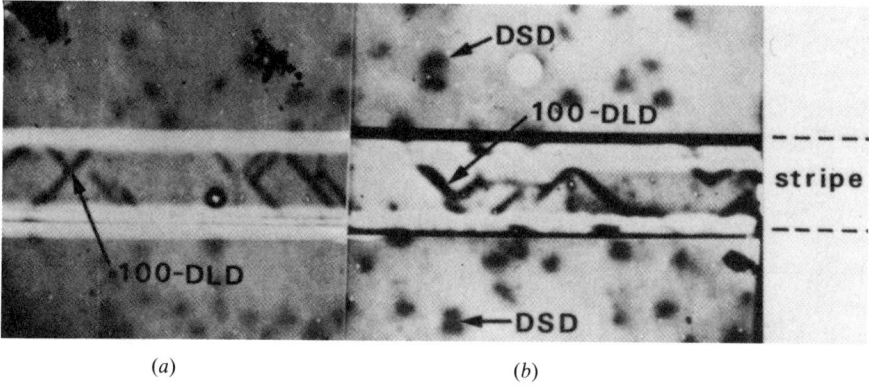

FIG. 22. Electron-beam-induced current images of degraded GaAs/GaAlAs lasers. Many dark spot defects are seen outside the stripe region. Inside the active strip only $\langle 100 \rangle$ dark line defects are seen in (a), whereas $\langle 100 \rangle$ and $\langle 110 \rangle$ dark line defects are seen in (b) (Ishida and Kamejima, 1979).

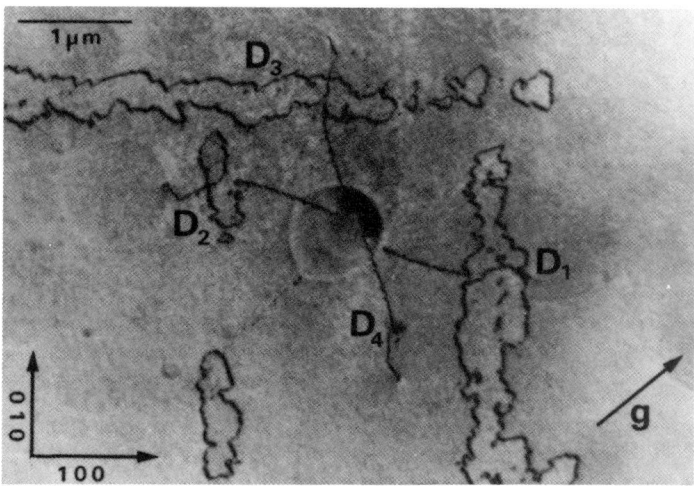

FIG. 23. Electron micrograph taken with a 1 MeV electron microscope shows a dislocation cluster consisting of four dislocations D_1, D_2, D_3, and D_4 and $\langle 100 \rangle$ dislocation dipoles developing from these dislocations; the operating reflection is 220 (Ishida and Kamejima, 1979).

transmission electron microscopy, Fig. 23. There are only two ways by which dislocations can multiply in a material: glide, and climb. The presence of irregularly shaped dislocations imply that climb has indeed occurred. Since the climb kinetics are extremely slow in these materials, Mahajan (1981) and Matsui (1983) have suggested that dislocation glide plays a major role in the evolution of DLDs and that the role of dislocation climb is minor.

It is well known that the occurrence of atomic ordering and phase separation in the metallic systems leads to hardening (Stoloff and Davies, 1966; Cahn, 1962, 1963, 1965; Kato et al., 1980; Kato, 1981; and Park et al., 1986). Several mechanisms have been proposed to rationalize the observed strengthening. First, the unit glide in the ordered materials leads to the formation of antiphase boundaries that cost energy. Second, phase separation produces regions that differ in composition. These regions could produce local strains and could have different shear moduli. Therefore, the gliding dislocations would have to flex around the locally strained regions. However, to produce microscopic strain, dislocations will have to move across the glide plane. We reckon for the reasons listed above that dislocation glide and multiplication will be difficult in the ordered and phase-separated III–V materials. Since dislocations cannot multiply easily during the operation of a light-emitting device consisting of a phase-separated and ordered active layer, the device exhibits enhanced degradation resistance.

VI. Summary

Over the last five years considerable progress has been made in understanding the deformation behavior of intrinsic, In-substituted, n- and p-type GaAs. The situation regarding other III–V and II–VI crystals, with the exception of CdTe, is very unsatisfactory. The deformation behavior of GaAs can be divided into three temperature regimes: (1) low (room temperature to 150°C), (2) middle (150–600°C) and (3) high temperature (650°C and above). In the low-temperature regime, the deformation studies have been carried out using the confining pressure to prevent the propagation of brittle cracks. The zinc-doped crystals are considerably harder than the Se-doped crystals. This has been rationalized in terms of the formation of microtwins in the Zn-doped crystals. These twins block slip dislocations gliding on intersecting slip systems. On the other hand, in the Se-doped crystals, decoupling occurs between the $30\beta/30\alpha$ partials constituting one of the screw segments, whereas the separation between the $30\alpha/30\beta$ is reduced at the opposite segment. This latter feature induces cross-slip and leads to reduced work hardening and thicker slip bands. The situation in the intermediate temperature range can also be rationalized in terms of the observed dislocation structure. In addition, the decrease in yield drop with increasing temperature can be interpreted on the basis of the Johnston–Gilman approach.

The isoelectronic substitutions in GaAs and CdTe crystals raise the athermal component of the yield stress. Following the work of Ehrenreich and Hirth (1985), the magnitude of the increase can be estimated by invoking that these substitutions produce tetrahedral units of different sizes within the material that act as weak obstacles to gliding dislocations. However, the Ehrenreich–Hirth model cannot explain strengthening due to dopants. This deficiency must stem from the fact that dopant-induced electronic effects may have a considerable influence on the mobility of dislocations, which must in turn affect the evolution of the microstructure, and thus the strength of the crystals.

We need to develop a better understanding of the photoplastic effects in compound semiconductors.

The occurrence of atomic ordering and surface phase separation in III–V ternary and quaternary layers strengthens them, i. e., dislocation glide and multiplication are difficult in these materials. This particular feature is responsible for the enhanced degradation resistance observed in light-emitting devices based on the InP/InGaAsP system.

Acknowledgments

The deformation work on CdTe crystals was supported at CMU by NVEOL, whereas the studies on atomic ordering and phase separation were sponsored by the Department of Energy (DE-FG02-87ER45329). The author is grateful to both the agencies for financial support. Also, the author acknowledges fruitful discussions with M. A. Shahid, R. S. Rai, J. P. Harbison, and V. G. Keramidas.

References

Androussi, Y., Vanderschaeve, G., and Lefebvre, A. (1989). *Phil. Mag. A* **59**, 1189.
Augarde, E., Mpaskoutas, M., Bellon, P., Chevalier, J. P., and Martin, G. P. (1989). *Inst. Phys. Conf. Ser.* #100, 155.
Beam, E. A. III, Mahajan, S. and Bonner, W. A. (1990). *Mats. Sci. Eng. (B)* **7**, 83.
Boivin, P., Rabier, J., and Garem, H. (1990a). *Phil. Mag. A* **61**, 647.
Boivin, P., Rabier, J., and Garem, H. (1990b). *Phil. Mag. A* **61**, 619.
Cahn, J. W. (1962). *Acta Met.* **10**, 179.
Cahn, J. W. (1963). *Acta Met.* **11**, 1275.
Cahn, J. W. (1965). *J. Chem Phys.* **42**, 93.
Chen, G. S., Jan, D. H., and Stringfellow, G. B. (1990). *Appl. Phys. Lett.* **53**, 2475.
Cockayne, B., Brown, G. T., and MacEwan, W. R. (1983). *J. Cryst. Growth* **64**, 48.
Depraetere, E., Vignaud, D., Farvacque, J. L., Sieber, B., and Lefebvre, A. (1990). *Phil. Mag. A* **61**, 893.
Dutt, B. V., Mahajan, S., Roedel, R. J., Schwartz, G. P., Miller, D. C., and Derick, L. (1981). *J. Electrochem. Soc.* **128**, 1573.
Ehrenreich, H., and Hirth, J. P. (1985). *Appl. Phys. Lett.* **46**, 668.
Escaig, B. (1968). *J. Phys., Paris* **29**, 225.
George, A., and Rabier, J. (1987). *Revue Phys. Appl.* **22**, 941.
Gerold, V., and Karnthaler, H. P. (1989). *Acta Met.* **37**, 2177.
Griggs, D. T., and Kennedy, G. C. (1956). *Am. J. Sci* **254**, 712.
Haasen, P. (1957). *Acta Met.* **5**, 598.
Haasen, P. (1964). *Disc. Faraday Soc.* **38**, 191.
Hirth, J., and Lothe, J. (1982). "Theory of Dislocations," pp. 373-379. John Wiley, New York.
Hornstra, J. (1958). *J. Phys. Chem. Solids* **5**, 129.
Hünfeld Conference (1979). *J. Physique Colloq.* **40**, C6.
Hutchinson, P. W., and Dobson, P. S. (1975). *Phil. Mag.* **32**, 745.
Ishida, K., and Kamejima, T. (1979). *J. Electron. Mater.* **8**, 57.
Jordan, A. S., Caruso, R., and von Neida, A. R. (1980). *Bell System Tech. J.* **59**, 593.
Kato, M. (1981). *Acta Met.* **29**, 79.
Kato, M., Mori, T., and Schwartz, L. H. (1980) *Acta Met.* **28**, 285.
Kondow, M., Kakibayashi, H., Tanaka, T., and Minagawa, S. (1989). *Phys. Rev. Lett.* **63**, 884.
Lefebvre, A., Francois, P., and DiPersio, J. (1985). *J. Phys. Lett., Paris* **46**, 1023.
Legoues, F. K., Kesan, V. P., Iyer, S. S., Tersoff, J., and Tromp, R. (1990). *Phys. Rev. Lett.* **64**, 2038.
Mahajan, S. (1975). *Met. Trans.* **6A**, 1877.
Mahajan, S. (1981). *In* "Defects in Semiconductors I," (J. Narayan and T. Y. Tan, eds.), p. 465. North-Holland, New York.
Mahajan, S. (1983). *Inst. Phys. Conf. Serv.* #67, p. 259.
Mahajan, S. (1989). *Progress in Mats. Sci.* **33**, 1.

Mahajan, S. (1991), to be published in *Proc. of 5th Brazilian School on Semiconductor Physics.*
Mahajan, S., and Chin, G. Y. (1973a) *Acta Met.* **21**, 173.
Mahajan, S., and Chin, G. Y. (1973b) *Acta Met.* **21**, 1353.
Mahajan, S., and Chin, A. K. (1981). *J. Crystl. Growth* **54**, 138.
Mahajan, S., Barry, D. E., and Eyre, B. L. (1970). *Phil. Mag.* **21**, 43.
Mahajan, S., Brasen, D., and Haasen, P. (1979a) *Acta. Met.* **27**, 1165.
Mahajan, S., Johnston, Jr., W. D., Pollack, M. A., and Nahory, R. E. (1979b) *Appl. Phys Lett.* **34**, 717.
Mahajan, S., Bonner, W. A., Chin, A. K., Miller, D. C., and Temkin, H. (1981). Bell Laboratories, unpublished results.
Mahajan, S., Shahid, M. A., and Laughlin, D. E. (1989). *Inst. Phys. Conf. Ser.* #*100*, p. 143.
Matsui, J. (1983). *In* "Defects in Semiconductors II," (S. Mahajan and J. W. Corbett, eds.), p. 285. North-Holland, New York.
McDevitt, T. L., Mahajan, S., Laughlin, D. E., Bonner, W. A., and Keramidas, V. G. (1990). *Mats. Res. Soc. Sym. Proc.* **198**, 609.
McDevitt, T. L., Mahajan, S., Laughlin, D. E., Bonner, W. A., and Keramidas, V. G. (1992). *Phys. Rev. B* **45**, 6614.
Meier, D. L., Greggi, J., Rohatgi, A., O'Keefe, T. W., Rai-Choudhury, P., Campbell, R. B., and Mahajan, S. (1985). *Proc. of the 18th IEEE Photovoltaic Specialists Conference*, p. 596.
Murgatroyd, I. J., Norman, A. G. and Booker, G. R. (1990). *J. Appl. Phys.* **67**, 2310.
Pande, C. S., and Hazzledine, P. M. (1971). *Phil. Mag.* **24**, 1039.
Park, K-H., LaSalle, J. C., and Schwartz, L. H. (1986). *Acta Met.* **34**, 1853.
Pashley, M. D., Haberern, K. W., Friday, W., Woodall, J. M., and Kirchner, P. D. (1988). *Phys. Rev. Letts.* **60**, 2176.
Pauling, L. (1960). "The Nature of the Chemical Bond," p. 246. Cornell University Press, Ithaca, New York.
Penning, P. (1958). *Philips Res. Rep.* **13**, 79.
Petroff, P. M., and Hartman, R. L. (1973). *Appl. Phys. Lett.* **23**, 469.
Philips, B. A., Norman, A. G., Seong, T. Y., Mahajan, S., and Booker, G. R. (1992). Submitted for publication.
Pirouz, P. (1987). *Scripta Met.* **21**, 1463.
Pirouz, P., and Hazzledine, P. M. (1991). *Scripta Met.* **25**, 1167.
Rabier, J., and Boivin, P. (1990). *Phil. Mag. A* **61**, 673.
Rabier, J., and Garem, H. (1985). *In* "Materials under Extreme Conditions V," MRS Europe (H. Ahlborn, H. Fredriksson, and E. Luscher, eds.), p. 177. Les Editions de Physique, Paris.
Rabier, J., and George, A. (1987). *Revue Phys. App..* **22**, 1327.
Rabier, J., Garem, H., Demenet, J. L., and Veyssiere, P. (1985). *Phil. Mag. A* **51**, L67.
Rai, R. S., Guruswamy, S., Faber, K. T., and Hirth, J. P. (1989). *Phil. Mag. A* **60**, 339.
Rai, R. S., Mahajan, S., Michel, D. J., Smith, H. H., McDevitt, S., and Johnson, C. J. (1991). *Mats. Sci. Eng. (B)* **10**, 219.
Seki, Y., Matsui, J., and Watanabe, H. (1976). *J. Appl. Phys.* **47**, 3374.
Seki, Y., Watanabe, H., and Matsui, J. (1978). *J. Appl. Phys.* **49**, 822.
Shahid, M. A., and Mahajan, S. (1988). *Phys. Rev. B* **38**, 1344.
Shahid, M. A., Mahajan, S., Laughlin, D. E., and Cox, H. M. (1987). *Phys. Rev. Letts.* **58**, 2567.
Sher, A., Chan, A. B., Spicer, W. E., and Shih, C. K. (1985). *J. Vac. Sci. & Tech.* **AS3**, 105.
Sleeswyk, A. W., and Verbraak, C. A. (1961). *Acta Met.* **9**, 917.
Srivastava, G. P., Martins, J. L., and Zunger, A. (1985). *Phys. Rev. B* **31**, 2561.
Stoloff, N. A. and Davies, R. G. (1966). *Progress in Mats. Science* **13**, 1.
Suzuki, T., Gomyo, A., and Iijima, S. (1988). *J. Crystal. Growth* **93**, 396.
Takeuchi, S., Maeda, K., and Nakagawa, K. (1983) *In* "Defects in Semiconductors II," (S. Mahajan and J. W. Corbett, eds.), p. 461. North-Holland, New York.

CHAPTER 6

Injection of Dislocations into Strained Multilayer Structures

John P. Hirth

DEPARTMENT OF MECHANICAL AND MATERIALS ENGINEERING
WASHINGTON STATE UNIVERSITY
PULLMAN, WASHINGTON

I.	INTRODUCTION	267
II.	DISLOCATION BEHAVIOR	268
III	INTERFACE EQUILIBRIUM.	271
	1. *Elastic Stresses*	271
	2. *Misfit Dislocation*	273
	3. *Equilibrium Arrays and Critical Thicknesses*	276
IV.	PARTIAL EQUILIBRIUM ARRAYS.	280
V.	DISLOCATION SPREADING	282
VI.	DISLOCATION INJECTION	284
	4. *Nucleation*.	284
	5. *Multiplication*	289
VIII.	SUMMARY	290
	ACKNOWLEDGMENTS	290
	REFERENCES	291

I. Introduction

The work of Guinier (1938) and Preston (1938) indicating elastic strains associated with GP precipitates established the concept of coherency strains connected with coherent interfaces, including those for isolated precipitates and those for planar interfaces of oriented overgrowths on layer structures. The theoretical work of Frank and Van der Merwe (1949) showed that the coherency strains could be relieved by the presence of misfit dislocations. Misfit dislocations were soon thereafter verified by etch pitting (Goss *et al.*, 1956) or by transmission electron microscopy (Delavignette *et al.*, 1961; Matthews, 1961). Since then, there have been a larger number of observations of misfit dislocations as reviewed by Matthews (1979) for early work.

With the demonstration of the potential of strained multilayer structures for electronic applications such as quantum well structures (Osbourn, 1983a and b) there has been renewed interest in the interface structure. The present work is restricted to oriented single or multilayers with planar interfaces — pertinent to the proposed electronic applications.

An oriented overgrowth layer is said to have an epitaxial or endotaxial orientation with respect to the substrate layer, and the interface is identified by the crystallographic habit places of the two adjoining crystals. If there is one-to-one atom matching at the interface, the interface is coherent or, in the recent alternate terminology in condensed matter physics, commensurate. If the strain is relieved by the introduction of misfit dislocations, the interface becomes semicoherent. The initial terminology (Osbourn, 1983a) for an alternating sequence of strained coherent layers was a strained superlattice, but this choice was unfortunate because of confusion with the crystallographic superlattices that occur for ordered compounds. The alternative description as a strained layer structure is followed here.

Thin strained layers tend to be stable in the coherent state, both thermodynamically and kinematically, while thicker layers become unstable with respect to misfit dislocation injection. The consideration of the critical instability condition is the major thrust of the present work. There are four aspects of the instability problem: (i) the equilibrium interface state, (ii) partial equilibrium states with incomplete dislocation arrays, (iii) the spreading of dislocations that thread through the layers, and (iv) the injection and spreading of dislocations by nucleation, multiplication, and motion. We briefly discuss the first three aspects of the problem and concentrate on the fourth, which is the one of most practical interest in limiting the performance of devices based on strained multilayers.

II. Dislocation Behavior

Figure 1 illustrates a coherent interface, an interface containing a misfit location, an interface containing a misorientation dislocation, and a *combined* dislocation with both misfit/misorientation character, so designated to distinguish it from a mixed screw/edge dislocation. As found by Gradmann (1964) and others (Matthews, 1979), many of the dislocations formed to relieve misfit have combined character and so contribute to misorientation as well. For simple cases such as two fcc crystals meeting at a plane of high symmetry, the nature of the dislocations, perfect or partial dislocations belonging to one of the other bulk lattice, is easy to comprehend. In less simple cases, two lattices are related by a CSL correspondence lattice

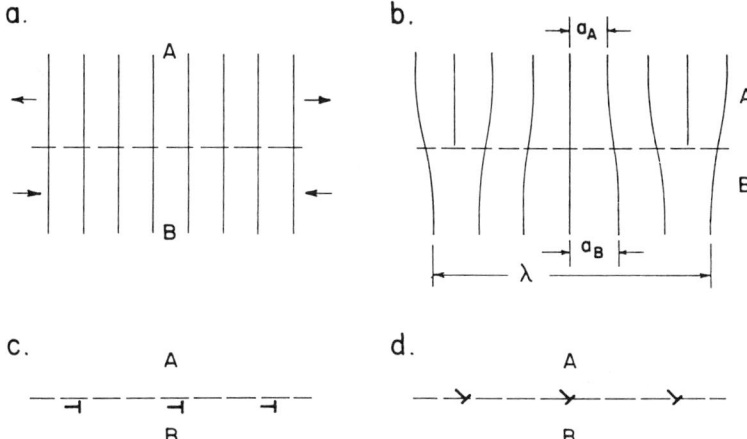

FIG. 1. Representation of (a) coherent interface, (b) relaxed crystals with misfit dislocations, (c) misorientation dislocations, and (d) combined dislocations.

(Bollman, 1979) (a lattice of crystal positions common to both bulk lattices). The reciprocal of this lattice is the DSC lattice, and the Burgers vectors of interface dislocations are perfect or partial IDs (interface dislocations) belonging to the DSC lattice; the perfect and partial Burgers vectors of the two bulk lattices also belong to the DSC lattice.

If the dislocation is not a perfect bulk dislocation, it will have an interface fault associated with it, as illustrated for the simple case of a twin boundary in Fig. 2. Ledges at interfaces, because of differences in lattice spacing in the two lattices, perforce have an associated dislocation called a step translational dislocation, with ID character. Fault translational dislocations, which are also IDs, separate interface regions of different fault nature.

There are a few other types of dislocations at interfaces. Continuous misorientations across an interface, e.g., at a first-order twin boundary, can be described in terms of intrinsic IDs (Hirth and Balluffi, 1973; Balluffi and Olson, 1985). The coherent strain field can be considered to arise from a continuous distribution of infinitesimal dislocations. Finally, small added elastic strains — associated with dilatations normal to faults, for example — can lead to small dislocations with translational character. In neither of the two latter two cases do the Burgers vectors belong to the DSC lattice. Detailed discussions of these dislocations are presented elsewhere (Hirth and Balluffi, 1973; Balluffi and Olson, 1985; Pond, in press; van der Merwe and Jesser, 1988). Here, our primary emphasis will be on misfit or combined dislocations with perfect bulk Burgers vectors.

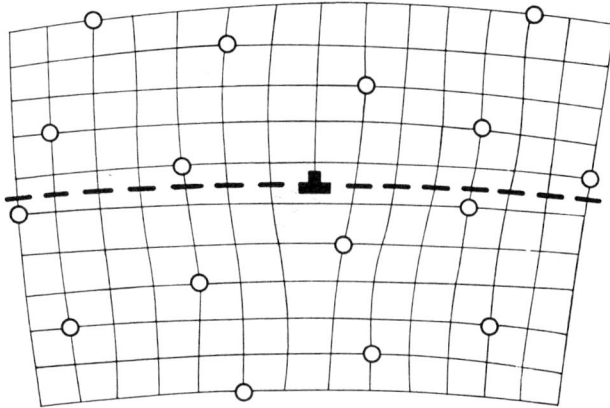

FIG. 2. Projection of simple cubic lattices along [00$\bar{1}$] parallel to a (310) twin boundary. Grid is the DSC lattice and balls are atom positions. Different atom configurations across twin plane on two sides of ID indicate different fault energies.

The original analysis of Frank and van der Merwe and continued analysis by the latter author (van der Merwe and Jesser, 1988; van der Merwe, 1950) are in terms of a description of the dislocations by the Frenkel–Kontorova model. Later analyses have been made in terms of the Volterra model (Matthews and Blakeslee, 1974, 1975, 1976). Other Peierls-type models are also possible, with either the standard sinusoidal form (Chapter 8 of Hirth and Lothe, 1982; van der Merwe, 1963) or with a parabolic form (Jesser and van der Merwe, 1988). We emphasize that, for single dislocations, the long-range field in all cases converges to the Volterra linear elastic result. All of these models give artificial descriptions of the near-core regions: None, for example, contain the core dilatation that is a nonlinear elastic feature of atomic simulations of dislocations (Sinclair et al., 1978). Also, there is no first-principles physical calculation of core structure, although atomic simulations have been made using an empirical potential for silicon (Dodson, 1987). Hence, we follow here the simple Volterra model for the dislocations. Since most calculations involve dislocation interactions, this model should be accurate when the dislocation spacing is large, say 10 or more atomic spacings, compared to core dimensions. For more closely spaced dislocations, core interactions and core relaxations could make the Volterra model physically unrealistic, in which case one of the other models could be superior.

Anisotropic elastic descriptions are available (Willis, 1970; Stroh, 1958; Barnett and Lothe, 1973) and have been used in some calculations. However, in most cases, the complexity of such calculations, involving numerical methods, makes the use of anisotropic elasticity cumbersome, and we shall restrict the present analysis to isotropic elasticity. A calculation (Hirth, 1986) of the thermodynamic critical thickness for the Ga–As–P multilayer system by both methods showed that the anisotropic result differed from the isotropic one by about 25 percent, so errors of this order can be anticipated in the use of isotropic elasticity.

Image forces associated with elastic inhomogeneity (Kamat et al., 1987) can also influence results, driving misfit dislocations off the interface and into the softer bulk material, for example. This effect becomes important when the elastic constants of the bulk crystals differ by more than about 50 percent. In most cases of interest for electronic applications, the elastic constants do not differ by this amount, and the inhomogeneity effect can be neglected with less resultant error than is produced by the neglect of anisotropic elastic effects. A counterexample is the Nb/Al_2O_3 case where the image effects drive misfit dislocations five atom spacings into the softer Nb crystal (Kamat et al., 1987; Mader, 1987).

Finally, diffusional effects influence the forces on dislocations. The graded concentrations produced near an interface by interdiffusion distribute and lower the coherency strain near the interface, and hence change the elastic force on misfit dislocations (Matthews, 1979; Hirth, 1964; Ball and Laird, 1977). Flux divergences at interfaces growing by diffusion and/or vacancy annihilation fluxes lead to osmotic forces (climb forces) on dislocations that can be large compared to elastic forces (van Loo et al., in press). These effects have not been considered quantitatively in any treatment of the injection or motion of misfit dislocations and represent an opportunity for further work.

III. Interface Equilibrium

1. Elastic Stresses

As a simple example of the thermodynamic limit for instability, we consider the example in Fig. 3 of two cubic crystals meeting on $\{1\ 0\ 0\}$ habit places.

In a multilayer structure the stresses are uniform except near free surfaces. For a layer in the multilayer structure of Fig. 3, coherency gives rise to equal biaxial stresses $\sigma_{11} = \sigma_{22} = \sigma$ and strains $\varepsilon_{11} = \varepsilon_{22} = \varepsilon$. The boundary conditions $\sigma_{33} = 0$ then leads to the result

$$\sigma = c\varepsilon, \qquad (1)$$

where

$$c = 2\mu(1 + v)/(1 - v) = 2\mu\kappa, \qquad (2)$$

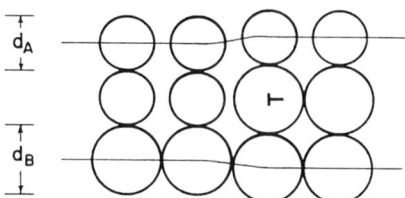

FIG. 3. Ledge translational dislocations at an interface ledge with Burgers vector equal to the difference in interplanar spacing Δd at the interface.

with μ the shear modulus, v Poisson's ratio, and κ an abbreviated Poisson factor. Since no net force can act on the system, the stresses partition in the two layers so that

$$\sigma_A h_A + \sigma_B h_B = 0, \qquad (3)$$

with h the layer thickness. The misfit strains must also add to remove the total misfit.

$$\varepsilon_0 \equiv \Delta c / \langle a \rangle = \varepsilon_A + \varepsilon_B, \qquad (4)$$

where a is the lattice parameter, $\Delta a = a_B - a_A$, and $\langle a \rangle \cong a$ since in all cases of interest $\varepsilon_0 < 0.01$. The above equations combine to give the result

$$\begin{aligned} \varepsilon_A &= \varepsilon_0 h_B c_B / (h_A c_A + h_B c_B), \\ \varepsilon_B &= \varepsilon_0 h_A c_A / (h_A c_A + h_B c_B). \end{aligned} \qquad (5)$$

At the free edge surfaces, the normal stresses must be released to satisfy the free surface boundary condition. This can be accomplished by imposing edge tractions to cancel the in-plane stresses, Fig. 4. The resulting stress field can be calculated by superposing results for line forces acting on half-spaces (Drory et al., 1988), Fig. 5. The result for the interface shear stress at the central interface among N layers is a sum that to first order gives

$$\sigma_{12} = 2c\varepsilon_0 h^2 / \pi (h^2 + x^2). \qquad (6)$$

This stress falls off with distance x into the interface, becomes negligibly small when $x > 4h$, and has a maximum value at the surface of

$$\sigma_{12} = 2c\varepsilon_0 / \pi. \qquad (7)$$

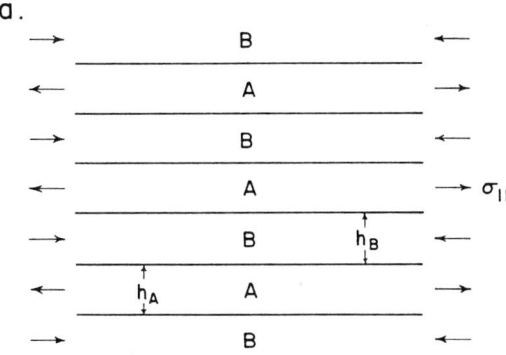

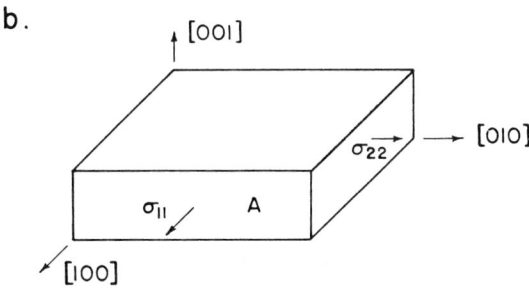

FIG. 4. (a) Alternating stresses in a strained layer structure in a fcc crystal, and (b) projection of a single A layer.

The peak interface shear is thus similar in magnitude to the uniform normal stress, Eq. (1).

For thin bilayers there would also be stresses associated with the relaxation of bending moments (Shinohara and Hirth, 1973), but these are negligible for the multilayer case. For multilayers coherently affixed to a substrate, the elastic field is also modified. For example, if a multilayer of A and B were deposited on a thick B substrate, the strain fields would be $\varepsilon_A = \varepsilon_0$, $\varepsilon_B = 0$ instead of Eq. (5).

2. MISFIT DISLOCATION

For one simple fcc $\{100\}$ case, the long-range strain field can be completely removed by edge misfit dislocations $\frac{1}{2}\langle 110 \rangle$ lying in the interface, an array

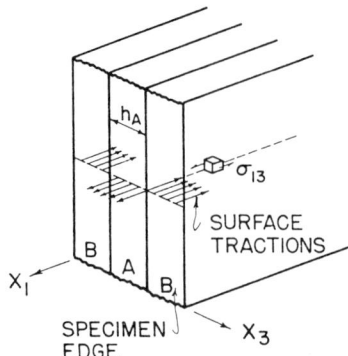

FIG. 5. Surface tractions at free edge of specimen and resulting shear stresses σ_{13}.

sometimes observed (Matthews, 1979). Counting of lattice planes for such an array shows that the misfit dislocation spacing is

$$\lambda \cong b\langle a\rangle/\Delta a = b/\varepsilon_0, \tag{8}$$

with ε_0 and $\langle a \rangle$ defined as in Eq. 4. If λ/b is close to an integer value, the small difference can be accommodated by a set of translational IDs with no net **b** when summed over the habit plane. If λ/b is not close to an integer value, then a sequence of alternating spacings λ_i is required such that $\langle \lambda_i \rangle = b/\varepsilon_0$. The latter complication becomes more important for small λ.

In determining the equilibrium arrays, one could work with global energies and convert the previous coherency strain fields to strain energies. As with many dislocation problems, such a procedure is cumbersome and a simpler method is to determine the Peach–Kochler thermodynamic force on the dislocation and integrate it to determine *local* interaction energies. In order to illustrate the procedures, which would involve lengthy numerical sums for the multilayer case, we consider the case of a single layer of thickness h_A between two B layers with thicknesses $h_B \gg h_A$ and with $c_A = c_B = c$. The misfit dislocation array can then be considered to arise by the reversible process illustrated in Fig. 6.

As shown in Fig. 6, a set of dislocation dipoles of infinitesimal separation and with dipole spacing λ is imagined to form. The dislocations of set II are then separated first on the plane $y = 0$ to a position $x = x_e$ and then separated at $x = x_e$ to a position $y = h_A$. Since the spacings of set II dislocations do not change in this process, there is no self work done by dislocations in set II in the overall process (equivalently, the Peach–Koehler force is zero among set II dislocations). However, the motion of the set II

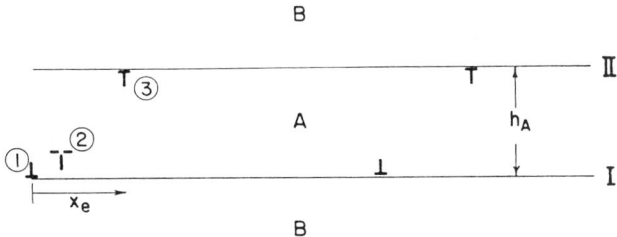

FIG. 6. Creation of set I and set II dislocations illustrated for one dipole: creation at 1 and separation via 2 to equilibrium position 3.

dislocations results in reversible work corresponding to the forces arising from set I dislocations and from the coherency stresses. The relevant components of the Peach–Koehler force are

$$\frac{F_x}{L} = -\sigma_{xy} b_x \tag{9}$$

and

$$\frac{F_y}{L} = \sigma_{xx} b, \tag{10}$$

where $b_x = b = \frac{1}{2}\langle 110 \rangle$.

In the evaluation of σ_{xy} and σ_{xx}, it is convenient to work with reduced distances $X = x/\lambda$ and $Y = y/\lambda$. The stresses of the set I dislocations in the configuration of Fig. 6 are well known (Chapter 19 of Hirth and Lothe, 1982) and have the components

$$\sigma_{xy} = \sigma_0 \sin 2\pi X (\cosh 2\pi Y - \cos 2\pi X - 2\pi Y \sinh 2\pi Y), \tag{11}$$

$$\sigma_{xx} = -\sigma_0 [2 \sinh 2\pi Y (\cosh 2\pi Y - \cos 2\pi X - 2\pi Y (\cosh 2\pi Y \cos 2\pi X - 1)], \tag{12}$$

where

$$\sigma_0 = \mu b / 2\lambda (1 - \nu)(\cosh 2\pi Y - \cos 2\pi X)^2. \tag{13}$$

The work of interaction is the same for each dislocation in set II. Hence, the forces of Eqs. (9) and (10) need only be integrated for one dislocation. The result is the interaction and coherency work per dislocation dipole pair and can be divided by two to get the work per misfit dislocation. Moreover, one should actually consider the perpendicular set of misfit dislocations of Fig. 5 to form at the same time. Analogous to the result for the biaxial coherency stress, added work must be done for set II dislocation to compensate for the displacement u_x produced by the cross-grid set of dislocations parallel to z, and this will increase the work by a factor involving v. Thus, the energy per dislocation using our work done by all thermodynamic forces is

$$\frac{W}{L} = \frac{1}{2}\int_0^{h_A}\frac{F_y}{L}dy + \frac{1}{L}\int_0^{x_e}\frac{F_x}{L}dx$$

$$= \frac{\lambda}{2}\int_0^{h_A/\lambda}(\sigma_{xx}^D + \sigma_{xx}^C)bdY - \frac{\lambda}{2}\int_0^{X_e}\sigma_{xy}^D bdX, \qquad (14)$$

where the superscripts D and C indicate, respectively, the dislocation fields of Eqs. (11) and (12) and the coherency fields of Eqs. (1) and (5). When h_A is less than a critical value h_c, W/L is positive and the coherently strained state is stable. When h_A is greater than h_c, W/L is negative and the misfit dislocation state is stable. We consider solutions to Eq. (14) first in the two limiting cases $h_c \ll \lambda$ and $x_c = \frac{1}{2}$, and then for the general case.

3. Equilibrium Arrays and Critical Thicknesses

The configurations in the two limiting cases are shown in Fig. 7. In the limit $h_c \ll \lambda$, the stress fields of Eqs. (11) to (13) reduce to those for a single edge dislocation and the equilibrium configuration is the lone dislocation dipole, 45° result (Hirth and Lothe, 1982). This is the case treated by Matthews (1979) and Blakeslee (1976), and we term it the MB result. The total energy from Eq. (14) is, with $x' = 1 - v$,

$$\frac{W}{L} = \frac{\mu b^2 \kappa'}{4\pi}\left[\ln\left(\frac{\sqrt{2}h_A}{r_o}\right) - \frac{4\pi h_A}{\lambda}\right], \qquad (15)$$

where $r_o \cong b/3$ is the core cutoff radius typical for covalent crystals (Hirth and Lothe, 1982), the outer cutoff radius is the dipole separation $\sqrt{2}h$, the first term in square brackets is the dislocation interaction energy, and the second term is the work done by the coherency stress in forming the dipole. Normalized to unit area of interface, the total energy is the above result

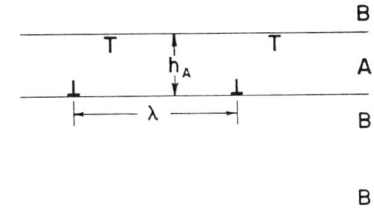

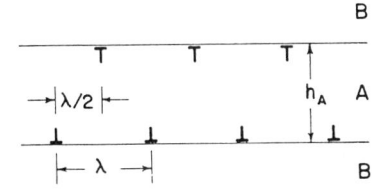

FIG. 7. Equilibrium arrays for h slightly greater than $h_c < \lambda$, (b) $h_c > \lambda$.

multiplied by the total lengths of dislocation in the square grid, $2/\lambda$, or

$$\frac{W_T}{L} = \frac{\mu b^2 \kappa'}{2\pi\lambda}\left[\ln\left(\frac{3\sqrt{2}h_A}{b}\right) - \frac{4\pi h_A}{\lambda}\right]. \tag{16}$$

Thus, the MB result for the critical thickness where $W_T/L = 0$ is

$$h_c/\ln(3\sqrt{2}h_c/b) = \lambda/4\pi = b/4\pi\varepsilon_o. \tag{17}$$

For a multilayer with $h_A = h_B = h$ in the MB limit, the dislocation interaction energy would be the same, but the coherency stress in the A layer would be one-half as large. The result would thus be of the form of Eqs. (15) and (16) with the last term in square brackets one-half as large. The critical thickness would be

$$h_c/\ln(3\sqrt{2}h_c/b) = \lambda/2\pi = b/2\pi\varepsilon_o. \tag{18}$$

For the single A layer, as h increases, the glide equilibrium position X_e approaches $\frac{1}{2}$ and equals $\frac{1}{2}$ for $h > 0.245\,\lambda$. In this limit, the test misfit dislocation configuration is that of Fig. 7b. In this limit, a convenient integration path for Eq. (14) is first to integrate over X to $X_c = \frac{1}{2}$, and then to integrate over Y to h_A/λ at $X = X_c$. For the first integral, σ_{xy}, Eq. (11), reduces to

$$\sigma_{xy} = \frac{\mu b}{2\lambda(1-\nu)} \frac{\sin 2\pi X}{(1-\cos 2\pi X)}. \tag{19}$$

For the second integral, σ_{xx}, Eq. (12), reduces to

$$\sigma_{xx} = -\frac{\mu b}{2\lambda(1-\nu)} \frac{(2\sinh 2\pi Y + 2\pi Y)}{(\cosh 2\pi Y + 1)}. \tag{20}$$

The resulting total energy is

$$\frac{W_T}{L} = \frac{\mu b^2 \kappa'}{4\pi\lambda} \{\ln[\cosh\alpha + 1] - \ln[\cosh\gamma - 1] + \alpha\tan(\alpha/2) - 1 - 4\alpha\}, \tag{21}$$

where $\alpha = 2\pi h_A/\lambda$ and $\gamma = 2\pi b/\lambda$. Thus, for this case of $X_e = \frac{1}{2}$, $Y_e = h_A/\lambda$, and for a given λ, the value of h_c, or equivalently α_c, is determined for the case $W_T/L = 0$, i.e.,

$$\ln[\cosh\alpha_c + 1] + \alpha_c\tanh(\alpha_c/2) - 4\alpha_c = \ln[\cosh\gamma - 1] + 1. \tag{22}$$

For the general case where $0 < X_c < \frac{1}{2}$, the resulting total energy is

$$\frac{W_T}{L} = \frac{\mu b^2 \kappa'}{4\pi\lambda} \{\ln[\alpha\sinh\alpha] - \ln[\cosh\gamma - 1] - 4\alpha\}. \tag{23}$$

Here, the glide equilibrium position X_e or equivality $\eta_e = 2\pi X_e$, for a given α_e, is determined from

$$\cos\eta_e = \cosh\alpha_e - \alpha_e\sinh\alpha_e \tag{24}$$

The critical value α_c is determined for the case $W_T/L = 0$ from the relation

$$\ln[\alpha_c\sinh\alpha_c] - 4\alpha_c = \ln[\cosh\gamma - 1]. \tag{25}$$

For the multilayer case, the fields of sets of dislocations such as those in Fig. 7 must be summed. The details are presented elsewhere (Feng) and are beyond the scope of this overview. However, we present the result for the sum of terms of the type of Eq. (21) to show the form of the result. In the multilayer case, this limiting form $X_e = \frac{1}{2}$ is more favored than for the single layer case and applies for almost all ranges of λ.

We define summed functions

$$p(m) = \ln \sinh(m\alpha/2) + (m\alpha/2)\coth(m\alpha/2),$$

$$q(m) = \ln \coth(m\alpha/2) - m\alpha/\sinh(m\alpha),$$

where m is an interger. The total energy per layer for an n multilayer is then

$$\frac{W_T}{L} = \frac{\mu b^2 \kappa'}{2\pi\lambda} \left\{ \frac{1}{(n+1)} \sum_{j=1}^{n} (-1)^{j+1} jp(n-j+1) \right.$$

$$+ \frac{1}{(n+1)} \sum_{j=1}^{(n+1)/2} (2j-1)q(n-2j+2)$$

$$\left. - \frac{1}{2}[\ln \sinh(\gamma/2) + 1] - \alpha \right\}. \tag{26}$$

A comparison of the results from Eqs. (16), (21), and (23) is given in Fig. 8 for a sample case. As can be seen, the multiple dislocation results differ from the M–B result of Eq. (16) when h_c/λ becomes small. This verifies the results of van der Merwe and Jesser (1988), although, as discussed in detail elsewhere (Feng), they treated a constrained-glide-equilibria case. However, we note that the M–B result does not apply except in the limit $\lambda \gg h$ *not* because they used the Volterra dislocation approximation as suggested in van der Merwe

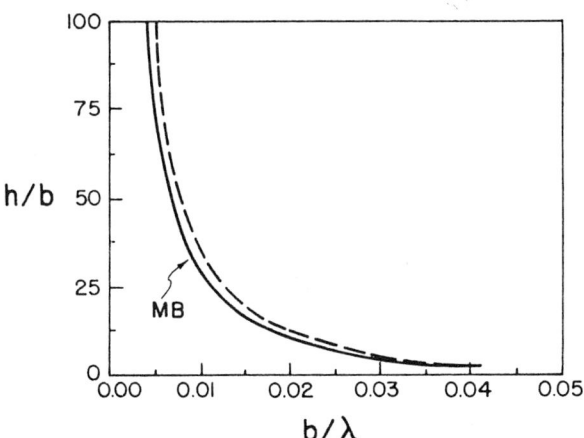

FIG. 8. Plot of h_A/λ versus b/λ for the single dipole case of Eq. (16) and for the multiple dipole case, Eqs. (21) and (23). Values of X_e are noted for the multiple dipole case.

and Jesser (1988), but because M-B neglected multiple dislocation interactions. Indeed, as mentioned previously, the parabolic core potential results for the dislocation strain field (van der Merwe and Jesser, 1988; Jesser and van der Merwe, 1988) should agree with the Volterra results except in the core vicinity. The form of Eq. (26) is similar to the parabolic potential results, although the specific sums and integrals differ. The Volterra and parabolic potential results are expected to differ only in the core term (reflected by b or γ in the present results) except in the limit γ or $h \to b$, where the elastic fields would differ.

IV. Partial Equilibrium Arrays

For Eqs. (21), (23), and (26) and their multilayer counterparts, W_T/L monotonically decreases with decreasing λ to the equilibrium value λ_e of Eq. (8) once $h > h_c$. Once the formation of misfit dislocations becomes thermodynamically favorable, then the formation of the equilibrium cross-grid of misfit dislocations becomes favorable also. For the pure misfit dislocation case, in the absence of frictional forces or other extrinsic effects, there is no constraint to the formation of the equilibrium misfit dislocation array.

However, there are several reasons why complete equilibrium might not be achieved once h exceeds h_c. First, if misfit is relieved by the combined dislocations of Fig. 1d instead of the misfit dislocations of Fig. 1b, dislocation interactions arise that influence the local equilibrium values of h and λ. For the (001) diamond cubic interface case, the combined dislocations could glide on the system $\frac{1}{2}[011](111)$, Fig. 9a, to create pure edge, combined dislocations with length $h' = (\sqrt{6}/2)h$. The interaction energy is then of the form in Eq. (15) but with h' replacing $\sqrt{2}h$. Only the misfit component $b_m = \frac{1}{2}[010] = (\sqrt{2}/2)b$ would relieve coherency stresses so that for the coherency portion of Eq. (15) b_m would replace b. Hence, Eqs. (16) and (17) would be modified to the form

$$\frac{W}{L} = \frac{\mu b^2 \kappa'}{2\pi\lambda}\left[\ln\left(\frac{3\sqrt{6}h}{2b}\right) - \frac{2\pi h\varepsilon_0}{b}\right] \tag{27}$$

and

$$h_c/\ln(3\sqrt{6}h_c/2b) = b/2\pi\varepsilon_0. \tag{28}$$

Hence, the critical thickness would be larger than for the misfit dislocations case.

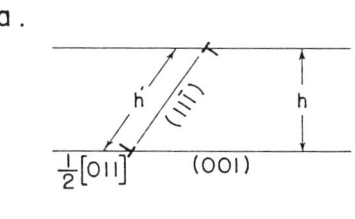

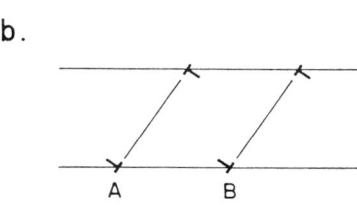

FIG. 9. Combined dislocations formed by glide on [011](11$\bar{1}$): (a) single dipole, (b) multiple dipole.

An analogous qualitative trend holds in the multilayer case, Fig. 9b. For this case there are added interactions for all dislocations, and each term in Eq. (26) would be modified. Just as for the single layer case, h_c would be increased for the multilayer case. In the configuration of Fig. 9b, the misorientation dislocation components of the dislocations, between A and B for example, are also repulsive. Thus, unlike the pure misfit dislocation case, for the combined dislocations W/L does not monotonically decrease with decreasing λ to X_e satisfying Eq. (8) once h exceeds h_c. Hence, there will be a constrained equilibrium value $\lambda_m > \lambda_e$ and a corresponding residual coherency strain present. As h increases, when $h > h_c$, λ_m approaches λ_e asymptotically and the coherency strain vanishes only in this limit.

In a similar manner, when h or $\lambda \to b$, core-core interactions between dislocations become important, modifying the interaction energy portion of the various expressions for W/L and hence also for h_c. The parabolic potential results (van der Merwe and Jesser, 1988), for example, contain this feature. There is little guideline to the form of the core interaction from atomistic calculations because of uncertainties in the interatomic potentials. However, available results (Sinclair et al., 1978) suggest that the very short-range (one of two atom distance) core-core interaction energy is attractive, while the longer range interaction is repulsive. The former effect would tend to increase h_c, while the latter would tend to increase λ_m with the consequences as discussed in the preceding paragraph.

Also analogously, the presence of a Peierls barrier or of frictional forces on dislocations arising from extrinsic defects in the crystals would contribute to the force balance or dislocations at local equilibrium. The effect again would be to tend to increase both h_c and λ_m.

V. Dislocation Spreading

In the growth of a multilayer structure, threading dislocations can grow into a growing layer and penetrate through it. Of interest is the critical condition for spreading of the dislocation into a dipole coating the interface, Fig. 10. As a specific example, we again consider a diamond cubic (001) interface and a glide dislocation on the system $\frac{1}{2}[011](111)$. The Peach–Koehler force on the segment D in Fig. 10b arising from the coherency stress is conveniently determined in coordinates fixed on the glide system, $\mathbf{i}' = [011]/\sqrt{2}$, $\mathbf{j}' = [111]/\sqrt{3}$, and $\mathbf{k}' = [211]/\sqrt{6}$. In these coordinates, $\mathbf{b} = b, 0, 0$; $\sigma'_{12} = (\sqrt{6}/6)\sigma_{11}$ and the dislocation sense vector ξ is parallel to $\mathbf{k}'$. The force on the dislocation is then the product of the Peach–Koehler force per unit length, $\sigma'_{12}b$, and the segment length $h' = (\sqrt{6}/2)h$. For the single layer

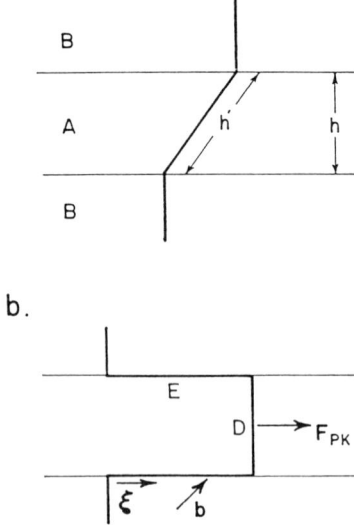

FIG. 10. View parallel to the (001) interface of the spreading of a $\frac{1}{2}[011]$ dislocation on $(11\bar{1})$: (a) view along the direction $[\bar{1}10]$ of spreading and parallel to the glide plane; (b) view along $[\bar{1}\bar{1}0]$, perpendicular to $\mathbf{b}$, to ξ, and to the direction of spreading.

6. INJECTION OF DISLOCATIONS INTO STRAINED MULTILAYER STRUCTURES

case treated in the earlier section where $\varepsilon_A = \varepsilon_0$, the resulting force is

$$F_{PK} = \mu b \kappa \varepsilon_0 h. \tag{29}$$

For the multilayer case where $\varepsilon_A = \varepsilon_B = \varepsilon_0/2$, the force is

$$F_{PK} = \mu b \kappa \varepsilon_0 h/2. \tag{30}$$

For the configuration of Fig. 10, the line tension force on the segment D, arising from the two dislocations on the interface, is

$$F_T = 2 \cdot \frac{\mu b^2 \kappa'}{4\pi} \ln\left(\frac{h'}{r_0}\right) = \frac{\mu b^2 \kappa'}{2\pi} \ln\left(\frac{3\sqrt{6h}}{2b}\right), \tag{31}$$

with the cutoff related to the the spacing h as discussed previously. Equating Eqs. (29) and (31) for the single layer case, we find the equation for the critical thickness to be identical to the energy result of Eq. (28). For $h > h_c$, F_{PK} dominates and the dislocation spreads, while it remains as a threading dislocation if $h < h_c$. The multilayer result follows from Eqs. (30) and (31).

Thus, the MB asymptotic energy result gives the exact answer for the critical thickness for spreading of an isolated threading dislocation in the simple line tension approximation. In the line tension approximation, the anisotropic elastic result is also simple enough in many cases to give an analytical result (Hirth, 1986) for h_c. For closely spaced threading dislocations, dislocation interactions would enter the force balance and the result would differ: The effect would appear roughly when the spacing between threading dislocations became less than h or λ_e.

For a threading dislocation reaching the free surface of a single layer on a thick substrate, Fig. 11, F_{PK} is still given by Eq. (29). However, there is now only one dislocation producing a line tension force, and the outer cutoff radius h' is replaced by the image dislocation spacing $2h$. Thus,

$$F_T = \frac{\mu b^2 \kappa'}{4\pi} \ln\left(\frac{2h}{r_0}\right) = \frac{\mu b^2 \kappa'}{4\pi} \ln\left(\frac{6h}{b}\right). \tag{32}$$

Thus, in this case the critical thickness is given by

$$h_c/\ln(6h_c/b) = b/45\pi\varepsilon_0. \tag{33}$$

Comparing Eqs. (28) and (33), we see that spreading is easier for the layer at a free surface. This is of interest in practical applications in that it implies that

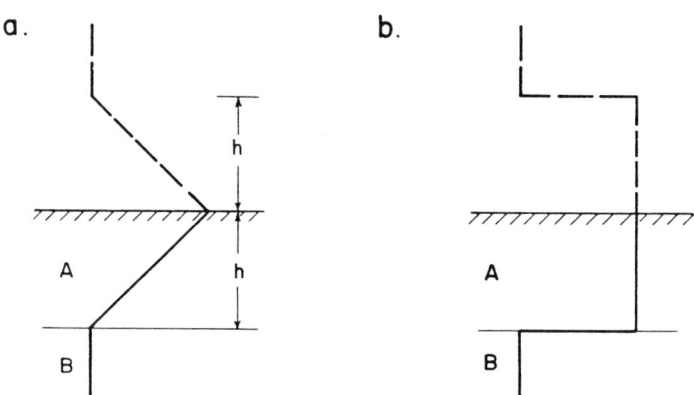

FIG. 11. Free surface version of Fig. 10. Image dislocation shown as dashed line.

substrate dislocations are more easily removed by sweeping for the first layer, favoring the growth of a perfect multilayer structure.

The line tension approximation is not exact because there is an interaction energy between segments D and E, for example, that comes into play in the initiation of spreading. The interaction energy can be determined for either internal layers or the free surface layer by standard methods (Chapter 8 of Hirth and Lothe, 1982) for dislocation segment interactions. However, at present, this extension of the force balance model has not been performed. Also, of course, forces associated with the Peierls barrier or with frictional effects arising from extrinsic defects, if present, would modify the force balance and the corresponding critical thickness. If these latter forces are large, dynamic effects could become important, as discussed in the next section.

VI. Dislocation Injection

4. NUCLEATION

For a perfect layer under conditions where $h > h_c$, there are possible kinematic constraints to the generation of dislocations. The first of these is dislocation nucleation. People and Bean (1985, 1986) draw attention to the possibility of such a constraint for strained layers. Their result led to an expression of the form $h_c/\ln(\text{const} \cdot h_c) \propto (1/\varepsilon_0^2)$, in contrast to Eq. (17) or its later versions. However, in their expression for the energy of dislocation they introduced a typical (constant) dislocation spacing of four or five lattice spacings (People and Bean, 1985 and 1986). Instead, it is more appropriate to introduce the spacing $\lambda \propto (1/\varepsilon_0)$, and then they would have recovered an expression for h_c analogous to the M–B result.

Nucleation of perfect dislocations loops in the bulk has been considered by Frank (1950), and that of perfect and partial dislocation half-loops at free surfaces by Hirth (1963). Matthews (1979) presented a partial version of the latter work, deriving an expression for the critical-sized loop and discussing it but not considering the nucleation rate in detail. Maree et al. (1987) have also reconsidered the problem, including the possibility of surface nucleation, but also focus on the loop energy and not the rate equations. Finally, nucleation at the edge of a multilayer (1986) has been considered in the near-surface limit where σ_{12}, Eq. (6), is constant.

Here, we apply the earlier treatment (Hirth, 1963) to nucleation in the interior of a strained layer where the complications of varying coherency stress fields are absent. The results qualitatively resemble those for the surface case. We first consider the nucleation of a loop of combined $\frac{1}{2}[011](111)$ dislocation within a strained multilayer for the example of a diamond cubic (001) interface, Fig. 12a. The energy of the loop of radius r is (Nabarro, 1952; p. 169 of Hirth and Lothe, 1982),

$$W_1 = \frac{\mu b^2 r}{4} \frac{(2-v)}{(1-v)} \left[\ln \frac{8r}{r_0} - 2 \right] = \frac{\mu b^2 r}{4} \frac{(2-v)}{(1-v)} \ln\left(\frac{24r}{be^2}\right). \tag{34}$$

The coherency strain energy released is given in terms of the resolved coherency stress $\sigma'_{12} = (\sqrt{6}/6)\sigma_{11}$ of Fig. 10b by

$$W_2 = -\sigma'_{12} b \pi r^2 = -(\sqrt{6}/6)\pi b \mu \kappa \varepsilon_0 r^2. \tag{35}$$

Maximizing the total energy $W_1 + W_2$ with respect to r, we find for the radius of the critical-sized loop that can grow with a decrease in energy

$$r_c = \frac{\sqrt{6}b}{8\pi\varepsilon_0} \frac{(2-v)}{(1+v)} \ln\left(\frac{24r_c}{be}\right), \tag{36}$$

and for the energy of the critical-sized loop

$$W_c = \frac{\mu b^2 r_c}{8} \frac{(2-v)}{(1-v)} \ln\left(\frac{24r_c}{be^3}\right). \tag{37}$$

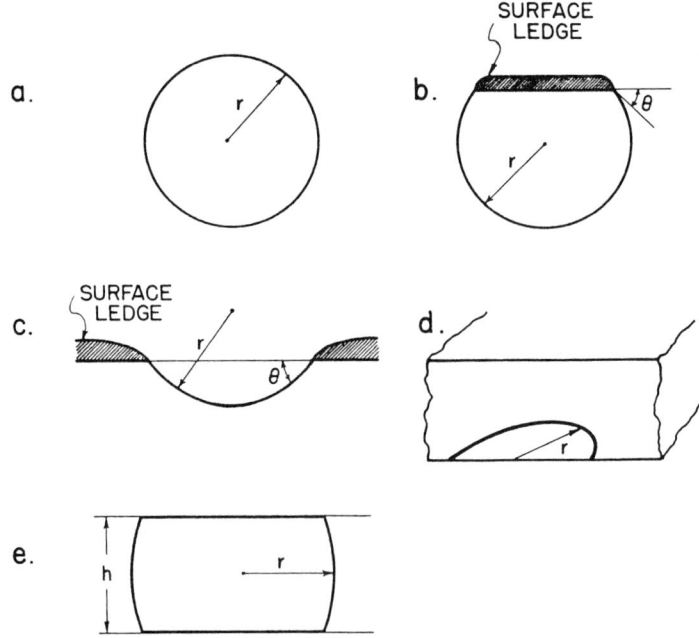

FIG. 12. Critical-sized glide dislocation nuclei for several cases with the Burgers vector in the plane of the loop: (a) nucleation within bulk A layer; (b) nucleation at a singular surface, requiring ledge creation; (c) nucleation at a vicinal surface, removing a ledge; (d) nucleation in the interface plane at a crystal edge, and (e) nucleation within an A layer for the case that $r_c > h'$.

Following standard rate theory (Feder *et al.*, 1986), the concentration of critical nuclei is

$$n^* = n_1 \exp(-W_c/kT), \tag{38}$$

where $n_1 \cong 1/b^3$ is the concentration of single atoms per unit volume. The nucleation rate is given by (Davis and Hirth, 1966).

$$J = Z\omega n^*, \tag{39}$$

where $Z \cong 0.1$ is the Zeldovich factor and

$$\omega = (8\pi r_c/b)v_D \tag{40}$$

is the frequency factor, with v_D the Debye frequency. Typically, $\omega \cong 10^{15}$ s^{-1} and $n_1 \cong 10^{22}$ cm^{-3}, so Eq. (39) becomes

$$J = 10^{36} \exp(-W_c/kT) \text{ cm}^{-1} \text{ s}^{-1} \qquad (41)$$

For nucleation to be important, J must equal a critical observable value J^*. Here, unlike the earlier treatments, the critical value is one in a thin multilayer in a few thousand seconds or one in a region viewed in an electron microscope: Either estimate gives $J^* \cong 10^{20}$ cm^{-1} s^{-1}. The condition that must satisfied for nucleation to be observable is given by Eq. (41) with J^* substituted for J or

$$W_c \leq 37kT. \qquad (42)$$

For the nucleation of a partial dislocation, there is an added energy term associated with the creation of a circle of stacking fault,

$$W_3 = \gamma \pi r^2, \qquad (43)$$

where γ is the stacking fault energy. Since Eqs. (35) and (43) have the same functional dependence on r, the same procedure as those for the perfect dislocation can be used if one replaces ε_0 in Eq. (35) by $\varepsilon_0' = \varepsilon_0 - \gamma/\mu b$. Unless γ is quite large, partial dislocation nucleation is favored relative to the perfect-dislocation case because b in Eq.(37) is smaller by a factor of 0.577.

To test the above cases, we use the same parameters as in a previous example (Ehrenreich and Hirth, 1985), typical of compound semiconductors, $\mu = 48$ GPa, $v = 0.23$, and $b = 0.400$ nm for the perfect dislocation with $\gamma = 40$ mJ/m^2. The solution of the transcendental Eq. (42) then gives the results in Table I for r_c and ε_0, the critical misfit to give appreciable

TABLE I

VALUES OF r_c/b AND ε_0^* FOR SEVERAL CASES OF DISLOCATION NUCLEATION

	Perfect Dislocation				Partial Dislocation			
	293 K		1,000 K		293 K		1,000 K	
	r_c/b	ε_0^*	r_c/b	ε_0^*	r_c/b	ε_0^*	r_c/b	ε_0^*
Internal loop	0.98	0.31	1.31	0.26	1.51	0.43	2.64	0.30
Singular surface	0.73	0.43	1.18	0.31	1.21	0.57	2.55	0.33
Vicinal surface	2.85	0.12	3.60	0.11	7.47	0.11	11.0	0.087

nucleation. As seen in the table, nucleation is essentially impossible at room temperature, because the theoretical strength of a perfect crystal ($\sim \mu/10$) (Kelly, 1973) would be reached before the critical nucleation strength. Even at 1,000 K, nucleation is only a marginal possibility.

Another possibility, treated earlier (Hirth, 1963) but neglected in later analyses (Matthews, 1979; Maree et al., 1987), is nucleation at the free surface of a growing layer. There are two possibilities: nucleation at a singular surface, Fig. 12b, and nucleation at a vicinal surface, Fig. 12c. For the vicinal surface one must supply the added energy of the surface ledge

$$W_4 = 2r\varepsilon, \qquad (44)$$

where ε is the ledge energy per unit length. However, only a portion of a loop need be created because of surface image interactions. The energies W_1, W_2, and W_3 are all reduced by a factor that is a function of the contact angle, θ, in turn determined by a balance of dislocation line tension and the surface ledge tension, approximately given by

$$\frac{\mu b^2}{2} \cos \theta = \varepsilon. \qquad (45)$$

For calculation purposes, the typical value of ε is taken as 0.69 nJ/m.

The result for the vicinal surface is similar, but the ledge is removed during nucleation, so W_4, of the same form as in Eq. (45), becomes negative. The cumbersome equations for this case are given elsewhere (Hirth, 1963; Kamat and Hirth, in press) and are not repeated here, but the results are also shown in Table I. Even in this more favorable case, nucleation is a likely process only for large misfits in the vicinal surface–partial-dislocation case. The other possibility of nucleation at an edge (Hirth and Evans, 1986), Fig. 12d, gives results quite close to those for the singular surface. A final possibility, discussed by Matthews (1979), occurs if $r_c > h'$, whereupon the critical configuration becomes that of Fig. 12c. The exact energy is available for this configuration (p. 113 in Hirth and Lothe, 1982). However, the values of r_c in Table I are so small that this case seems unlikely, so we do not proceed with the analysis.

In summary of the nucleation results, dislocation nucleation for the semiconductor case is only likely for the partial dislocation–vicinal surface case, and otherwise should not occur. The physical basis for this result in comparison with the easier nucleation results for metal is the typical larger magnitude of **b** for the semiconductor case, a factor that strongly influences W_c in Eq. (37). This difficulty in nucleation is a favorable factor in the stability of strained layer structure. Yet the presence of dislocations, presumably injected by nucleation, is often observed (Matthews, 1979; Radzinski et al., 1988). We can only conclude that nucleation is augmented by extrinsic stresses, externally imposed or associated with defects. The resolved shear stress components τ of such defects would modify Eq. (35) to

$$W_2 = -(\sigma'_{12} + \tau)b\pi r^2. \tag{46}$$

With stresses τ present, the critical misfit would be changed from the values in Table I to values ε'_0 given by

$$\varepsilon'_0 = \varepsilon_0 - \sqrt{6\tau/\mu\kappa}. \tag{47}$$

If τ were of the order of the coherency stresses, as would be the case for microcracks, surface asperities, disclinations (Pond, in press), or inclusions, the critical misfit could be appreciably reduced.

5. MULTIPLICATION

The nucleation rate is so dependent on ε_0 and τ that it changes from a negligible value to an enormous value over a small range of stress. Contrariwise, the rate of dislocation multiplication and the dislocation velocity are much less stress-dependent. For example, under phonon or electron damping control, the dislocation velocity is linear in stress. To take account of this weaker strain dependence, Dodson and Tsao (1987) have introduced the useful concept of excess stress to drive the dislocation multiplication processes. If the critical coherency stress for thermodynamic stability is σ_0 when ε_0 satisfies Eq. (18), for example, the driving force for dislocation motion would be zero when σ equalled σ_0. If the coherency stress σ were greater than σ_0, the term σ_0 would act as a drag stress and the driving force would be related to some function of the excess stress ($\sigma - \sigma_0$).

In general, the critical stress for nucleation greatly exceeds σ_0, so that if nucleation is the constraint to dislocation injection, multiplication and motion will be rapid and the excess stress considerations will be unimportant in controlling dislocation generation. If a few dislocations are already present in the structure, however, multiplication can occur at stresses well below those required for nucleation. Under these conditions, the concept of a critical excess stress to produce an appreciable rate of dislocation generation becomes important. Dodson and Tsao have shown that this leads to a kinetically controlled, required, critical thickness h_k greater than the thermodynamic value. The excess stress monotonically increases with increasing strain rate or dislocation generation rate. Hence, h_k is a kinetic factor such that h_K for a given dislocation content decreases with increasing time period of consideration.

VII. Summary

Equilibrium arrays of misfit dislocations in layer structures are treated in the Volterra dislocation model. Long-range interactions lead to deviations from the early dipole results for the equilibrium critical layer thickness and have the important consequence that asymmetrical arrays are favored for the regime of a small ratio of layer thickness to misfit dislocation spacing. Once the critical thickness is reached, the energy decreases monotonically with dislocation spacing until the equilibrium spacing is reached so that there is no thermodynamic constraint to achieving the equilibrium array. Several cases of dislocation spreading are considered.

The nucleation of dislocations is considered. Nucleation in the absence of extrinsic forces of defects is found to be unlikely except for the case of a partial dislocation nucleating at a vicinal surface in a material with low stacking-fault energy. The nucleation constraint leads to a lower bound in misfit below which nucleation is unlikely, a factor that should be of practical importance in the use of strained layer structure in applications. The interrelation of nucleation and the overstress needed for dislocation generation is considered.

Acknowledgments

The author is grateful for the support of this research by DARPA through ONR Contract No. 0014-86-K-0753 with the University of California, Santa Barbara, with a subagreement with Washington State University.

References

Ball, C. A. B., and Laird, C. (1977). *Thin Solid Films* **41**, 307.
Balluffi, R. W., and Olson, G. B. (1985). *Metall. Trans.* **16A**, 529.
Barnett, D. M. and Lothe, J. (1973). *Phys. Norvegica* **7**, 13.
Bollman, W. (1970). "Crystal Defects and Crystal Interfaces." Springer-Verlag, Berlin.
Davis, T. L., and Hirth, J. P. (1966). *J. Appl. Phys.* **37**, 2112.
Delavignette, P., Tournier, J., and Amelinckx, S. (1961). *Philos. Mag.* **6**, 1419.
Dodson, B. W. (1987). *Phys. Rev.* **B35**, 2795, 5558.
Dodson, B. W., and Tsao, J. Y. (1987). *Appl. Phys. Lett.* **51**, 1325.
Drory, M. D., Thouless, M. D., and Evans, A. G. (1988). *Acta Metall.* **36**, 2019.
Ehrenreich, H., and Hirth, J. P. (1985). *Appl. Phys. Lett.* **46**, 668.
Feder, J., Russell, K. C., Lothe, J., and Pound, G. M. (1966). *Adv. Phys.* **15**, 111.
Feng, X. Ph.D. dissertation, Washington State University, Pullman, Washington.
Frank, F. C. (1950). *In* "Symposium on Plastic Deformation of Crystalline Solids," p. 89. Carnegie Institute of Technology, Pittsburgh.
Frank, F. C., and van der Merwe, J. H. (1949). *Proc. Roy. Soc. (London)* **A198**, 216.
Goss, A. J., Benson, K. E., and Pfann, W. G. (1956). *Acta Metall.* **4**, 332.
Gradmann, U. (1964). *Ann. Phys. Lpz.* **13**, 213.
Guinier, A. (1938). *Compt. Rend.* **206**, 1641.
Hirth, J. P. (1963). *In* "The Relation between the Structure and Mechanical Properties of Alloys," p. 217. H. M. Stationary Off., London.
Hirth, J. P. (1964). *In* "Single Crystal Films" (Francombe, M., and Sato, H., eds.), p. 173. Pergamon, Oxford.
Hirth, J. P. (1986). *S. Afr. J. Phys.* **9**, 72.
Hirth, J. P., and Balluffi, R. W. (1973). *Acta Metall.* **21**, 929.
Hirth, J. P., and Evans, A. G. (1986). *J. Appl. Phys.* **60**, 2372.
Hirth, J. P., and Lothe, J. (1982). "Theory of Dislocations," 2nd edition. Wiley, New York.
Jesser, W. A., and van der Merwe, J. H. (1988). *J. Appl. Phys.* **63**, 1059, 1928.
Kamat, S. V., and Hirth, J. P. (1990). *J. Appl. Phys.* **67**, 6844.
Kamat, S. V., Hirth, J. P., and Carnahan, B. (1987). *Scripta Metall.* **21**, 1587.
Kelly, A. (1973). "Strong Solids," 2nd edition. Clarendon, Oxford.
Mader, W. (1987). *Mat. Res. Soc. Proc.* **82**, 403.
Maree, P. M. J., Barbour, J. C., van der Veen, J. F., Kavanagh, K. L., Bulle-Lieuwma, C. W. T., and Viegers, M. P. A. (1987). *J. Appl. Phys.* **62**, 4413.
Matthews, J. W. (1961). *Philos. Mag.* **6**, 1347.
Matthews, J. W. (1979). *In* "Dislocations in Solids" (Nabarro, F. R. N., ed.), Vol. 2, p. 461. North-Holland, Amsterdam.
Matthews, J. W., and Blakeslee, A. E. (1974). *J. Cryst. Growth* **27**, 118.
Matthews, J. W., and Blakeslee, A. E. (1975). *J. Cryst. Growth* **29**, 273.
Matthews, J. W., and Blakeslee, A. E. (1976). *J. Cryst. Growth* **32**, 265.
Nabarro, F. R. N. (1952). *Adv. Phys.* **1**, 332.
Osbourn, G. C. (1983a). *J. Vac. Sci.* **B1**, 379.
Osbourn, G. C. (1983b). *Phys. Rev.* **B27**, 5126.
People, R., and Bean, J. C. (1985). *Appl. Phys. Lett.* **47**, 322.
People, R., and Bean, J. C. (1986). *Appl. Phys. Lett.* **49**, 229.
Pond, R. C. *In* "Dislocations in Solids" (Nabarro, F. R. N., ed.), Vol. 8, North-Holland, Amsterdam. In press.
Preston, G. D. (1938). *Proc. Roy. Soc. (London)* **A167**, 526.

Radzinski, Z. J., Jiang, B. L., Rozgonyi, G. A., Humphreys, T. P., Hamaguchi, N., and Bedair, S. M. (1988). *J. Appl. Phys.* **64**, 2328.
Shinohara, K., and Hirth, J. P. (1973). *Philos. Mag.* **27**, 883.
Sinclair, J. E., Gehlen, P. C., Hoagland, R. G., and Hirth, J. P. (1978). *J. Appl. Phys.* **49**, 3890.
Stroh, A. N. (1958). *Philos. Mag.* **3**, 625.
van der Merwe, J. H. (1950). *Proc. Phys. Soc. (London)* **A63**, 616.
van der Merwe, J. H. (1963). *J. Appl. Phys.* **34**, 123.
van der Merwe, J. H., and Jesser, W. A. (1988). *J. Appl. Phys.* **64**, 4968.
van Loo, F. J. J., Pieraggi, B., and Rapp, R. A. (1990). *Acta Metall.* **38**, 1769.
Willis, J. R. (1970). *Philos. Mag.* **21**, 931.

CHAPTER 7

Critical Technologies for the Micromachining of Silicon

Don L. Kendall, Charles B. Fleddermann and Kevin J. Malloy

UNIVERSITY OF NEW MEXICO
THE CENTER FOR HIGH TECHNOLOGY MATERIALS
ALBUQUERQUE, NEW MEXICO

I INTRODUCTION	293
II. WET CHEMICAL ETCHING OF SINGLE-CRYSTAL SILICON	295
1. *Why is the (111) Surface So Slow to Etch?*	296
2. *What about Ethylene Diamine–Based Etchants?*	298
3. *The Electrochemical Model for Anisotropic Etching*	300
4. *Etch Stopping in Basic Solutions*.	301
5. *Porous Silicon for Micromachining and Chemical Factories*	302
6. *Micromirrors on Silicon*.	303
7. *Full Three-Dimensional Circuits and Signal-Induced Feedback Treatment* . .	307
III. DRY ETCHING TECHNIQUES FOR MICROMACHINING	311
8. *Dry Etching Techniques*.	311
9. *Applications of Plasma Etching in Micromachining*	313
10. *Critical Issues in Plasma Etching of Microstructures*	314
IV. BONDING AND MICROMACHINING.	317
11. *Introduction*	317
12. *Bonding Techniques*	318
13. *The Si Surface and the Physics of Bonding*	321
14. *Applications of Bonding*.	328
15. *Compound Semiconductors and Other Materials*	330
16. *Critical Questions*	332
V. CONCLUSIONS	333
ACKNOWLEDGMENTS	333
REFERENCES	333

I. Introduction

Micromachining is a term generally applied to the wide range of three-dimensional (3-D) structures that can be fabricated using techniques originally developed for the microelectronics industry. Principally, micromachining revolves around single-crystal or polycrystalline Si, although associated materials such as oxides, nitrides, glasses, polymeric materials, and metals are also necessary and used. Other semiconductors such as the III-V compound semiconductors are also the subject of investigations.

In this work we introduce the most important technologies for Si micromachining. We discuss wet-chemical and plasma etching, as well as bonding methods. Rather than attempting to review all applications of these rather broad technologies, we focus on the mechanisms and critical unanswered questions. We introduce related methods for compound semiconductors where appropriate. Other than semiconductor bonding, we do not describe additive processes. Such processes include the selective deposition of single-crystal or polycrystalline Si, the selective deposition of metals such as tungsten, and an interesting technology for producing deep three-dimensional structures called lithographic galvanformung abformung (LIGA) (Ehrfeld et al., 1988). LIGA uses x-ray lithography of very thick resists followed by electroplating of metals to build up tall 3-D structures. Also not described and potentially of importance for micromachining, well-focused laser beams have been used to locally stimulate the etching of Si and the etching and deposition of other materials (Ehrlich and Tsao, 1989; Bloomstein and Ehrlich, 1991).

In particular, the various proposed mechanisms for wet-chemical anisotropic etching are discussed in depth. This includes several very recent topics involving innovative electrochemical thinning, as well as porous Si. New data are presented on the chemical formation of micromirrors on Si and GaAs surfaces. The Multiple Use Substrate (MU-STRATE) is discussed in some detail as a possible means of reaching into the terabit/cm^3 range of logic element density. A method (signal-induced feedback treatment) of analog e-beam testing and actively modifying up to a billion devices in a few minutes is presented as a way to use the MU-STRATE. The plasma etching discussion focuses on the ultimate importance of very high anisotropy, especially on the possibility of attaining this with purely chemical plasmas by using the crystallographic dependence of etch rates of different Si surfaces. Finally, the various wafer bonding techniques and physical mechanisms are discussed, especially as they relate to micromachining applications.

We do not reiterate the myriad of interesting applications that have already been discussed in several review articles (e.g., Petersen, 1982; Wise and Najafi, 1991; de Rooij, 1991). Those seriously interested in the field of micromachining should surely study these articles, as well as those related to the mechanical microgears and pressure sensors that are based on the etching of polycrystalline Si and sacrificial oxides (Fujita and Gabriel, 1991; Muller, 1990; Guckel *et al.*, 1990; Monk *et al.*, 1991). The recent work using hinged structures to greatly increase the vertical scale is also of interest (Pister *et al.*, 1991), but will not be discussed further.

II. Wet Chemical Etching of Single-Crystal Silicon

The existence of selective and anisotropic etches for Si and SiO_2 has formed the basis of many of the initial micromachining investigations. To briefly summarize, we will consider the mechanisms operative in two classes of etchants: the KOH-based etchants and the ethylene diamine-based etchants. Both types show highly anisotropic behavior, etching (111) Si planes at a much slower rate than other planes. Both etchants show a doping dependence, with etch rates slowing considerably as the Si becomes more than 3×10^{19} cm^{-3} p-type with boron doping. Considerably more etchant systems exhibit isotropic etching properties, and the reader is referred to any number of standard textbooks on Si IC processing for those details. Furthermore, the existence of etches selective for SiO_2 over single-crystal Si (HF) or polycrystalline Si over single-crystal Si (e.g., by HCl vapor in a CVD reactor) is also crucial to many micromachining processes without being the subject of our discussion here.

Some of us thought in the seventies that the ability to etch very narrow vertical grooves in (110) Si was going to change everything we did (Kendall, 1975, 1979, 1990; Kendall and deGuel, 1985; Bassous, 1978; Bean, 1978). The "vertical structure revolution" implied never quite materialized. It may yet happen, but on a much larger technology base and in different forms and not necessarily (though possibly) on the (110) orientation. For example, many of the preceding functions have been performed on other crystal orientations, especially the (100). The plasma technologies were developed in this same time period, and these are not generally limited by crystal orientation restrictions, except under special conditions to be discussed later.

We will begin by discussing current models of the highly unusual etching behavior of Si in alkaline solutions. The fundamentals of the etching processes will set the limits of future technological advancements. The most critical questions in the anisotropic etching of Si revolve around the extremely slow etching of the (111) surfaces compared to all others, and whether we will ever be able to obtain much higher etch ratios than 400/1 using *any* technology. We will return to this.

1. WHY IS THE (111) SURFACE SO SLOW TO ETCH?

This question has baffled workers ever since it was first shown that the (100) and (110) surfaces were attacked in KOH : H_2O at rates of about 200 and 400 times faster, respectively, than the (111) surface (Kendall, 1975). Kendall's "working hypothesis" (1979) suggested that the (111) surface oxidized in a KOH : H_2O solution at a rate much greater than did the other crystal surfaces, thereby blocking or passivating the (111) against etching. Thus, the dissolution of the continually forming oxide became the rate-limiting step for the etching of the (111) face of Si. All the other crystal orientations were thought to oxidize more slowly in alkaline solutions and never to become passivated against etching. This simple idea was consistent with the observed quite similar etch rates of both (111) Si and SiO_2 at temperatures around 85°C, so the hypothesis seemed to be reasonable. However, it would predict 110 : 111 etch rate ratios of greater than 1,000 at 27°C, instead of the observed 400 : 1 or lower ratios at this lower temperature. Another reason for reserving judgement on this model is that the (110) has been found to be the fastest-oxidizing surface under some conditions, rather than the (111), the latter of which is the fastest during high-T dry or wet oxidation (Kendall, 1990). Furthermore, Palik and coworkers, using ellipsometry (Palik *et al.*, 1985), have not been able to detect any large difference in the oxide thicknesses between (100) and (111) wafers that were freshly etched in KOH : H_2O. In fact, the 400 : 1 ratios have also been called into question by Kendall and de Guel (1985), and also by Clark *et al.* (1987), with 200 : 1 being given for the highest 110 : 111 etch rate ratio. We will return to this after discussing an alternate etching mechanism.

A second model for the very slow etching of the (111) surface is based on a type of molecular steric hindrance of this crystal surface (Kendall, 1990; Glembocki *et al.*, 1991). It is premised partially on the simple observation that the etch rate of Si has a distinct maximum at about 22 wt. % KOH concentration and falls nearly to zero at very high KOH concentrations near the solubility limit of KOH in water. Thus, for almost any reasonable chemical reaction — for example,

$$Si + 2OH^- + 2H_2O \Rightarrow Si(OH)_2(O^-)_2 + 2H_2, \qquad (1)$$

proposed by Palik *et al.* (1985) — it is apparent that it is the product of the OH^- ion and the *free* water molecule concentration (both squared) that enters into the reaction to form the etch products on the right-hand side of (1). Now the OH^- concentration increases continuously with increasing

KOH in H_2O, but the *free* H_2O *decreases* with increasing KOH. The latter is understandable in terms of the complexing of OH^- by three to five water molecules. As the OH^- concentration increases, more and more free water molecules are complexed, and the chemical activity of the water decreases almost to zero. Using published values for the chemical activity of water in KOH solutions, this model explains (quantitatively) the maximum etch rate at 22 wt. % and accounts in general for the decrease of the etch rate at high KOH concentration for the (100) and the (110).

Thus, we imagine a pellet of KOH dropping into a deep container of already saturated $KOH:H_2O$. It cannot dissolve, since there is no free water to "etch" it. The same occurs for a Si "pellet" in a saturated solution.

Now in this model there is an interesting way to explain why the etch rate of the (111) might fall even faster than those of the other crystal planes. Picture, if you will, an OH^- ion that is complexed in a sort of tripod arrangement with three water molecules forming the legs. These legs straddle each dangling bond on every Si atom on the (111) surface. The OH^- ion at the top of the tripod is blocked off from the surface so it cannot react, even if there were a few stray free water molecules. This makes an attractive explanation for the very low etch rate of the (111) at almost any KOH concentration. It also explains why it drops proportionally faster than the rate on other planes at high KOH concentrations, since the (111) is affected by both the diminishing free water *and* the increasing complex concentration. The other planes are not passivated nearly so effectively by the tripod complexes, since there is no way to block off all of the Si on the other non-threefold symmetric surfaces. The net result is an etch-ratio 110:111 that increases almost linearly from about 1.0 (i.e., isotropic etching) at very low KOH concentrations to perhaps 500:1 approaching the solubility limit of KOH in water. However, in that earlier work, Kendall and de Guel (1985) suggested that this maximum ratio could conceivably be as small as 200:1 if the *full* oxide lip correction were taken into account in the lateral undercutting measurement. Clark *et al.* (1987) later found a maximum etch ratio of about 200, but their measurements of the (111) etch rate were taken by measuring the undercutting of circular openings on the (100) rather than long straight openings on the (110).

The apparent maximum etch ratio of about 400 in a solution of 44 wt. % $KOH:H_2O$ was obtained by careful measurements of the lateral undercutting of the Si under a precisely aligned long and narrow opening in a thermally grown SiO_2 layer on a (110) wafer (Kendall, 1975). The alignment had to be within about 0.1° of the intersection of one of the perpendicular {111} planes to the top (110) surface in order to obtain such high etch ratios.

When misaligned from this intersection by some angle θ, the undercutting, u, was given by the simple linear relation

$$\frac{u}{D} = \frac{\theta}{40°},\qquad(2)$$

where D is the depth of etching in the center of a wide groove, and θ is the misalignment in degrees from the perfect perpendicular {111} intersection. As mentioned earlier, these maximum etch ratios could be as small as 200 : 1 if the oxide lip was assumed to be receding at the full oxide etch rate. Clark *et al.* (1987) came to a similar conclusion by using a Si_3N_4 mask deposited over a 500 Å "stress relief" thermal oxide. However, they could not be sure that the etch rate on the (111) was not enhanced by the strain introduced by the nitride–oxide combination, or by the thin underlying oxide itself dissolving at the same rate as the apparent rate on the (111). Thus, an unambiguous measurement of the *unstressed* (111) may await an inert (non-etching) mask that introduces no stress to the Si interface. One approach might be to deposit a stress-free plasma nitride. Another would be to etch a perfectly aligned and flat chem-mechanically polished (111) wafer on both sides in a $KOH:H_2O$ solution, and then to measure the full slice thickness by IR spectrophotometry. The mask stress problem, incidentally, is much like the anisotropic etching of vertical grooves in (100) InP along one of the perpendicular (110) planes. Oxide or nitride masking gave only about a 20 : 1 ratio due to interface stresses (Coldren *et al.*, 1983), but an etch mask of almost stress-free lattice matched InGaAsP was later found that gave a (100) : (110) etch ratio of better than 50 : 1 in pure HCl solutions.

2. What about Ethylene Diamine-Based Etchants?

While on the topic of anisotropic etch mechanisms, it is of considerable interest to consider what happens in the other major anisotropic etching solutions, namely those solutions containing ethylenediamine, or "EDA." These solutions often also contain pyrocatechol in order to speed up the etch rate, in which case the mixture is called "EDP." The behaviors of the $EDA:H_2O$ and the EDP mixtures are similar from the standpoint of relative etch rates on various crystal planes (Abu-Zeid *et al.*, 1985). They are often used, in spite of their much greater toxicity, because they have a much lower etch rate on the masking silicon oxide at high etching temperatures. However, these $EDA:H_2O$ mixtures also give much higher etch rates on the (111) plane than the $KOH:H_2O$ solutions. Furthermore, the (110) etches at about half the rate of the (100), rather than vice versa for the KOH mixtures.

It appears that the EDA molecule "fouls" the passivation mechanism on the (111) to some degree, perhaps by competing with the OH : $3H_2O$ "tripods" for the dangling bonds on this surface. Thus, (111) surfaces etch markedly faster in the EDA mixtures than in KOH : H_2O, since the passivation is only partial. On the other hand, the EDA seems to help passivate the (110) surface relative to the (100) surface, thereby decreasing the etch rate of the (110) surface as compared to KOH mixtures. The result is an etch ratio of about 200/100/5 for the 100 : 110 : 111 in EDA solutions, instead of the 200/400/1 seen for these planes in KOH : H_2O mixtures. The etch rates of the (100) are about the same in the two mixtures at 85°C, so the etch rates are normalized to 200 in these etch ratios. The relative ratios are shown graphically in the KOH and EDA systems in Fig. 1, with the (111) rates multiplied by 10 to make them more visible. The difference is quite striking.

These results with EDA : H_2O are particularly surprising when it is realized that the EDA molecule, $H_2NCH_2CH_2NH_2$, in a water mixture grabs H^+ ions from two HOH molecules to realise two OH^- ions to make a strong basic solution that is expected to be similar to KOH : H_2O in many respects. On this basis one would expect the etching behavior of Si in EDA : H_2O to be quite similar to that of KOH : H_2O mixtures, since both etchants should operate by reaction (1). However, the EDA : $2H^+$ ion-complex evidently intervenes rather strongly on the (111) and (110) surfaces, enhancing the etch rate of the (111) and suppressing the (110) rate. The elongated EDA molecule (or ion complex) appears to demonstrate some sort of passivation of the

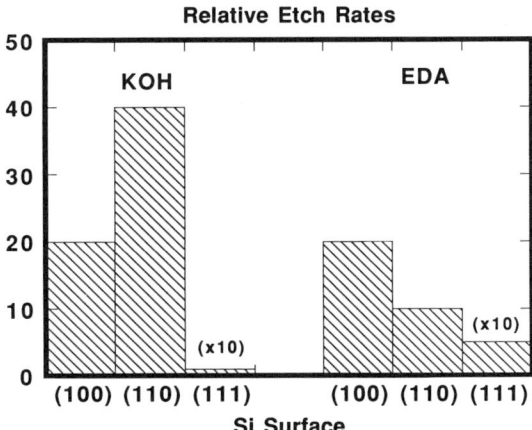

FIG. 1. Comparison of etch rates on Si using KOH : water and ethylene diamine : water, showing the large decrease in the rates on the (110) and the increase on the (111) using the EDA mixture. The (100) rate is much less affected.

twofold-symmetric (110) surface. If this is a valid argument, one might be able to select some other aliphatic base or additive molecule that would make the (110) dissolve more slowly than the (111). This would be a significant discovery, since it would make it possible to etch vertical grooves on the (100) surface, thereby making a number of three-dimensional ICs possible on a surface with low surface state density. As mentioned earlier, such a vertical etch capability has already been shown for the (100) surface of InP by arranging for the (110) etch rate to be slower than the relatively fast rate of the (111B) plane. In that case the (111A) is still a very slow-etching plane, but it doesn't interfere with *one* of the two sets of orthogonal perpendicular {110} planes on the (110) surface.

3. The Electrochemical Model For Anisotropic Etching

Another model for the anisotropic etching of Si is based on the electrochemistry occurring at the interface between different crystal orientations and the basic solution (Seidel et al., 1990). They proposed that the first event in the dissolution "is an oxidation step where four hydroxide ions react with one surface Si atom, leading to the injection of four electrons into the conduction band." In the reduction step that follows, "the injected electrons react with water molecules to form new hydroxide ions and hydrogen." They also proposed that "the anisotropic behavior is due to small differences of the energy levels of the back bond surface states as a function of the crystal orientation."

At high concentrations of KOH in water, Seidel et al. (1990) showed the etch rate decreases with the fourth power of the water concentration. This latter observation has a similar net result to the hydration model discussed earlier, which suggested that the etch rate decreased at high KOH along with the chemical activity of the water, which was dictated by the hydration complexes of the water with the hydroxide ions.

We believe the hydration model is somewhat more amenable to physical modeling and visualization. The relative merits of the two models or other models will ultimately rest on their success in predicting and modifying the anisotropic behavior of different crystal surfaces in the presence of different chemicals. At present, both models can be adjusted to give good fits to the experimental results for the metal hydroxides. The hydration model is quite simple and rather fundamental and is supported by good predictions based on independent measurements of the chemical activity of the OH^- ion and water. The electrochemical model also has merit, but it probably needs a more precise geometrical explanation of the "energy levels of the back bond surface states" before it can be applied to explain the extreme anisotropy of etching. The steric (geometrical) arguments of the hydration model are attractive intuitively for explaining the anisotropy, but there is as yet no direct evidence for such a passivation effect on the (111) plane (see experimental suggestions in Kendall, 1990).

4. ETCH STOPPING IN BASIC SOLUTIONS

Another very interesting effect that may have a possible electrochemical explanation is the extremely slow etching of Si at hole concentrations greater than about 3×10^{19} cm^{-3}. Raley et al. (1984) were the first to show that the etch rate of Si in EDP solutions decreased as the fourth power of the hole concentration. They suggested that the transfer of charge (probably injected electrons) across the semiconductor interface played an essential role in the oxidation-reduction reaction. Seidel et al. (1990) clarified this concept and also presented extensive new data showing that low concentrations of KOH : H_2O gave results that were quite similar to the fourth-power dependence obtained in EDP solutions.

The other electrochemical process that is used for an etch stop involves a positive connection with an n-type region of a Si wafer that contains a pn junction. The counter electrode is usually Pt and is biased negative. Thus, the pn junction is reverse biased around 1.0 volt in such a way that no current can flow to the p-Si, so the p-Si is not passivated and thereby etches in the basic solution at a rate similar to that of an unbiased wafer. In a typical experiment, the p-type substrate is dissolved until it reaches an n-epitaxial layer. When the positively biased n-region is exposed to the etching solution, passivation occurs and the etching stops. This process is widely used in the production of thin Si membranes for pressure sensors, as well as for many other applications. See, for example, Jackson et al. (1981), Palik et al. (1982), and Kloeck et al. (1989).

A recent innovation by Wang et al. (1991) in the etch stopping process uses a different biasing scheme from that described above. They connect the p-type Si to the positive terminal and allow the n-type parts of the wafer to be exposed to the KOH : H_2O etching solution and the negatively biased counterelectrode. However, the voltage of about 2.0 V is only applied for one second, and then the voltage is removed for 30 seconds. The p-Si anodizes to a thickness of about 7 Å during this pulse, while the n-type Si develops an anodic film of less than 0.5 Å. The latter provides essentially no etch stopping of the n-regions during the subsequent 30 s etching phase, but the p-regions are fully stopped for over 80 s for this particular pulse. Thus, the n-Si dissolves and the p-Si is stopped from etching if such a pulsing procedure is used. This allows a great deal of process flexibility and does not require a low-leakage reverse-biased *pn* junction, which is sometimes difficult to obtain in aqueous solutions.

A final comment is in order concerning a novel way of producing very thin Si membranes (Lee, 1991). This is a variant of an earlier electrochemical method that used anodization in a solution of dilute HF : H_2O (van Dijk and de Johnge, 1970). However, in the new method a thin, high-resistivity *n*-type or "intrinsic" region is first produced in a uniformly doped *p*-type crystal by Li drift (more precisely by *diffusion*) by soaking in a LiCl : H_2O solution for a few minutes near room temperature. The wafer is subsequently anodized in a 5% HF solution to produce membranes in the range of 0.5 to 1.0 μm thickness. The Li is subsequently removed from the membrane by boiling in DI water. It would seem feasible to the present authors to use LiOH as the anodizing ambient, and therefore to form the high-resistivity region before the thinning process is undertaken in the same solution. There is clearly much work to be done before such techniques can be used in commercial processes, but they do invite some interesting proposals. For example, would patterned light on the sample during the initial diffusion process result in a new form of lithography? Also, is there an analogue of this type of diffusion, albeit at much lower levels, when using KOH or NaOH electrolytes? The question of possible device instabilities arises here.

5. Porous Silicon for Micromachining and Chemical Factories

Porous Si forms when a Si anode is electrochemically treated in a solution of HF and H_2O (and sometimes ethanol). The pores can be varied in diameter and spacing over a range of about 1 nm to several hundred nanometers at different hole and electron concentrations and by using different current densities. Porous Si can be oxidized or nitrided at relatively low temperatures and is the basis of the FIPOS process — that is, fully insulated porous oxidized silicon. A brief survey of recent proposals to use porous Si as the basis for a variety of micromaching applications is presented by Smith *et al.* (1991) and references cited therein. Kelly *et al.* (1991) also give specific details on how to thin Si with a precision of better than 0.3% using a controlled current–time sequence followed by etching of the porous Si in NaOH : H_2O. These same workers (Guilinger *et al.*, 1991) have also shown that oxidized porous polycrystalline Si makes a very sensitive detector of water vapor, since the dielectric constant of the porous matrix increases by more than one magnitude in the presence of water vapor at 600 ppm by volume. Along this line, we have not seen much work on the optical properties of thin porous Si used as antireflecting films, except Prasad *et al.*, 1982. We have produced such "films" in the laboratory, but have not attempted to measure any effects of atmospheric gases or humidity on the reflectivity.

Kendall (1990) has also shown that pores can be introduced into thin (<1 μm) and tall (>200 μm) vertical Si studs. It is believed that these will eventually be used as semipermeable membranes for a wide variety of applications. Complex interdigitated vertical stud membranes may someday form the basis "for salt water purification, fuel cells, voltage aided ion separation, and for other separation and reaction schemes. This will require a great enlargement of our understanding of anisotropic etching, cavity sealing techniques, capillary and chemical effects in pores as small as 1 or 2 nm, protein and other pore coating schemes, as well as the physical properties of the pores that can now be produced" (Kendall, 1990).

In that same paper it was also proposed to use the unusual properties of misalignment steps on the walls of anisotropically etched holes to make a sort of chiral molecule filter and possibly a new type of air foil — the latter was something like a flying carpet. These are speculative concepts, but appear to be within the reach of today's technology.

6. Micromirrors on Silicon

The seed for the micromirror work was first published in 1988, from which we quote the opening sentences:

> During the thinning of Si slices with KOH : water solutions, shallow circular depressions are often observed in the midst of flat well-polished regions. Rather good real images of structures on the ceiling of the laboratory are sometimes seen in the optical microscope when the sample is tilted slightly and the focusing plane is a millimeter or so above the slice surface (Kendall *et al.*, 1988).

The general behavior of the circular depressions was described in that work as a function of the etching depth. Nevertheless, there were several aspects of the micromirrors that we did not understand. For example, the first question was how could a square inverted pyramidal pit become a circular depression after thinning the {100} fourfold symmetric slice? Also, why did the growth of the depression start out linearly and then convert to a decidedly sublinear relation? Also, how could these depressions be so close to spherical? Why was the surface quality inside the boundaries of the depression so much smoother than the surrounding relatively flat {100} regions? How smooth could the surfaces be on an atomic basis? What is the controlling etch rate and how could it be modified? What was the attainable range of *f*/

numbers? How small could they be made, and how large? Can the top edges be smoothed? Is the process diffusion-limited or surface reaction–limited?

We have now partially answered the first question, as well as several of the simpler questions. For example, we have shown that the formation is primarily reaction rate–limited. We have also shown that it is the {411} etch rate that controls the initial process, but we still have not developed a model to explain the nearly spherical surfaces and circular depressions that develop after the etching depth is greater than about seven times the original pinhole (or inverted pyramidal pit) width dimension, d_0. The preliminary data on the progression of the etched depression in 30 wt. % KOH : water is shown in Fig. 2. The first V-shaped segment in the figure shows the original circular opening in the oxide mask after a preliminary etching step in KOH : H_2O. The circular opening allows an inverted pyramid to form (Bassous, 1978; Bean, 1978), and then the etching essentially stops at the four slow-etching {111} planes. These very slow-etching planes at 54.74° to the top {100} are apparent in Fig. 2, although they are distorted by the expansion of the vertical scale of the Alphastep measurements. Then the oxide is removed, the thinning procedure begins, and the square depression becomes more and more circular, reaching an essentially circular shape (within 5%) when the etching depth is about seven times the original diameter. After this, the depression remains circular for all succeeding time as far as we have been able to determine.

Figure 2 represents data at 40°C from widely disparate starting values for d_0, ranging from 8.5 μm to 232 μm. The striking concordance of the data over such a wide range of d_0 strongly suggests that the dissolution process is limited by the etch rates on the different crystal orientations around the {100} planes, and that diffusion limitation is only very slightly involved. This concordance is expected to be even more precise at higher KOH. The increasing diffusion limitation at KOH less than 20 wt. % should be useful in modifying the f/number and focal length at a given etch depth, as will secondary treatments in isotropic etches, chemi-mechanical polishing, and plasma etching.

We now have related the curvatures of the depressions in a general way to the etch rates between the (100) and the fast-etching (411). This orientation dependence around the (100) was shown in Weirauch's early paper (1975). Recent work by Mayer et al. (1990) has given the ratio of the (411) : (100) rates at different KOH concentrations. By a simple geometrical analysis of Fig. 2, we get about 1.41 for this ratio at 30 wt. %, which is in good agreement with Mayer et al., 1990, who gets a ratio of 1.4 at the same composition. Mayer also gets 1.6 at 15% and 1.27 at 50%. A complete understanding of the progression of the shapes of Fig. 2 will require a three-dimensional analysis

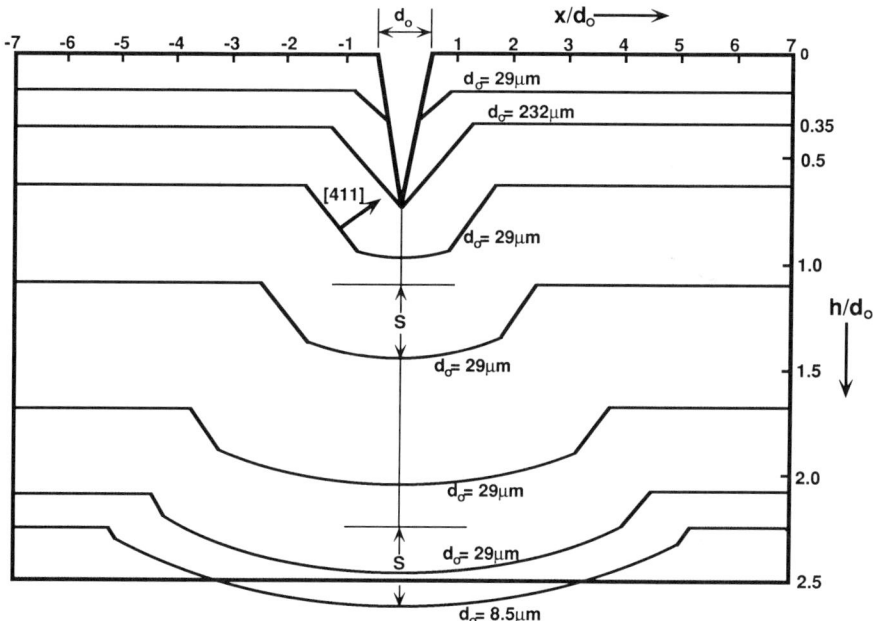

FIG. 2. Normalized progression of micromirrors in silicon using 30 wt. % KOH : H$_2$O at 40°C. The original opening diameter or lateral distance across the top of the inverted pyramid after the pre-etch in the same solution is d_0. The depth of etching after the oxide mask is removed is h. Note that the vertical scale is 3.8 times bigger than the horizontal, so the actual curvature is much weaker than shown. The actual angle of the [411] with respect to the top (100) surface is only 19.47°. The V-groove at the top consists of {111} planes that subtend 54.74° with the (100).

similar to that of Shaw's (1979) two-dimensional predictions of etching near a passivating mask edge, as well as incorporation of the expected diffusion limitation contribution at low KOH. The variation of the etch ratio, from 1.6 down to 1.27, gives one a significant range of control over the limiting sagitta s of Fig. 2. We note that the focal length, f, of a shallow circular depression can be approximated by the expression

$$f = \frac{D^2}{16s}, \qquad (3)$$

where D is the final diameter of the depression and s is the "sag" or sagitta of the curve.

a. Why Are the Micromirrors So Smooth?

The micromirrors are markedly smoother than the surrounding {100} top surface. We earlier estimated this smoothness to be "generally smaller than 40 nm" in 30% KOH (Kendall et al., 1988). However, we have now remeasured this in a sample that was etched in one continuous thinning step to a depth of 122 μm, and we find that the typical steps are less than 5 nm high and separated by 3 to 5 μm near the center of a relatively large micromirror of 550 μm diameter. The smoothness is expected to be even better than 5 nm in 50 to 60% KOH, but at considerable sacrifice in the etch rate to produce a given structure.

We believe the micromirrors are smoother than their flat surroundings because the etching steps always start at the center of the micromirror and flow continuously outward from the center. They do not build up high and low spots due to the normal stochastic processes that occur on large flat surfaces, since the center of the depressions is the slowest-etching plane and the steps should not have much tendency to move toward the center, but only outward and upward. At some point in the growth of the depression, there should be a "roughening transition" when the center of the depression gets large enough and flat enough to support step motion in both inward and outward directions. The outward motion of steps may also be at the heart of the unexpected tendency of the depressions to grow toward circles with almost spherical curvatures.

b. Additives and Other Etchants

Preliminary studies of alcohol additives to KOH gave rather poor micromirror formation, but we have no understanding as to why this is so. In general, it appears that the strong organic bases such as hydrazine and ethylenediamine: water have quite a different behavior from the KOH and NaOH systems (see Fig. 1). This manifests itself in stronger undercutting of the {111} walls using organic additives or the organic bases, but less etching on the exposed convex corners of mesa structures (Kendall, 1990). We have not yet attempted to make micromirrors with these organic base solutions.

Other chemical solutions are of interest here, especially if they are able to change the general form of the micromirrors. For example, two groups have recently studied the tetramethyl ammonium hydroxide (TMAH) system as an anisotropic etchant for Si (Tabata et al., 1991; Schnakenberg et al., 1991). The big advantage over KOH is the much lower oxide mask etch rate, and presumably the much lower heavy-metal contamination level. However, the etch rates are somewhat slower than in KOH, and the surfaces are covered with hillocks below 20 wt. % of TMAH. Higher concentrations are much smoother, but also much slower. Other hydroxides such as CsOH and NH$_4$OH are also of interest for micromirror formation (Schnakenberg et al., 1990), but have not yet been investigated for this application.

c. Cylindrical III–V Micromirrors

There have been several studies of the anisotropic etching of GaAs. None of these have discussed the issue of thinning a flat surface with a preexisting pyramidal pit or V-shaped groove. However, we predicted in Kendall et al. (1988) that "based on the concepts of this work and the analytical methods of Shaw (1979), it may indeed be possible to etch long cylindrically shaped grooves in the III–V compounds. The critical requirements appear to be a two stage etching process, a soft minimum in the etch rate on the crystal surface being etched, and equal rates in the two lateral directions perpendicular to the long grooves."

De Guel (1990) undertook a somewhat detailed study of the GaAs system with several different anisotropic etching solutions. She did this to provide special substrates for MOCVD deposition of III–V heterostructures. She demonstrated that cylindrical grooves could indeed be produced on the {100} surface with properties similar to the results shown in Fig. 2 on Si {100}. Her best results were obtained when the starting groove was V-shaped with the slow-etching Ga surfaces forming the walls. After removal of the nitride etch mask and thinning of a sample with an approximately 10 μm wide V-shaped groove to a depth of 60 μm, the lateral width of the cylinder was 188 μm and the sagitta s was about 3.7 μm. The etching was done at 25°C in 8 min in H$_2$SO$_4$: H$_2$O$_2$: H$_2$O in a volume ratio of 1 : 8 : 1 (MacFadyen, 1983).

The results for grooves running perpendicular to these same grooves gave very poor surfaces, since they did not meet the criteria of the statement in quotes — that is, the fast-etching As {111} surfaces interfered with the process. Finally, it was also shown that spherical surfaces could not be attained on a GaAs surface using this technique, as was predicted earlier (1988) on the basis of the Ga-face and As-face difference for the four walls of an etch pit on (100) GaAs.

7. FULL THREE-DIMENSIONAL CIRCUITS AND SIGNAL-INDUCED FEEDBACK TREATMENT

There is immense potential for improvement of our present IC circuit density. We presently use less than 1 % of the volume of a typical Si chip for

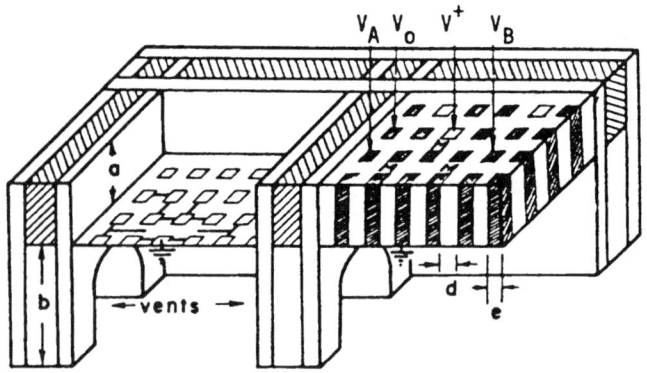

FIG. 3. Perspective view of the multiple use substrate or MU-STRATE, showing different voltage probes (e-beams or other) on one side and interconnects on the opposite side. The M designates Molecular arrangements on the front side when used for Molecular Electronic Device arrays.

our active elements, namely the top few microns of a 500 μm thick chip. The remainder of the chip is primarily a support structure. One of the most promising proposals to break this inefficient usage of semiconductor real estate is to conduct current from the top of the slice to the bottom via very small (0.1 μm) square (or parallelogram-shaped) conductors. The proposed wafer thickness is about 1 to 5 μm, made in portions of a thicker wafer (Kendall, 1983, 1985). The fabrication of such small feedthroughs requires a complicated but technically feasible procedure, as will be outlined later. The structure is presented in Fig. 3 to show the *kind* of three-dimensional configuration that may be needed to reach into the gigabit/chip or perhaps even the terabit/chip range of element density. It will be noted that the basic wafer shown in Fig. 3 has the following properties:

1. Beam accessible from both top and bottom.
2. Very thin, yet self-supporting.
3. Matable with substrates and packages from both sides.
4. Interlaced with closely spaced highly conductive isolated paths from top to bottom.
5. Capable of discretionary interconnections from either or both sides.
6. Readily coolable.
7. Equipped with special pads to provide different voltage levels and clocking to different subsystems.

This particular structure was originally created to serve the unique needs of mating a slurry of miscellaneous molecules on one side of the structure with directed discretionary interconnects on the other side of the thin structure. The idea was to use electron beam testing to identify particular logic functions in the random assemblage of molecules, then to effectively "freeze" the good ones by not heating them in a subsequent e-beam heating process. This heating procedure used the analog picture generated by the assemblage (or its negative) to feed back at a higher current to the assemblage itself. This process was repeated until a large fraction of the assemblage operated with the desired logic function(s). To perform this massive adjustment of billions of localized elements would require many years if done individually in series, but it could be performed in a few minutes or hours if done by projecting back the testing information onto the array, where it is noted that the scan time for a given test is a fraction of a second. This localized knowledge was then used to wire the circuit on the back side of the slice, also with an analog feedback process. This "signal-induced feedback treatment" or SIFT process (Kendall, 1972) allows a sort of directed natural selection and may be similar to how the brain organizes itself to perform particular logic functions.

We do not necessarily expect to be performing such "SIFTing" operations in the near future on molecular-based logic, but we might try something similar on simpler applications. For example, in the 1972 patent we showed how to rapidly fine-adjust the dark current of one million diodes in a Si vidicon array by a SIFT process of e-beam heating intermittently at 600°C and e-beam testing near room temperature. This utilized the well-known oxygen donor reaction of Czochralski crystals. Threshold adjustment of MOS circuits is another real possibility for the SIFT process, as is beta adjustment of special bipolar circuits. The possibilities are endless.

The fabrication of the structure of Fig. 3 on (110) Si is complicated by several effects, even though the deep etching capability for long and deep grooves is now well established. First, small holes that run vertically through Si cannot be produced because of the competing (111) slow-etching planes that limit the depth of etching to about 30% of the widest dimension of the groove (Kendall, 1979). This means that a square hole of 0.1 μm width would stop etching after progressing only 0.03 μm. Barth and Shlichta (1987) found that they could avoid this problem by "spoiling" the (111) competing (nonperpendicular) planes by laser etching, and then cleaning up the laser-etched hole by subsequent etching in a $KOH : H_2O$ solution. However, the smallest holes that one could etch with this latter technique would probably be of the order of the wavelength of the light, say 0.4 μm, and to produce a billion in a 1 cm^2 area would not be too cost-effective. Nevertheless, one might consider doing them all at once with a very intense crossed-laser interference scheme.

The second scheme that comes to mind is to etch crossed grooves in the form of screenwire, leaving vertical studs of highly conducting Si of the desired 0.1 μm width. This is not feasible because of the severe corner etching that will occur on such studs. In fact, the groove widths (and starting vertical stud widths) would have to be more than 1 μm to have 0.1 μm of remaining stud at the top surface after etching to a depth of, say, 5 μm (see Mayer et al., 1990, for a discussion of convex corners on (100) mesas). The corner compensation used by Mayer et al., (1990) would use far too much area for this application. Certain organic additives are helpful in reducing this exposed corner etching, but not nearly as much as needed here.

We will now discuss a possible fabrication sequence that could indeed be used to produce such a structure. It involves the use of (110) Si to produce very narrow vertical grooves, subsequent filling of these grooves with single-crystal Si, re-etching a nearly orthogonal set of narrow grooves, refilling the grooves again with single-crystal Si, removing the regions that have not been etched and refilled with a special etchant, filling these holes with a good conductor, and finally thinning and otherwise shaping the structure in appropriate regions so that the top and bottom of the wafer can be accessed with multiple types of beams, and also so that the thin wafers can be bonded and stacked into massive, thermally efficient three-dimensional arrays. An experienced process engineer will recognize that all of these steps are possible, but collectively they present a sizable challenge. If the deep-groove etching process could be done with adequate aspect ratio using some form of plasma etching, numerous steps could be saved, and the structure could probably be made using another crystal orientation or material — for example, (100) Si, if low surface-state density MOS circuits were to be fabricated. Summarizing the steps for the (110) process, we have the following:

1. Anisotropic chemical etch narrow grooves in p^+ (but not p^{++})Si to depth a along the y-direction (Wise and Najafi, 1991; Seidel et al., 1990).
2. Fill narrow grooves with single-crystal n or p Si using chemical vapor deposition (CVD) (Smeltzer, 1975).
3. Repeat steps 1 and 2 along x-direction.
4. Remove p^+ regions using concentration-dependent etching (CDE) (Muraoka et al., 1973).
5. Oxidize or otherwise coat walls with dielectric.
6. Refill narrow holes with metal to form narrow feedthroughs.
7. Thin the required regions to create access to beams of various types.

This is clearly not a simple process, but the preceding structure offers the possibility of reaching logic function densities of $10^9/\text{cm}^2$ on each level and a volume density of $10^{12}/\text{cm}^3$ by stacking the levels every 10 μm. Furthermore, the structure would have multiple uses that could be determined after the basic structure is formed. We call this the multiple use substrate, or "MU-STRATE." (We could also call it the "MUSIC" for multiple use substrate for integrated circuits.)

III. Dry Etching Techniques for Micromachining

In order to pattern silicon into the extremely small electronic devices required for integrated circuits, plasma etching techniques were developed by the semiconductor electronics industry. Using plasma etching, highly directional etch profiles can be obtained, allowing many devices to be packed into a small surface area on a semiconductor substrate. Plasma etching has already been used in a variety of ways in fabricating microgears, microactuators, and sensors. This section will offer an overview of the application of plasma techniques to silicon microstructures, with a focus on plasma techniques used to achieve anisotropic etch profiles.

8. DRY ETCHING TECHNIQUES

Before reviewing the application of plasma etching to the fabrication of microstructures in silicon, it is important to briefly summarize the types of plasma-based etching processes and their characteristics. Reviews on plasma etching are found in Fonash (1985) and Coburn (1982).

Plasma etching is a term that is used in various ways. In a very general sense, it is used to describe the use of plasmas for etching materials. It is also used in a more specific sense that will be used here: etching in a plasma where the substrate is not biased, and thus the etching comes about mostly through chemical reactions of plasma-produced radicals with the substrate, forming volatile compounds that are pumped away by a vacuum system.

In contrast, ion milling utilizes noble-gas atoms (most frequently argon) that are ionized in a plasma and are accelerated to a substrate by various means including substrate bias. Here the etching is effected by sputtering, the physical removal of material due to the impact of energetic species on the surface. Since the plasma species in this case are noble gases, no chemical reactions take place at the substrate.

Etch rates can be increased substantially through combining these two techniques in a process called reactive ion etching (RIE). Here the plasma is used to produce both chemically reactive species and ions. Etching takes place through the combined actions of energetic ions accelerated to the substrate, which damage the surface and remove material through sputtering, and through the chemical reactivity (perhaps increased by sputter damage) of the radicals produced in the plasma. RIE generally produces the fastest etch rates and allows for highly directional etches (to be discussed in the next section). A related dry etching technique called reactive ion beam etching (RIBE) also utilizes a combination of sputtering and chemical etching to realize high etch rates. However, in RIBE the reactive ions are formed in an ion gun and are extracted by means of biased grids producing a beam of ionized radicals that impinge on the substrate.

Ashing is a generic term used for plasma etching of organic films, especially photoresists. Ashing is achieved using a plasma to dissociate molecular oxygen into reactive atomic oxygen radicals. These radicals react with the organic film to form carbon dioxide and water vapor, which are then pumped out of the system.

A variety of laser-based techniques have also been employed for etching materials. These utilize the large flux of photons produced by a laser to dissociate molecules, forming radicals that react with the substrate. Alternatively, the photons may excite electron–hole pairs within the material to be etched; the electrons or holes interact with the ambient gas at the surface to produce volatile species that are then removed. Lasers can also act as localized heating sources to selectively heat the substrate to a point where the process gas can be broken down pyrolytically, forming reactive radicals that etch the substrate. Finally, it is possible to use high-energy lasers to remove material by ablation. A review of the various laser-etching processes can be found in Ashby (1987).

The characteristics of a plasma-etched structure depend on the particular plasma etch technique used and can differ greatly from wet etching results. Of primary importance in dry etching is the directionality of the etch: Is it isotropic (the etch rate is the same in all directions, and hence significant mask undercutting can take place), or is the etch anisotropic (highly directional)? For many applications it is desirable to have a very anisotropic etch with sharply etched features and little or no undercutting of the etch mask. Alternatively, isotropic processes might be useful for removing sacrificial layers underneath other important layers, or where tight registration tolerances are not required. In the absence of a bias or some other means for directing the energy of radicals and ions produced in a plasma, etching will generally be isotropic. Anisotropic dry etching is achieved by utilizing the directed energy of ions from the plasma to remove material. Processes yielding anisotropic etch profiles include RIE, RIBE, and ion milling.

In many cases it is important to etch away one material while leaving other materials untouched. This is referred to as etch selectivity. Selectivity of etching processes is enhanced through the use of chemical reactants that only attack the layers to be etched. For example, high selectivity is achieved in etching of GaAs compared to AlGaAs by using etching gases containing both chlorine and fluorine (such as CCl_2F_2) that freely etch Ga and As, but when exposed to AlGaAs form involatile aluminum fluorides that then act as an etch stop. An example of very poor selectivity would be the etching of SiO_2 layers on silicon substrates using NF_3 to open holes for introducing dopants into the silicon. NF_3 etches silicon at a somewhat faster rate than silicon dioxide. When the oxide layer is etched through, the etch rate increases as the silicon is etched, leading to problems of process control. In general, ion milling has very poor selectivity because nearly all materials have similar ion sputter rates. Perhaps the most crucial selectivity is between the layers used as etch masks (typically photoresists) and the material being etched, which limits the etch depth that can be obtained. This is especially important for the fabrication of microstructures where very deep etches are frequently required. Selectivity can generally be enhanced through appropriate choices of etching gases that will react with some materials, but have limited reactivities with others. For example, RIE can be highly selective by appropriate choice of etch gas.

9. Applications of Plasma Etching in Micromachining

The most widely used techniques for micromachining silicon are still based on wet chemistry. Nevertheless, some workers have made extensive use of plasma-based techniques, and this work will be described here. A review of fabrication techniques for microstructures with a good description of applications of plasma etching for micromachining is provided by Delapierre (1989).

Plasma etching techniques have been most commonly used in the fabrication of micromotors, gears, turbines, etc. Mehregany et al. (1987) reported the fabrication of microgears and turbines using RIE of silicon in CCl_2F_2/SF_6 achieving 50 μm etch depths with surface smoothness of ± 1 um and less than 5% undercutting of the aluminum mask. The backing substrate was removed using SF_6 etching in a magnetron system. Mehregany et al. (1988) also formed gears, turbine blades, and tongs 40 to 50 μm thick and 300 to 2,400 μm in diameter using RIE with CHF_3 to etch silicon dioxide, and a polysilicon RIE step using a $Cl_2/CFCl_3/Ar$ gas mixture. Later work from the same group showed more extensive use of plasma-based etching techniques for fabricating variable-capacitance side-drive electric micromotors (Mehregany et al., 1990a, b). This work involved using RIE with CCl_4 to pattern polysilicon, RIE of aluminum masks using Cl_2, O_2 RIE of polyimide, and CCl_4 RIE to partially release bearings, followed by an isotropic etch using SF_6 to completely free the bearings. They describe the use of plasma etching for low-temperature oxide (LTO) layers, but note that the endpoint detection is poor using this process, and so finish the etch using buffered oxide etch (BOE) in which endpoint detection is more readily facilitated. These same micromachining techniques are also applied to the fabrication of electrohydrodynamic pumps (Bart et al., 1990). Fan et al. (1988) used two different plasma etching methods to form pin joints, gears, cranks, springs, and slides in phosphorus-doped polysilicon. In general, RIE with CCl_4 was used, but when thinner layers were to be etched and some mild undercutting of the mask could be tolerated, isotropic etching in SF_6 was employed. Reithmuller and Benecke (1987) fabricated silicon microactuators using ion milling to pattern a Cr/Au plating layer.

Plasma etching has been used in a variety of other applications. Schmidt et al. (1988) fabricated floating-element shear stress sensors using a parallel plate rf reactor. O_2 was used to etch polyimide, and SF_6 was used for silicon etching. They found that for a 30 μm thick polyimide layer, the mask was undercut by 6 μm during etching in this reactor. Allen et al. (1990) micromachined silicon plates suspended by polyimide beams for use as mirrors in fiber-optic applications. Using isotropic etching with SF_6, the silicon backing substrate was removed to release the plates and beams. Akamine et al. (1990) formed integrated scanning tunneling microscopes (STM) by forming a tunneling tip, actuator, and counterelectrode on the same substrate. The actuators were cantilevers, which were released from the silicon substrate with an SF_6/C_2ClF_5 etch, which does not affect the ZnO layers in the device. Polyimide was also removed with an O_2 plasma etch. Jerman (1990) used plasma etching to micromachine corrugated silicon diaphragms. Depending on the application, either isotropic or anisotropic etching could be used. He reported etched groove depths from less than a micron up to 50 μm.

10. Critical Issues in Plasma Etching of Microstructures

a. Anisotropy

As mentioned earlier, anisotropic profiles are generally achieved in plasma etching using ion milling or RIE. Perhaps the most difficult structures to form are deep trenches for capacitors. Aspect ratios as high as 11 for capacitor trenches on silicon have been reported by Herb et al. (1987). They also discuss how the shape of the bottom of trenches can be controlled by changes in the etching parameters such as pressure and process gas. Vertical etch rates in capacitor trenches seem to be limited by conductance of reactive gases into the trench, and removal of the etch by-products out of the trench (Coburn and Winters, 1990). Concave sidewalls are reported under certain conditions by many authors; these are thought to be caused by ion scattering in the plasma sheath, or by deflection of ions by insulating masks that have become charged by the plasma (Arikado et al., 1988). Chin et al. (1985) found that the vertical etch rate decreases as the aspect ratio of the trenches becomes larger. Similar problems will likely become important as the size of micromachines decreases.

The sharp etch profiles achieved in RIE are often aided by formation of chemically impervious films on sidewalls. For example, during RIE of silicon dioxide using CF_4, a Si_xOF_y layer forms on the etched sidewalls (Thomas and Maa, 1983). This film does not readily react with fluorine, so the sidewalls are protected from etching. At the bottom of an etched well, however, the carbon film is sputtered off by the action of ions, exposing fresh material to etching by fluorine radicals. In this manner, highly anisotropic etching of silicon may be achieved. Clearly, further work will be required before reliable etching of narrow features will become routine in micromachining.

An alternative route to achieving narrow structures in silicon might be through the use of purely chemical plasmas. It has been shown that by an appropriate choice of etching gas, certain crystal planes of silicon etch faster than others. For example, chlorinated gases seem to show some difference in etch rate for different crystal planes, whereas fluorinated gases etch all crystal surfaces equally fast (Herb et al., 1987). The best study of this phenomena is from Kinoshita and Jinno (1977), who measured the ratio of etch rates for the silicon (100), (110), and (111) planes in $CCl_4 : O_2$ plasmas and found a ratio of $30 : 15 : 1$ for these surfaces. Clearly this indicates some possibility for anisotropic etching using chlorinated gases. Limited studies of this have been performed by Carlile et al. (1988), who found that the angle of sidewalls in trenches after etching in chlorinated plasma was determined by the substrate temperature; Wohl, Mattes and Weisheit (1988), who used CF_2Cl_2/O_2 plasmas with added helium showing that the addition of helium increases the anisotropy on (100) silicon surfaces and leads to cleaner sidewalls; and Uetake et al. (1990), who studied etching of polysilicon in chlorinated discharges with added nitrogen and found some anisotropy enhancement over fluorinated discharges.

The use of such purely chemical plasma etching to achieve highly anisotropic etches needs much more study to determine the ultimate aspect ratios that can be obtained. A good starting point might be to study etching of very narrow trenches on (110) silicon. This would tell us whether plasma etching of Si could ultimately produce some of the very high aspect ratios of $100 : 1$ or greater that are possible with the hydroxide wet chemical etchants discussed previously. Such a dry etching technology might make the (110) a viable crystal orientation for MOS and other applications. Perhaps even more fruitful would be the development of very high anisotropy vertical etching using the (100) surface. As mentioned earlier with respect to Fig. 1, this would probably require fast etching in the (100) direction and very slow etching of the (110) inside narrow grooves. This is a dubious possibility, but one that could have very significant ramifications if achieved.

With respect to the preceding, we need only mention the immediate impact on chip "real estate" that a 100 : 1 aspect ratio narrow groove would have on present-day trench capacitors of CMOS and other VLSI circuits, especially if such ratios could be produced on 0.5 μm width grooves with 50 μm depths. New vertical isolation schemes would also proliferate. An extreme example of this would be a relatively straightforward approach to the MU-STRATE structure discussed in Fig. 3. In this case, deep (5 to 10 μm) and narrow (0.1 μm) holes would be etched into a Si wafer in addition to the wider grooves of Fig. 3 for interwafer connections, and then the grooves would be filled with tungsten or other metal that can be deposited in high–aspect ratio holes.

b. Surface Damage

Surfaces exposed to a plasma generally exhibit damage. This damage typically takes three forms: damage due to ion bombardment, incorporation of impurities into the material, and formation of impurity layers on the surface (Fonash, 1985). All of these damage mechanisms can affect the electrical properties of the material and should lead to differences in the frictional and hardness properties of the surface. Dieleman and Sanders (1984) found that silicon exposed to argon ions in the range from 400 to 2000 eV had a 40 to 100 Å thick amorphized layer at the surface. Singh *et al.* (1983) found substantial incorporation of iron and chromium, presumably from the chamber walls during ion-beam etching. Coburn and Kay (1979) studied the formation of polymer films in CHF_3 discharges. Most of these studies focus on the resulting degradation in the electrical properties of materials. Thus far, very little work has been done to study the effects of plasma-induced damage during etching of microstructures. Understanding changes in frictional and wear properties of materials exposed to plasmas will be essential to further progress in fabricating microstructures.

Although surface damage can be quite severe, post-etching treatments can be used to return the surface to near-original quality. Damage can be removed by annealing, brief chemical etches, or using isotropic plasma etches (which would be expected to have fewer energetic ions impinging on surfaces). Most intriguing is the possibility of using laser-assisted etching processes, which should generate fewer ions and hence lead to less ion-induced damage (Fonash, 1985).

IV. Bonding and Micromachining

11. INTRODUCTION

Even though one can trace the roots of direct bonding of Si wafers back several decades, its applicability to semiconductor micromachining has been realized only recently. With that relevance has come a wealth of new applications, as bonding replaces more cumbersome technologies and permits the machining of structures never before possible.

The observation that some materials with smooth surfaces form a "permanent" bond merely by being placed in contact has intrigued scientists as it is rediscovered in many forms. Opticians are well aware (Smart and Ramsay, 1964; Smith, 1965; Smith and Gussenhoven, 1965) of the technique called "optical contacting" and the ability to make specialized beam-splitting prisms and achromatic doublets without glues or other adhesives. Vacuum technologists have experienced vacuum welding and used similar principles for glass-to-metal seals. However, these techniques have remained essentially empirical until their relevance to semiconductor micromachining became apparent. Recent studies of this phenomenon in silicon-wafer bonding has shed some light on the mechanisms common to contact bonding of all materials. Because of the importance of these techniques to micromachining, we'll try to take a broad perspective by emphasizing the fundamentals.

A number of physical phenomena contribute to the attachment of two planar materials. Fundamentally, the driving force for bonding results from the difference between the free energy of two surfaces and the free energy of the bonded configuration. Physical mechanisms — principally hydrogen bonding forces, but possibly including van der Waals forces — provide the energetic reduction in many circumstances. If the surfaces do not possess either the flatness or the surface bonding configurations necessary, externally applied electrostatic forces across the gap in materials can supplement these attachment forces. Annealing can encourage the formation of permanent chemical bonds between the two surfaces, or the externally applied electric field can assist in electrochemical reactions at the interface. These phenomena, in combination or separately, can ultimately result in bonds indistinguishable from the bulk material or from those formed by the deposition of a dissimilar material on a substrate. It is not surprising that most of the materials useful in silicon micromachined devices have been successfully bonded. However, while a great deal of insight exists as to the mechanisms of bonding in certain circumstances, no general strategy or understanding has yet been put forth, and many bondable pairs of materials remain unexplored, limiting the usefulness of this technique. Our goals in this section are to spell out as much as is known about bonding relevant to the micromachining of silicon, and to point out potential applications to micromachining of other semiconductors.

12. Bonding Techniques

a. Field-Assisted Bonding

Besides the somewhat apocryphal reports of optical contacting (Smart and Ramsay, 1964; Smith, 1965), the first well-characterized bonding technique was reported by Wallis and Pomerantz (1969) using electric fields coupled with low-temperature (300–600°C) annealing. They called their technique for making glass-to-metal seals "field-assisted" bonding; closely related techniques have been referred to as "electrostatic" or "anodic" bonding. Wallis and Pomerantz (1969) reported successful field-assisted bonding of many combinations of borosilicate glasses, including Pyrex™ (7740), soda lime (0800), potash soda (0120), aluminosilicate (1720), fused silica, and fiber optics, with metals including tantalum, titanium, Kovar, and the semiconductors silicon, germanium, and gallium arsenide. They also reported bonds between glass and thermally grown SiO_2. Metals with large thermal expansion mismatches with glass, such as aluminum, nickel, chromium, iron, and boron, could be bonded in thin-film form. The procedure is summarized in Fig. 4 for a Si–glass bond. Several physical mechanisms are relevant. Principally, in order for a significant electric field to occur in the gap between the Si and glass, a space charge or polarization region must develop in the

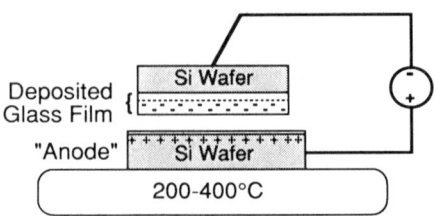

Fig. 4. Field-assisted glass–metal sealing, after Wallis and Pomerantz (1969). The polarization region develops in the glass because of the transport of positive ions and the relative immobility of the negatively charged defect sites left behind. The electrostatic force across the gap helps bring the two surfaces into intimate contact.

glass. It has been well established that electric fields in glass cause the drift of positively charged tramp alkali (usually fluxing ions such as Na^+), intermediates, or related impurities (Proctor and Sutton, 1960; Sutton, 1964a, b; DeNee, 1969; Wallis, 1970). Heating the glass to around 400°C enhances the mobility of these impurities and reduces the time required for the space charge region to form. These mobile impurities, in turn, leave behind relatively immobile negatively charged defects, giving rise to the electrostatic field of the observed polarity (Fig. 4.) Studies (Wallis, 1970; Borom, 1973; Gossink, 1978) show this space charge region exists for up to several micrometers into the glass. The resulting electrostatic forces across an air gap of about 10 nm effectively contribute only an extra 10^{-3} eV/chemical bond to the eventual attachment energy. However, as Anthony (1983) pointed out, these electro-static forces, coupled with viscous flow or elastic and inelastic deformation of the interface, are sufficient to achieve intimate contact of surfaces in the presence of some amount of surface roughness. The conclusion drawn is that the thinner the material to be bonded, the lower the force necessary for the required mechanical deformation.

The mechanical stresses arising during either bonding or subsequent heat treatments impose severe limits on the usefulness of the bond. Others have investigated bonding minimizing the mechanical deformation of Si–glass bonds by use of thermal-expansion-matching glasses such as alkaline-earth aluminosilicates (1729) (Spangler and Wise, 1987), or by using lower softening-temperature glasses (7570) (Esashi et al., 1990), permitting a room-temperature bonding process. The substrate does not need to be a glass, as successful anodic bonding using sputtered glass interlayers (Brooks et al., 1972; Esashi et al., 1990) has been established (Fig. 4). The electrostatics of $Si–SiO_2$ and glass systems have been described by Snow et al. (1965) and Snow and Dumesnil (1966).

The resulting chemical reactions occurring at the interface to complete the bonding process are less well understood. DeNee (1969) and Borom (1973) postulated some specific electrochemical reactions occurring between the glass and metal oxides (presumably including native SiO_2 on Si), including electrochemical dissolution of high points at the interface and transport of reactants. The net result is a smooth transition from glass through metal oxides to the metal. The role of the temperature in these reactions is also not understood. Joule heating at the interface due to the time-decaying current observed for field-assisted bonding has been shown by Kanda et al. (1990) to be unimportant in the glass–Si system.

Bonding can be selectively inhibited by a thin layer of metal predeposited on the glass surface (Roylance and Angell, 1979), which, when grounded, can also serve to electrically protect any circuitry. On the other hand, by following up on the understanding of bonding in the SiO_2 system, progress has been made in the bonding of Si_3N_4 (Albrecht et al., 1990). For this system, a steam anneal of the Si_3N_4 prior to field-assisted bonding presumably results in an oxide layer at the surface of the Si_3N_4 that could then form the field-assisted bond with a 7740 Pyrex substrate at the relatively low temperature of 475°C. Harendt et al. (1991a) found slightly lower bond strengths when one or both of the bond surfaces was Si_3N_4. Finally, indium–tin oxide films were reported bonded to glass by Esashi et al. (1990) and, by treatment of the surface with a hydrogen plasma, GaAs was reported to be bonded, using field-assisted techniques, to a glass substrate at 180–500°C in an N_2 ambient (Huang et al., 1990).

While the complexity associated with applying an electric field during bonding limits the usefulness of field-assisted bonding, we shall also see that direct bonding of Si (without an E field) requires considerably higher temperatures. Therefore, intermediate approaches have been developed to address these shortcomings. Already mentioned was the report by Esashi et al. (1990), describing field-assisted bonding at room temperature using lower softening temperature glasses. Related techniques requiring *no* electric field have also been developed. An 8 μm layer of epoxy resin (Hamaguchi et al., 1984) has been used to glue Si to quartz. There is also no need here to limit one of the bonding pairs to a glass. Glass can be used as an intermediary layer to assist the bonding between two similar or dissimilar materials. Successful electric field–free bonding between Si wafers using sputtered glass interlayers (Brooks et al., 1972) or using processing-compatible spin-on glasses for Si–quartz and Si–Si_3N_4 bonds (Kimura et al., 1983; Yamada et al., 1987a, b) all allow bonding at relatively low temperatures without electric fields. In a recent study, Field and Muller (1990) successfully bonded two Si wafers using a borosilicate glass as an intermediate layer. The glass was deposited on Si by solid-source boron doping during the oxidation of Si. Bonding (without an electric field) took place at 450°C. Interestingly, the temperature required for bonding phosphorus-containing glasses was reported to be much higher.

As something of a prelude to Si direct bonding, Frye et al. (1986) reported high-temperature (above 1,000°C) field-assisted bonding between two Si wafers without a glass interlayer. The wafers did have combinations of thermally grown and native SiO_2 layers present, however. While, in retrospect, the electrostatic fields were clearly not necessary for bonding, Frye et al. (1986) did detect charge transport reminiscent of the transport occurring in glass films during field-assisted bonding. This indicates that other species are involved besides the "tramp" alkali present in glass films. Frye et al. (1986) speculate that H^+ and OH^- are the important species in their experiment, fully consistent with current understanding of the direct bonding process at SiO_2 interfaces.

b. Direct Silicon Bonding

Given the studies of field-assisted bonding and the extensive experience with related techniques, it now seems quite obvious that Si wafers should be bondable by any of several approaches. The simplest technique, relying on both physical and chemical mechanisms together with heat treatments, is called silicon direct bonding by most (Shimbo et al., 1986), fusion bonding by some (Barth, 1990), and occasionally thermal bonding. It appears to have been first described in a patent (U.S. Patent No. 3,288,656) issued in 1966 to T. Nakamura of NEC (as mentioned by Yamada et al. 1987a and Barth, 1990). The first published description was by Lasky (1986), followed closely by Shimbo et al. (1986). This technique eliminates the use of the electric field to assist in the bonding and relies first on achieving intimate contact between Si microelectronic wafers and second on the chemistry of silica and water. Disadvantages, as already discussed, include the high temperature necessary to achieve bulklike bonds. However, even the lower bond strength associated with near-room temperature bonding can serve useful purposes: It permits wafers to be spun together, or enables a wafer to be used as a removable cap for the protection of a particularly vulnerable surface (Matzke, private communication).

13. THE Si SURFACE AND THE PHYSICS OF BONDING

The key to successful direct bonding of Si is achieving the proper chemical state of the surface. Grundner and Jacob (1986) published the definitive study of Si surfaces with the following general conclusions. Every Si surface exposed to an oxidizing ambient has at least 1.5 nm of native SiO_2 on the surface. It is also known that SiO_2 is terminated with a layer of Si atoms (Iler, 1974). Given proper preparation, those Si atoms can, in turn, be terminated in silanol (Si–OH) bonds (Fig. 5a). Such a surface exhibits hydrophilic behavior; deionized (DI) water wets or forms a low contact-angle droplet on the surface. Furthermore, there may also be several monolayers of water molecules present on such a surface. A hydrophilic surface may be prepared by any number of oxidizing etches such as the standard RCA clean ($H_2O : H_2O_2 : NH_4OH$; 5 : 1 : 1, followed by $H_2O : H_2O_2 : HCl$, 6 : 1 : 1), the so-called piranha etch ($H_2O_2 : H_2SO_4$), or hot HNO_3, all of which leave the surface terminated with OH and include some amount of bound water molecules.

The silicon surface type can be determined by examining the contact angle of water on the surface. For the hydrophilic surface, water will wet the

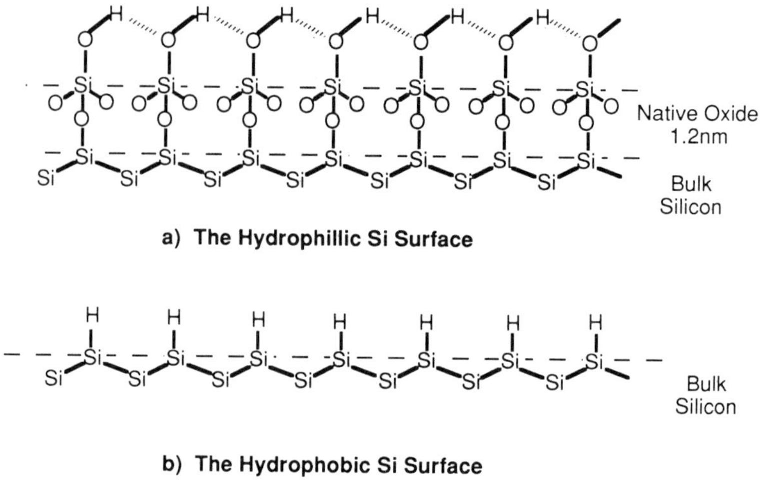

a) The Hydrophillic Si Surface

b) The Hydrophobic Si Surface

FIG. 5. The typical surface structure of (a) hydrophilic and (b) hydrophobic Si surfaces. The hydrophilic surface can have several additional layers of adsorbed water molecules.

surface, giving a very low contact angle. Studies show a hydrophobic surface also exists, where water beads up with a large contact angle. This contrasting Si surface is formed when the SiO_2 is completely removed and the resulting Si dangling bonds are saturated by hydrogen (Fig. 5b). Such is the case when the Si wafer is dipped in HF (Burrows et al., 1989; Yablonovitch et al., 1986). It is possible to saturate almost completely the surface with hydrogen.

a. Mechanisms and the Three Stages of Bonding

Almost all reports of Si wafer bonding have stressed the importance of a hydrophilic surface. Only one series of reports of weaker bonding between hydrophobic surfaces exists (Bengtsson and Engström, 1989, 1990), but the mechanism was attributed either to partial conversion to a hydrophilic surface (Bengtsson and Engström, 1989) or to evacuated bubbles (Stengl et al., 1989b). Since the normal hydrophilic Si surfaces involved in bonding have at least 1.4 nm of native SiO_2 on the surface, bonding is controlled by the surface chemistry of SiO_2, as pointed out by Lasky (1986). Silicon dioxide surfaces are terminated by Si atoms with chemisorbed –OH radicals, resulting in a layer of silanol groups (Fig. 5a). In addition, several layers of water molecules may adsorb on this surface under normal conditions. These considerations, therefore, are the basis of the models for direct bonding that have been formulated. From studies of bond strength (see below), three stages of bonding are deduced as described by Maszara et al. (1988) and elaborated upon by the study of Stengl et al. (1989a):

(1) Initial contact at room temperature (Fig. 6a). Surface strengths, as measured by crack propagation, reach $170 \text{ erg/cm}^2 = 0.17 \text{ J/m}^2$. Assuming 5×10^{14} bonding sites/cm^2 available on the SiO$_2$ surface (estimates range from 3.73 to 8×10^{14} (Stengl et al., 1989a; Michalske and Fuller, 1985), bond strengths approach 0.21 eV/bond, consistent with a hydrogen-bonding mechanism between silanol (Si–OH) groups on opposing surfaces. This is also the upper limit for a van der Waals bonding mechanism. However, the chemical reactions hypothesized for the rapid increase in bond strength suggest a hydrogen-bonding mechanism is more likely at this stage. Stengl et al. (1989a) suggest a slightly different mechanism than Maszara et al., involving the presence of water clusters at the interface. Hydrogen bonding between these water clusters is then the stage I bonding mechanism.

(2) Intermediate-temperature anneals at 300–1,000°C (Fig. 6b). The model of Maszara et al. (1988) hypothesizes Si–OH : HO–Si (silanol) converts to

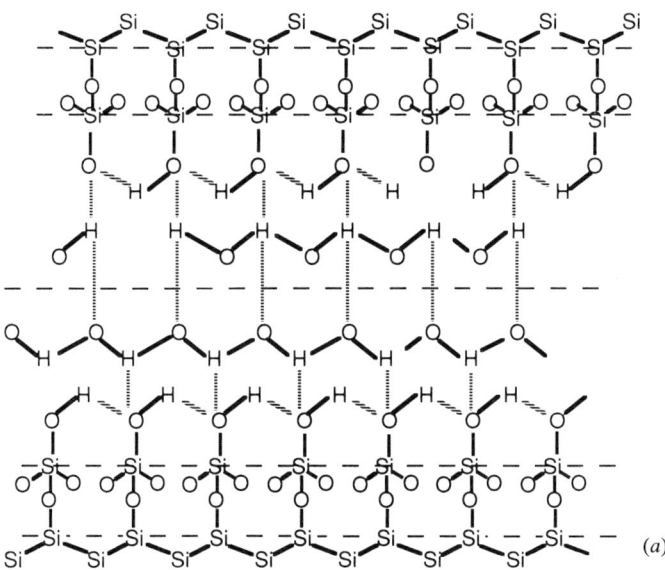

FIG. 6. The three stages of the Si direct bonding process as first discussed by Maszra et al. (1988) and by Stengl et al. (1989a). We follow Stengl et al. (1989a) more closely than Maszara et al. (1988) here. (a) Bonding through hydrogen bonds between surface water molecules. (b) Bonding through silanol groups. (c) Formation of siloxane bonds in the final configuration.

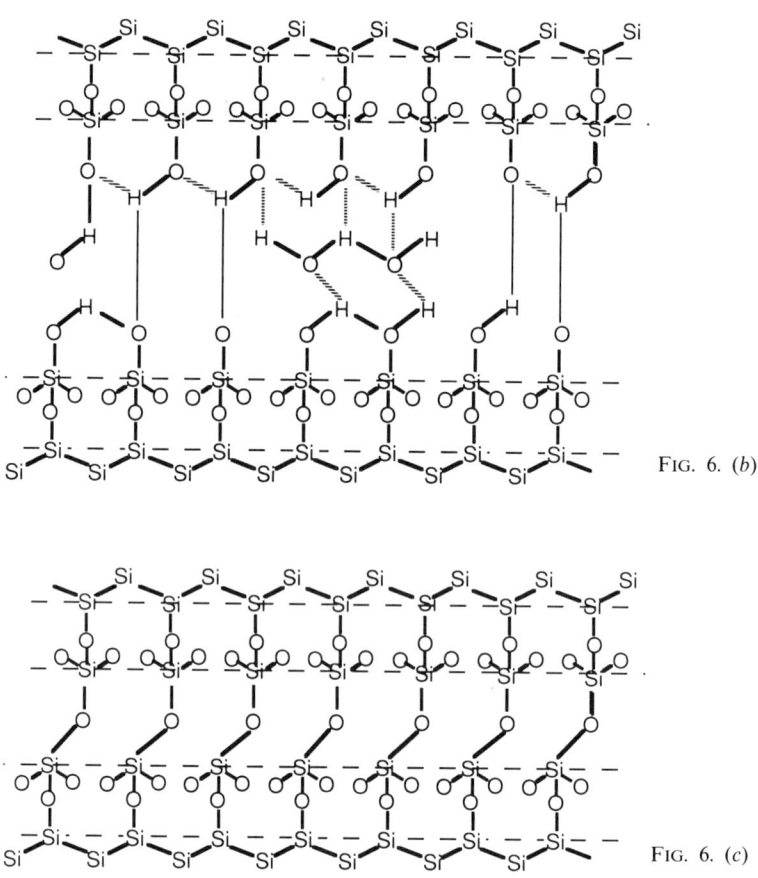

Fig. 6. (b)

Fig. 6. (c)

Si-O-Si (siloxane) through water elimination. Bond strength increases to 700 erg/cm^2 or 0.87 eV/bond. However, Stengl *et al.* (1989a) believe this is the stage where the water molecules coalesce and hydrogen-bonding across the silanol groups reaches completion. They base this conclusion partially on their and Ohashi *et al.*'s (1987) observation of interfacial water bubbles up to 800°C, although current thought is that these bubbles arise from residual hydrocarbon contamination (see below). Bond strength is independent of annealing time for these first stages, indicating that component diffusion is not involved.

(3) High-temperature annealing above 1,000°C (Fig. 6c). Maszara *et al.* (1988) postulate that the viscous flow of SiO$_2$ seals all remaining voids in this regime. Bond strength becomes dependent on time and averages

2.5 J/m² or 3.2 eV/bond. This is close to the energy of an Si–O bond (3.8 eV, Pauling, 1964) or an Si–Si bond (2.32 eV, Harrison, 1980). Stengl *et al.* (1989) suggest this stage involves the conversion of hydrogen-bonded silanol groups to siloxane bonds. Lasky had suggested the eventual dissolution of oxygen into the Si lattice. We'll later describe more detailed studies (Ahn *et al.*, 1989) showing the circumstances under which this oxygen dissolution occurs.

An interesting, potentially testable prediction of Stengl *et al.* (1989a) is that the distance between the wafers decreases by a factor of two after each reaction step, with the ultimate distance of 0.16 nm set by the Si–O bond length.

b. Measurement of Bond Strength

By far the most common technique for measuring bond strength is the crack propagation method described in Maszara *et al.* (1988) using Griffith's crack theory as elaborated upon by Gillis and Gilman (1964). Briefly stated, this method involves forcing a knife edge between the bonded wafers. The surface energy (bond strength) is proportional to y^2/L^4, where y is the thickness of the knife edge and L is the length of the resulting crack. Several groups report similar bond strengths for similar bonding conditions (Lehmann *et al.*, 1989; Harendt *et al.*, 1990, 1991). Other techniques include pressurized membrane burst tests (Shimbo *et al.*, 1986) and helium leak tests of enclosed cavities.

As bonding can occur between Si wafers with nothing but native oxide layers present, it is possible for the remaining oxygen to dissolve completely into the bulk Si, resulting in a Si–Si bonded interface. Ahn *et al* (1989) investigated the circumstance under which this occurs. Both float-zone and Czochralski substrates were investigated, and the authors found that the dissolution of the oxide layers depends on the oxygen interstitial concentration in the substrate. Float-zone substrates, with low oxygen concentrations, were capable of dissolving 3 nm of interfacial oxide after annealing at 1,150°C for 10 days. However, Czochralski Si, with a larger concentration of interstitial oxygen, actually showed growth in the interfacial oxide layer thickness. Accompanying these changes were disintegration of the layer into roughly spherical oxide precipitates at the interface. Similar effects occurred for thicker SiO_2 layers. If the wafers were misoriented, an amorphous layer remained at the interface, probably contributing to strain relaxation. Others have shown high-resolution TEM micrographs of the bonded interface (Shimbo *et al.*, 1986; Black *et al.*, 1987).

c. Techniques for Bonding and Particulate Contamination

Techniques for observing voids and bubbles during the bonding process and at bonded interfaces include IR microscopy or imaging (Black *et al.*, 1988; Stengl *et al.*, 1988; Mitani *et al.*, 1990; Harendt *et al.*, 1990; Bengtsson and Engström, 1990; Harendt *et al.*, 1991b, x-ray topography (Yamada *et al.*, 1991; Mitani *et al.*, 1990; Okabayashi *et al.*, 1990) acoustic microscopy (Black *et al.*, 1988; Okabayashi *et al.*, 1990), and optical projection (Okabayashi *et al.*, 1990). Several kinds of particles may be present in bonding (Shimbo *et al.*, 1986; Black *et al.*, 1988). The most complete study of interfacial bubbles/voids is by Mitani *et al.* (1990). They rule out surface roughness as a cause of interfacial voids in commercial Si wafers because the bonding energy should be sufficient to overcome wafer distortion energy (this becomes less important as one of the bonded pair becomes thinner — a thinner wafer can be distorted more than a thicker wafer with the same amount of energy). Instead, the three primary causes of bubble formation are trapped air, particulates, and surface contamination. Trapped air can be eliminated by initiating bonding from the center of the wafer out. Particulates can be eliminated either by sufficiently stringent cleanroom discipline or by techniques such as the micro-cleanroom spinner reported by Stengl *et al.* (1988). In this apparatus, the wafers to be bonded are mounted horizontally in a rack separated by Teflon spacers. The whole rack is subjected to the hydrophilization treatment described previously, and then flushed with DI water for several minutes at sufficient flow to inhibit the formation of interface voids. After flushing, a transparent cover is fitted to the rack, and the rack is mounted on a spinner. The transparent cover allows the wafers to be slightly heated by an infrared lamp during spin-drying. Bonding then is initiated by depressing the center of the top wafer to contact the bottom and removing the spacers. After removing the pair from the rack, bonding is completed by an appropriate annealing. As Stengl *et al.* (1988) point out, this enclosed flushing, spin-drying, and contacting eliminates the need for the stringent large-scale cleanroom conditions necessary to avoid particulate contamination.

The final source of interface voids and bubbles is surface contamination, particularly from adsorbed hydrocarbons (Mitani *et al.*, 1990). Even in wafer pairs showing no interface voids at room temperature, this group found two types of bubbles forming after 200–600°C anneals: up to 15 mm diameter bubbles visible through IR imaging, and small 1 mm bubbles only detectable through x-ray topography. Both types of voids apparently dissolve into the bulk Si with high-temperature ($>1,000°C$) annealing. Previously, it had been suggested (see Stengl *et al.*, 1989a) that these bubbles arose from interfacial water molecules during the intermediate stage of bond formation. However, variability with wafer supplier and other factors led Mitani et al. (1990) to suggest that these voids arise from hydrocarbons adsorbed upon the surface (Grundner and Jacob, 1986) during storage, transport, or handling. While proper wet chemical treatment could eliminate the large bubbles, only a 600–800°C anneal prior to initiation bonding was successful in reducing the smaller voids occurring after bonding.

d. Electrical Properties of the Interface

Assessment of the electrical properties of the bond began with Lasky's first report (1986) and continue. Two types of interfaces may be obtained; if thick oxide layers are used for bonding, MOS-like structures result (Lasky, 1986; Bengtsson and Engström, 1989, 1990; Black et al., 1988; Maszara et al., 1988). Capacitance–voltage studies of MOS capacitors indicate the best bonded interfaces have about 10^{11} cm^{-2} electrically active acceptor (negatively charged) defects at the bonded oxide interface, about one defect in every 10^4 bonds. When surfaces covered with thinner native oxide are studied, transport normal to the interface is usually investigated. The situation then depends on how much of the interfacial oxide layer remains after bonding, and on the doping of the adjacent Si. If the oxide remains intact, the thin 2.5 nm layer serves as a tunneling barrier to current flow (Stengl et al., 1989b). If the oxide at least partially disintegrates, the interfacial defects may give rise to a grain boundary-like barrier, provided both sides have the same type of doping (Bengtsson and Engström, 1989, 1990). Alternatively, the oxide could contribute a space charge generation–recombination current if the bond occurs in a *pn* junction depletion region (Bengtsson and Engström, 1989). In each case, the results imply the existence of approximately 10^{11} cm^{-2} electrically active acceptors at the bond interface. However, dopant diffusion occurring at the high annealing temperatures usually displaces the *pn* junction from the actual bond. The effect of the bond then depends on the doping level of the surrounding level: Heavy doping makes the barrier undetectable (Shimbo et al., 1986), while lower doping levels give rise to the aforementioned grain boundary effect (Bengtsson and Engström, 1989).

An interesting aside to the electrical properties of bonded interfaces has arisen in the work of Bengtsson and Engström (1989, 1990). They report bonding of Si wafers with hydrophobic (H-terminated) surfaces (described earlier). Such bonds were described as having smaller adhesive forces than hydrophilically bonded surfaces, but showed considerably reduced electrical defect density. This is consistent with what might be expected from hydrogen passivation of the dangling Si bonds. Bengtsson and Engström (1989) offered residual hydrophilic bonds as the origin of the bonding of the hydrophobic surfaces, consistent with the presence of some electrical defects.

14. Applications of Bonding

In keeping with the emphasis of this chapter on the techniques for the micromachining of Si, we'll only briefly discuss the applications of silicon direct bonding. Two principal applications of direct bonding have emerged:

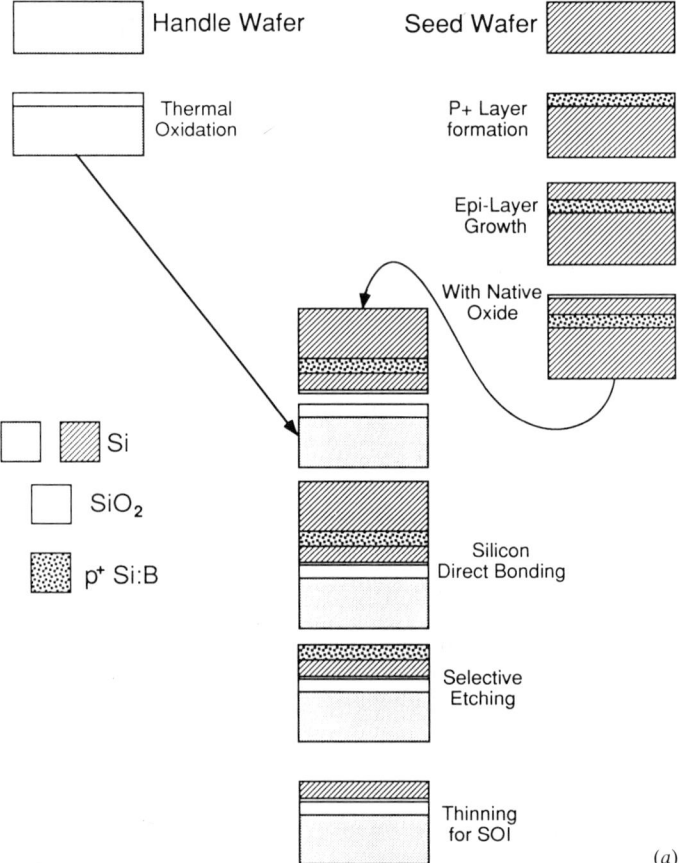

FIG. 7. Applications of silicon direct bonding, (a) Bond and etch back Si on insulator. (b) Comparison of sacrificial layer versus direct bonding for membrane formation (after Barth, 1990).

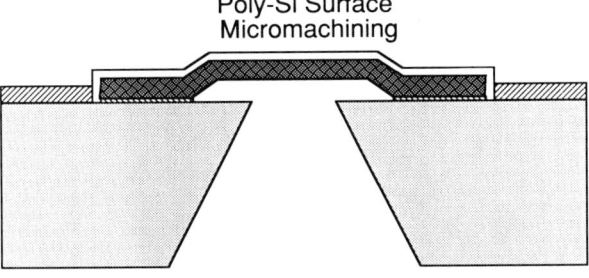

Poly-Si Surface Micromachining

- Nonplanar surface
- Long Etch Times for Sacrificial Spacer layers
- Back-side through-hole etching
- Residual Strains present
- Thermally Mismatched materials

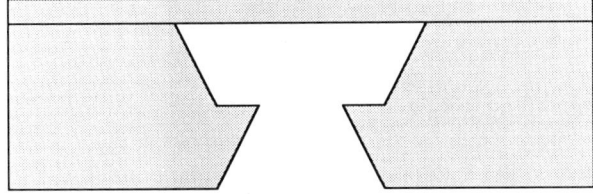

Si Direct Bonding Micromachining

- Planar surface
- Shorter Etch Times
- Top and Back Side Hole Etchings
- No Residual Strains
- Thermal Expansion Match

INTERFACE VOID FORMATION

FIG. 7. (*b*)

silicon-on-insulator (SOI) for high-speed radiation-resistant devices, and membranes for pressure sensors. Both require that one of the wafers be thinned by some etching or machining technique after bonding in order to reach the final configuration.

Silicon-on-insulator was the application first proposed by Lasky (1986). It involves attaching a device layer of Si to an insulating substrate. The substrate, referred to by Lasky (1986) as the handle wafer, is typically a thermally oxidized Si wafer with the oxide providing the dielectric insulation. As shown in Fig. 7a, the bonding proceeds as described earlier, with the attached "seed" wafer thinned to a thickness of typically 1 μm. Of the several different approaches developed for thinning the seed wafer, most involve preparing an etch-stop layer in the seed. The simplest approach is to bury a heavily boron-doped p^+-Si layer in the seed, remove the seed to the etch stop with a combination of nonselective and selective etches described previously, and then selectively remove the etch-stop layer. This approach has become known as "bond and etch back SOI" or BESOI.

Heavy doping of Si introduces a considerable number of defects such as dislocations and results in some boron diffusion into the device layer. Recently, other etch-stop layers have been reported, such as $Si_{1-x}Ge_x$ alloys using strain-sensitive etches (Godbey et al., 1990). However, the lattice mismatch occurring in such layers also could result in dislocation generation. Finally, an approach involving numerically controlled machining or polishing has also been reported (Yamada et al., 1991), although submicrometer polishing precision has yet to be demonstrated. Barth (1990) offers some cogent comments on the technological suitability of BESOI in comparison with other SOI technologies.

The other principal application of direct silicon bonding is in the preparation of membranes for accelerometers or pressure sensors. In this case, direct bonding and etch-back can replace the complex sequences of depositions and etching of sacrificial layers. Barth (1990) gives a tabular comparison of surface micromachining based on sacrificial layer techniques and direct bonding techniques for manufacture of such devices (Fig. 7b). Direct wafer bonding produces stress-free membranes without complex deposition and etching processing steps (Guckel et al., 1988), and hence has the potential of becoming the technology of choice for such structures.

15. Compound Semiconductors and Other Materials

Bonding of GaAs to Si enables researchers to marry optoelectronic devices based on GaAs with VLSI circuits based on Si. However, until recently, relatively few published studies have appeared on the subject. The most complete (Lehmann et al., 1989) describes bonding GaAs and InP to Si using the techniques of Si direct bonding. The only modification of the Si–Si bonding process was limiting the annealing step to low temperature ($<150°C$) to prevent loss of the group V element from the compound semiconductor. Bonding strengths were slightly below the strength of equivalently treated Si–Si wafer bonds and on the order of 0.2 Jm^{-2}. This is large enough to permit thinning of the GaAs by grinding or polishing to reduce stresses induced by thermal mismatch.

Another approach to bonding of GaAs has recently been developed by Yablonovitch et al. (1987, 1990). As shown in Fig. 8a and Fig. 8b, it is a two-

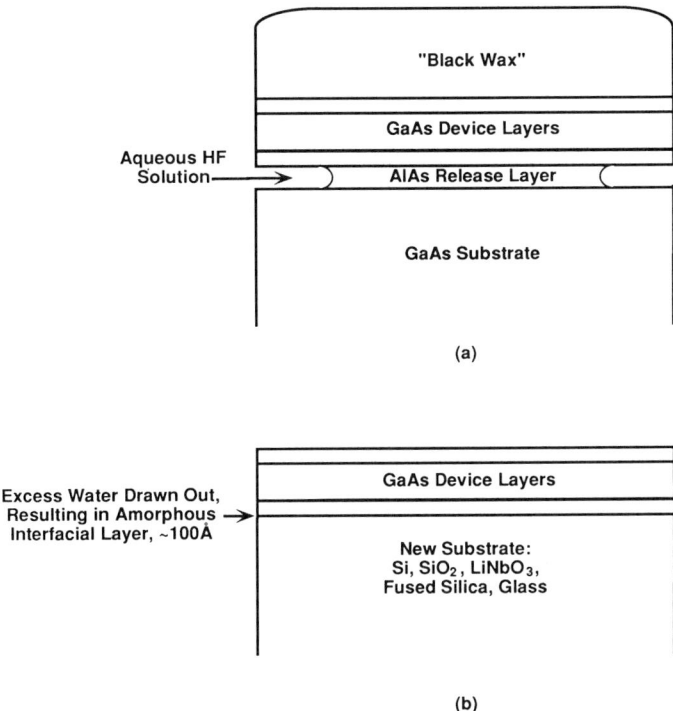

FIG. 8. (a) The epitaxial liftoff and (b) van der Waals bonding process for compound semiconductors as developed at Bellcore by E. Yablonovitch (1987, 1990).

step process that first uses the extreme selectivity in etch rates that HF exhibits for AlAs over GaAs. Since AlAs and GaAs are essentially lattice-matched, no strain or defects arise from the etch-stop or etch-release layer as is the situation for Si. Wax is used to place a slight compressive strain on the layer to be lifted off so HF can successfully undercut the whole layer. An analysis offered by Yablonovitch *et al.* (1987) argues that the thinner the etch release layer, the more successful the undercut, and layers as large as 2 cm × 4 cm were lifted off using AlAs release layers 50 nm thick (Yablonovitch *et al.*, 1990). By using strained, but not dislocated, layers of AlAs in InP-based structures, liftoff could also be obtained for those materials.

The second step involves bonding the film to the new substrate (Fig. 8b). Glass, sapphire, diamond, $LiNbO_3$, InP, and Si were all used successfully as substrates, with the only criteria being smoothness and a hydrophilic surface (as defined by a wetting angle of less than 90°). Once the film is released, it is manipulated into position over the substrate, and pressure is applied to initiate the bonding process. After most of the water is removed from the interface by pressure, the remaining water eventually disappears, leaving the film bonded to the new substrate. As is known from Si bonding studies, the thinness of the GaAs film (as little as 20 nm) minimizes the film stress necessary for conforming with the microscopically rough substrate. While no overt annealing is necessary, some baking appears to help eliminate the excess water (Yablonovitch et al., 1990), and given the proper substrate, the bonded combination survives some processing up to 725°C. The strength of the bond was not discussed other than to say the bond survived the "fingernail test" (Yablonovitch et al., 1987).

The exact nature of the bonding mechanism is in doubt. Yablonovitch et al. (1987, 1990) call the process "van der Waals (VDW) bonding due to the presumed role of interfacial forces." Smith and Gussenhoven (1965) also postulated a van de Waals mechanism for optical contacting when they observed bonding in a vacuum, although they did not appear to make any attempt to remove water from their optical surfaces. However, van der Waals bonding describes a rather specific type of bonding mechanism (see Kittel, 1976, for instance). In light of the crucial role of water in both VDW bonding and Si direct bonding, and the understanding of hydrogen bonding in silicon direct bonding, it seems probable that these two bonding techniques share some underlying physical mechanisms. GaAs has a more complicated native oxide structure than Si; several types of Ga oxides are possible, and As oxides dissolve in water. It seems quite likely that residual water molecules would remain adsorbed to the surface and could contribute to the formation of hydrogen-bond links with the substrate. High-resolution TEM micrographs of the interface between VDW-bonded GaAs and Si or GaAs substrates shows the formation of an amorphous interlayer of presumably mixed semiconductor oxide composition (Yablonovitch et al., 1990), but the resulting interfacial roughness might be greater than that in silicon direct bonding.

16. CRITICAL QUESTIONS

Clearly, the techniques for bonding of both GaAs and Si are still evolving. The incomplete understanding of the mechanisms of bonding has not inhibited use of bonding in micromachining applications. As applications widen, a natural synergy between the fundamental and the practical will continue to unfold. Since the review presented here will probably reflect the state of the art for but a short time, some speculation on critical questions in semiconductor bonding appears warranted.

In order to develop new combinations of bondable materials, we must move away from reliance on the physics and chemistry of the $Si-SiO_2$ system. Furthermore, the necessity for a high-temperature anneal limits the usefulness of Si direct bonding. While no other single materials combination possesses the range of useful properties of Si–SiO$_2$, an important opportunity for expanding both the processing parameters of bonding and the range of bondable materials combinations lies in the use of other solid-phase reactions for bonding. For instance, one report investigated Al, Ge, and Pt interlayers for the bonding of Si (Bhagat and Hicks, 1987), and another investigated solid-phase reactions between compound semiconductors (Lo et al., 1991). In particular, silicides probably offer new potential as interlayers for Si-compatible processing (e.g., Pt forms a silicide). The rich variety of known solid-phase reactions offer the ability to tailor such reactions to a number of processing conditions. We note that in GaAs, Pd has recently been used as such an interlayer (Yablonovitch et al., 1991). The multicomponent phase diagrams between compound semiconductors and metals greatly complicate the interface metallurgy, but Pd appears to form a Pd$_4$GaAs intermetallic compound that is thermodynamically stable.

V. Conclusions

Our imagination and engineering skills are both humbled and challenged as we reach into the many domains of future applications, whether in electronics, biology, botany, chemistry, or physics, or in some presently inconceivable combination of them all. The field of micromachining in all of its permutations elicits a vision that is hard to ignore — namely that we will hardly recognize the worlds of technology and science in another generation. The immense breadth of both micromachining techniques and applications has restricted our discussion here, but nevertheless, this work and some of the concepts and references cited herein must surely form a platform which we can begin to glimpse some of those worlds.

Acknowledgments

The authors thank R. Smart, E. Yablonovitch, and G. R. de Guel for helpful discussions.

References

Abu-Zeid, M. M., Kendall, D. L., de Guel, G. R., and Galeazzi, R. (1985). *Extended Abstracts 85-1, Electrochem. Soc. Toronto*, Abstract 275.
Ahn, K.-Y., Stengl, R., Tan, T.-Y., Smith, P., and Gösele, U. (1989). *J. Appl. Phys.* **65**, 561.

Akamine, S., Albrecht, T. R., Zdeblick, M. J., and Quate, C. F. (1990). *Sensors and Actuators* **A21–A23**, 964.
Albrecht, T. R., Akamine, S., Carver, T. E., and Quate, C. F. (1990). *J. Vac. Sci. Technol.* **A8**, 3386.
Allen, M. G., Scheidl, M., Smith, R. L., and Nokolich, A. D. (1990). *Sensors and Actuators* **A21–A23**, 211.
Anthony, Thomas R. (1983). *J. Appl. Phys.* **54**, 2419.
Arikado, T., Horioika, K., Sekine, M., Okano, H., and Horiike, Y. (1988). *Jpn. J. Appl. Phys.* **27**, 95.
Ashby, C. I. H. (1987). *AIP Conf. Proc.* #*160*, 586.
Bart, S. F., Tavrow, L. S., Mehregany, M., and Lang, J. H. (1990). *Sensors and Actuators* **A21–A23**, 193.
Barth, P. W. (1990). *Sensors and Actuators* **A21–A23**, 919.
Barth, P. W., and Shlichta, P. J. (1987). *NASA Tech. Briefs*, p. 72.
Bassous, E., (1978). *IEEE Trans. on Electron Devices* **ED–25**, 1178.
Bean, K. E. (1978). *IEEE Trans. on Electron Devices* **ED–25**, 1185.
Bengtsson, S., and Engström, O. (1989). *J. Appl. Phys.* **66**, 1231.
Bengtsson, S., and Engström, O. (1990). *J. Electrochem. Soc.* **137**, 2297.
Bhagat, J. K., and Hicks, D. B. (1987). *J. Appl. Phys.* **61**, 3118.
Black, R. D., Arthur, S. D., Gilmore, R. S., Lewis, N., Hall, E. L., and Lillquist, R. D. (1988). *J. Appl. Phys.* **63**, 2773.
Bloomstein, T. M., and Ehrlich, D. J. (1991). *Proc. of Transducer '91, San Francisco*, pp. 507–511.
Borom, M. P. (1973). *J. Amer. Ceram. Soc.* **56**, 254.
Brooks, A. D., Donovan, R. P., and Hardesty, C. A. (1972). *J. Electrochem. Soc.* **119**, 545.
Burrows, V. A., Chabal, Y. J., Higashi, G. S., Raghavachari, K., and Christman, S. B. (1988). *Appl. Phys. Lett.* **53**, 998.
Carlile, R. N., Liang, V. C., Palusinski, O. A., and Smedi, M. M. (1988). *J. Electrochem. Soc.* **135**, 2058.
Chin, D., Dhong, S. H., and Long, G. J. (1985). *J. Electrochem. Soc.* **132**, 1705.
Clark, L. D., Jr., Lund, J. L., and Edell, D. J. (1987). *Proc. IEEE Micro Robots and Teleoperators Workshop*, pp. 1–6.
Coburn, J. W. (1982). *American Vacuum Society Monograph Series* (N. Rey Whetten, ed.). American Vacuum Society.
Coburn, J. W., and Kay, E. (1979). *Solid State Technology* **22**(4), 117.
Coburn, J. W., and Winters, H. F. (1990). *Appl. Phys. Lett.* **55**, 2730.
Coldren, L. A., Furuya, K., and Miller, B. I. (1983). *J. Electrochem. Soc.* **130**, 1918.
de Guel, G. R. (1990). University of New Mexico, unpublished data.
de Rooij, N. (1991). *Proc. of Transducers '91, San Francisco*, pp. 8–13.
Delapierre, G. (1989). *Sensors and Actuators* **17**, 123.
DeNee, P. B. (1969). *J. Appl. Phys.* **40**, 5396.
Dieleman, J. and Sanders, F. H. M. (1984). *Solid State Technology* **27**(4), 191.
Ehrfeld, W., Goetz, F., Munchmeyer, D., Schleb, W., and Schmidt, D. (1988). *Technical Digest IEEE Solid State Sensor and Actuator Workshop, Hilton Head, South Carolina*.
Ehrlich, D. J., and Tsao, J. Y., Eds. (1989). "Laser Microfabrication: Thin Film Processes and Lithography," Academic Press, New York.
Esashi, M., Nakano, A., Shoji, S., and Hebiguchi, H. (1990). *Sensors and Actuators* **A21–A23**, 931.
Fan, L.-S., Tai, Y.-C., and Muller, R. S. (1988). *IEEE Trans. on Electron Devices* **35**, 724.
Field, L. A., and Muller, R. S. (1990). *Sensors and Actuators* **A21–A23**, 935.
Fonash, S. J. (1985). *Solid State Technology* **28**(1), 150.
Frye, R. C., Griffith, J. E., and Wong, Y. H. (1986). *J. Electrochem. Soc.* **133**, 1673.

Fujita, H., and Gabriel, K. J. (1991). *Proc. of Transducers '91, San Francisco*, pp. 14–20.
Gillis, P. P., and Gilman, J. J. (1964) *J. Appl. Phys.* **35**, 647.
Glembocki, O. J., Palik, E. D., de Guel, G. R., and Kendall, D. L. (1991). *J. Electrochem. Soc.* **138**, 1055.
Godbey, D. J., Twigg, M. E., Hughes, H. L., Palkuti, L. J., Leonov, P., and Wang, J. J. (1990). *J. Electrochem. Soc.* **137**, 3219.
Gossink, R. G. (1978). *J. Amer. Ceram. Soc.* **61**, 539.
Grundner, M., and Jacob, H. (1986). *Appl. Phys.* **A39**, 73.
Guckel, H., Burns, D. W., Visser, C. C. G., Tilmans, H. A. C., and Deroo, D. (1988). *IEEE Trans. Electron Devices* **ED-35**, 800.
Guckel, H., Sniegowski, J. J., Christenson, T. R., and Raissi, F. (1990). *Sensors and Actuators* **A21–A23**, 346.
Guilinger, T. R., Kelly, M. J., Granstaff, V. E., Peterson, D. W., Sweet, J. N., and Tuck, M. R. (1991). *Extended Abstracts 91–2, Electrochem. Soc. Phoenix, Arizona*, Abstract 354.
Hamaguchi, T., Endo, N., Kimura, M., and Ishitani, A. (1984). *Jpn. J. Appl. Phys.* **23**, L815.
Harendt, C., Graf. H.-G., Penteker, E., and Höfflinger, B. (1990). *Sensors and Actuators* **A21–A23**, 927.
Harendt, C., Höfflinger, B., Graf, H.-G., and Penteker, E. (1991a). *Sensors and Actuators* **A25–A27**, 87.
Harendt, C., Hunt, C. E., Apple, W., Graf, H.-G., Höfflinger, B., and Penteker, E. (1991b). *J. Electron. Mater.* **20**, 267.
Harrison, W. A. (1980). "Electronic Structure and the Properties of Solids," p. 171. Freeman, San Francisco.
Herb, G. K., Reiger, D.J., and Shields, K. (1987). *Solid State Technol.* **30**(10), 109.
Huang, Q.-A., Lu, S.-H., and Tong, Q.-Y. (1990). *Sensors and Actuators* **A21–A23**, 40.
Iler, R. K. (1974). "The Chemistry of Silica," Wiley, New York.
Jackson, T. N., Tishler, and Wise, K. D. (1981). *IEEE Electron Device Lett.* **EDL-2**, 44.
Jerman, J. H. (1990). *Sensors and Actuators* **A21–A23**, 988.
Kanda, Y., Matsuda, K., Murayama, C., and Sugaya, J. (1990). *Sensors and Actuators* **A121–A23**, 939.
Kelly, M. J., Guilinger, T. R., Stevenson, J. O. (1991). *Extended Abstracts 91–2, Electrochem. Soc. Phoenix*, Abstract 355.
Kendall, D. L. (1972). U.S. Patent 3,674,995.
Kendall, D. L. (1975). *Appl. Phys. Lett.* **26**, 195.
Kendall, D. L. (1979). *Annual Rev. of Mater. Sci.* **9**, 373.
Kendall, D. L. (1983). *Proc. of the Midwest Symposium on Circuits and Systems, Puebla, Mexico*, pp. 536–539. Also in "Proc. of the Workshop on Molecular Electronic Devices II," (F. L. Carter, ed.). Naval Research Lab, Washington, D.C., 1983.
Kendall, D. L. (1990). *J. Vac. Sci. Technol.* **A8**, 3598.
Kendall, D. L., and de Guel, G. R. (1985) "Micromachining and Micropackaging of Transducers" (Fung, C. D., Cheung, P. W., Ko, W. H. and Fleming, D. G., eds.), pp. 107–124. Elsevier Science, Amsterdam.
Kendall, D. L., de Guel, G. R., Guel-Sandoval, S., Garcia, E. J., and Allen, T. A. (1988). *Appl. Phys. Lett.* **52**, 836.
Kimura, M., Egami, K., Kanamori, M., and Hamaguchi, T. (1983). *Appl. Phys. Lett.* **43**, 263.
Kinoshita, H., and Jinno, K. (1977). *Jpn. J. Appl. Phys.* **16**, 381.
Kittel, C. (1976). "Introduction to Solid State Physics," 5th Ed., pp 78–79. Wiley, New York.
Kloeck, B., Collins, S. D., de Rooij, N. F., and Smith, R. L. (1989). *IEEE Trans. on Electron Devices* **ED-36**, 663.
Lasky, J. B. (1986). *Appl. Phys. Lett.* **48**, 78.

Lee, K. (1991). *Extended Abstracts 91–2, Electrochem. Soc. Phoenix*, Abstract 352.
Lehmann, V., Mitani, K., Stengl, R., Mii, T., and Gösele, U. (1989). *Jpn. J. Appl. Phys.* **28**, 2141.
Lo, Y. H., Bhat, R., Hwang, D. M., Koza, M. A., and Lee, T. P. (1991). *Appl. Phys. Lett.* **58**, 1961.
MacFadyen, D. N. (1983). *J. Electrochem. Soc.* **130**, 1934.
Maszara, W. P., Goetz, G., Caviglia, A., and McKitterick, J. B. (1988). *J. Appl. Phys.* **64**, 4943.
Mayer, G. K., Offereins, H. L., Sandmaier, H. and Kuhl, K. (1990). *J. Electrochem. Soc.* **137**, 12.
Mehregany, M., Gabriel, K. J., and Trimmer, W. S. N. (1987). *Sensors and Actuators* **12**, 341.
Mehregany, M., Gabriel, K. J., and Trimmer, W. S. N. (1988). *IEEE Trans. on Electron Devices* **35**, 719.
Mehregany, M., Bart, S. F., Tavrow, L. S., Lang, J. H., and Senturia, S. D. (1990a). *J. Vac. Sci. Technol.* **A8**, 3614.
Mehregany, M., Bart, S. F., Tavrow, L. S., Lang, J. H., Senturia, S. D., and Schlecht, M. F. (1990b). *Sensors and Actuators* **A21–A23**, 173.
Michalske, T. A., and Fuller, E. R., Jr. (1985) *J. Amer. Ceram. Soc.* **68**, 586.
Mitani, K., Lehmann, V., and Gösele, U. (1990). *IEEE Solid State Sensors Workshop, Tech. Digest, Hilton Head, South Carolina*, p. 74.
Monk, D. J., Soane, D. S., and Howe, R. T. (1991). *Proc. of Transducers '91, San Francisco*, pp. 647–650.
Muller, (1990). *Sensors and Actuators* **A21–A23**, 1–8.
Muraoka, H., Ohhashi, T., and Sumimoto, Y. (1973). "Semiconductor Silicon 1973," pp. 327–338. *Electrochem. Soc.*
Ohashi, H., Furukawa, K., Atsuta, M., Nakagawa, A., and Imamvia, K. (1987). *Proc. IEEE International Electron Devices Meeting, '87*, Washington D.C., pp. 678–681.
Okabayashi, O., Shirotori, H., Sakurazawa, H., Kanda, E., Yokoyama, T., and Kawashima, M. (1990). *J. Crystal Growth* **103**, 456.
Palik, E. D., Faust, J. W., Gray, H. F., Greene, R. F. (1982) *J. Electrochem. Soc.* **129**, 2051.
Palik, E. D., Bermudez, V. M., and Glembocki, O. J. (1985) *J. Electrochem. Soc.* **132**, 871.
Pauling, L. (1964). "The Nature of the Chemical Bond," 3rd Ed., p. 85. Cornell, Ithaca, New York.
Petersen, K. E. (1982). *Proc. IEEE* **70**, 420.
Pister, K. S. J., Judy, M. W., Burgett, S. R., and Fearing, R. S. (1991). *Transducers '91, San Francisco*, late paper.
Prasad, A., Balakrishnan, S., Jain, S. K., and Jain, G. C. (1982). *J. Electrochem. Soc.* **129**, 596.
Proctor, T. M., and Sutton, P. M. (1960). *J. Amer. Ceram. Soc.* **43**, 173.
Raley, N. F., Sugiyama, Y., van Duzer, T. (1984). *J. Electrochem. Soc.* **131**, 161.
Reithmuller, W., and Benecke, W. (1988). *IEEE Trans. on Electron Devices* **35**, 758.
Roylance, L. M., and Angell, J. B. (1979). *IEEE Trans. on Electron. Devices* **ED–26**, 1911.
Schmidt, M. A., Howe, R. T., Senturia, S. D., and Haritonitis, J. H. (1988). *IEEE Trans. on Electron Devices* **35**, 750.
Schnakenberg, U., Benecke, W., and Lochel, B. (1990). *Sensors and Actuators* **A21–A23**, 1031.
Schnakenberg, U., Benecke, W., and Lange, P. (1991). *Proc. of Transducers '91, San Francisco*, pp. 815–818.
Seidel, H., Csepregi, L., Heuberger, A., and Baumgaertel, H. (1990). *J. Electrochem. Soc.* **137**, 3612 and 3626.
Shaw, D. W. (1979). *J. Crystal Growth* **47**, 509.
Shimbo, M., Furukawa, K., Fukuda, K., and Tanzawa, K. (1986). *J. Appl. Phys.* **60**, 2987.
Singh, R., Fonash, S. J., Ashok, S., Caplan, P. J., Shappirio, J., Hage-Ali, M., and Ponpon, J. (1983). *J. Vac. Sci. Technol.* **A1**, 334.
Smart, R., and Ramsay, J. V. (1964) *J. Scientific Instruments* **41**, 514.
Smeltzer, R. K. (1975). *J. Electrochem. Soc.* **122**, 1666.

Smith, H. I. (1965). *J. Acoustical Soc. Am.* **37**, 928.
Smith, H. I., and Gussenhoven, M. S. (1965). *J. Appl. Phys.* **36**, 2326(c).
Smith, R. L., Fulmer, J., and Collins, S. D. (1991). *Extended Abstracts 91-2, Electrochem. Soc., Phoenix,* Abstract 353.
Snow, E. H., Grove, A. S., Deal, B. D., and Sah, C. T. (1965) *J. Appl. Phys.* **36**, 1664.
Snow, E. H., and Dumesnil, M. E. (1966). *J. Appl. Phys.* **37**, 2123.
Spangler, L. J., and Wise, K. D. (1987). *IEEE Electron Device Letters* **EDL-8**, 137.
Stengl, R., Ahn, K.-Y., and Gösele, U. (1988). *Jpn. J. Appl. Phys.* **27**, L2364.
Stengl, R., Ahn, K.-Y., Mii, T., Yang, W. S., and Gösele, U. (1989b). *Jpn. J. Appl. Phys.* **28**, 2405.
Stengl, R., Tan, T., an Gösele, U. (1989a). *Jpn. J. Appl. Phys.* **28**, 1735.
Sutton, P. M. (1964a). *J. Amer. Ceram. Soc.* **47**, 189.
Sutton, P. M. (1964b). *J. Amer. Ceram. Soc.* **47**, 219.
Tabata, O., Asahi, R., Funabashi, H., and Sugiyama, S. (1991). *Proc. of Transducers '91, San Francisco,* pp. 811-814.
Thomas, J. H., and Maa, T. S. (1983). *Appl. Phys. Lett.* **43**, 859.
Uetake, H., Matsuura, T., Ohmi, T., Murota, J., Fukuda, K., and Mikoshiba, N. (1990). *Appl. Phys. Lett.* **57**, 596.
van Dijk, H. J. A., and de Johnge, (1970). *J. Electrochem. Soc.* **117**, 2556.
Wallis, G. (1970). *J. Amer. Ceram. Soc.* **53**, 563.
Wallis, G., and Pomerantz, D. I. (1969). *J. Appl. Phys.* **40**, 3946.
Wang, S. S., McNeil, V. M., Schmidt, M. A. (1991). *Extended Abstracts 91-2, Electrochem. Soc. Phoenix,* Abstract 356. Also in *Proc. of Transducers '91, San Francisco,* pp. 819-822.
Weirauch, D. F. (1975). *J. Appl. Phys.* **46**, 1478.
Wise, K. D., and Najafi, K. (1991). *Science* **254**, 1335. Also in *Proc. of Transducers '91,* pp. 2-7.
Wohl, G., Mattes, M., and Weisheit, A. (1988). *Vacuum* **38**, 1011.
Yablonovitch, E., Allara, D. L., Chang, C. C., Gmitter, T., and Bright, T. B. (1986). *Phys. Rev. Lett.* **57**, 294.
Yablonovitch, E., Gmitter, T. J., Harbison, J. P., and Bhat, R. (1987). *Appl. Phys. Lett.* **51**, 2222.
Yablonovitch, E., Hwang, D. M., Gmitter, T. J., Florez, L. T., and Harbison, J. P. (1990). *Appl. Phys. Lett.* **56**, 2419.
Yablonovitch, E., Sands, T., Hwang, D. M., Schnitzer, I., Gmitter, T. J., Shastry, S. K. Hill, D. S., and Fan, J. C. C. (1991). *Appl. Phys. Lett.* **59**, 791.
Yamada, A., Kawasaki, T., and Kawashima, M. (1987a). *Electron. Lett.* **23**, 39.
Yamada, A., Kawasaki, T., and Kawashima, M. (1987b). *Electron. Lett.* **23**, 315.
Yamada, A., Jiang, B.-L., Rozgonyi, G. A., Shirotori, H., Okabayashi, O., and Kawashima, M. (1991). *J. Electrochem. Soc.* **138**, 2468.

CHAPTER 8

Processing and Semiconductor Thermoelastic Behavior

Ikuo Matsuba and Kinji Mokuya*

SYSTEMS DEVELOPMENT LABORATORY, HITACHI, LTD.
KAWASAKI-SHI, JAPAN

I. INTRODUCTION	339
1. General Considerations	339
II. THERMOELASTIC MODEL OF DISLOCATIONS	342
2. Wafer Temperature Model	342
3. Steady Temperature Profile	345
4. Transient Temperature Behavior	346
5. Thermal Stress Model	348
III. PREDICTION OF DEFECT ONSET	355
6. Estimation of Critical Stress	355
7. Comparison with Experiments	358
8. Dependence of Critical Stress on Strain Rate	359
IV. APPLICATION OF THERMOELASTIC MODEL	361
9. Optimization of Heat Processing	361
V. CONCLUSIONS	365
ACKNOWLEDGMENTS	366
REFERENCES	366
LIST OF SYMBOLS	367

I. Introduction

1. GENERAL CONSIDERATIONS

For many years the standard technique has been used for annealing and growing of silicon oxide films in the fabrication of large-scale integrated circuits (Sze, 1981). The silicon wafers are held vertically on a virtreous silica boat designed so that they sit in approximately an axially symmetric position in the silica tube of the diffusion furnace, as shown in Fig. 1. Besides cost and

---

* Present address: Takasaki Works, Hitachi, Ltd. 111 Nishiyokote-machi Takasaki-shi, 370-11 Japan

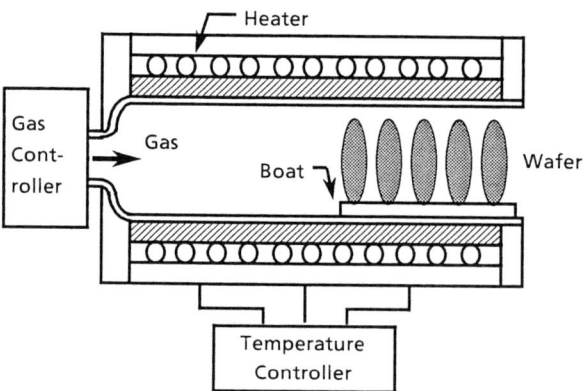

FIG. 1. Schematic diagram of diffusion furnace. The wafers on the boat are inserted into the diffusion furnace with a loading speed V.

reliability, a major advantage of this technique over other methods is uniformity of film thickness within a wafer and from wafer to wafer. The two most interesting variables are defects and film thickness, which depend on a number of process variables. The most important ones are temperature distribution in wafers and inlet gas partial pressure. It is a good approximation that a convective layer of gas ends at the surface of the cylinder that envelops the wafers, and that the gas is stagnant between them because the space between wafers is very small compared with the wafer diameter. Thus, the gas does not affect the wafer temperature. Hereafter, we are only concerned with the temperature distribution in the wafers. When a row of regularly spaced circular wafers is transferred from room temperature to the high temperature of a furnace, a transient temperature gradient is generated in each wafer. High-temperature processing of wafers often produces sufficient thermal stresses to generate defects such as slip and dislocations that reduce fabrication yield.

The wafer temperature distribution can not be measured directly, and only the tube wall temperature is routinely recorded by thermocouples. Therefore, it is difficult to determine how to control the boat loading speed into the furnace. A too-high boat speed will generate a thermal stress in wafers and will often exceed the critical stress level that is associated with permanent deformation, while, on the other hand, a very low speed requires excessive residence time in the furnace. Determining how the desired uniform temperature distribution has been achieved is even more of a problem. In current practice, the tube wall temperature and boat withdrawal time are based on certain rules of thumb and operator experience. Since, nowadays, the minimum line width of semiconductor devices is scaled down to 0.5 μm, it becomes more difficult to determine how to operate the processes. In the near future, it will be possible to apply on-line control to the furnace by measuring the dioxide film thickness or wafer temperature, directly.

Historically, theoretical analysis of wafer temperature was pioneered by Hu (1969), who studied a concentric row of wafers in a vacuum to obtain the wafer surface radial temperature distribution and the changes in the wafer shape due to thermal stress. He considered only the radiant losses of the wafer itself, the radiant exchanges between wafers, and the heat diffusion within the wafer. He ignored the reactor tube wall which otherwise appears as an external heat source. Aiming to improve on that and develop a practical process simulator, we (Matsuba et al., 1984) have proposed a model in which the temperature distribution of the wafers standing still in a furnace is in a steady state. This model was extended to simulate the transient behavior of wafer temperature in a subsequent paper (Matsuba et al., 1985).

Investigations of defect onset due to thermal stresses induced during transient periods in a furnace have also been reported (Morizane and Gleim, 1969). However, in these studies the temperature distribution over the wafer surface is assumed to be parabolic. In practice, the transient temperature distribution occurring when the wafers are inserted into a furnace does not have polar symmetry, but has circumferential dependence due to the boat jig supporting the wafers. It is necessary to consider this radiative heat exchange between the wafers and the boat (Leroy and Plaugonven, 1980). The characteristics of the transient wafer temperature distributions (Mokuya et al. 1985) depend on the heat treatment process conditions, namely, furnace temperature T_f, loading speed V, and wafer spacing H. For the purpose of predicting defect onset in actual cases, it is important to evaluate the thermal stresses corresponding to the parameters of these processes (Mokuya et al., 1986; Matsuba et al., 1986; Mokuya and Matsuba, 1989). Additionally, the critical stress of the target (target means the objective to be evaluated by the proposed method) wafer, namely, the threshold of defect onset, is considered to drift depending on the fabrication process. This is because of the different heat cycling and different initial strength of the wafers. Although the dependence of the critical stress on oxygen concentration has been studied in experiments (Mokuya and Matsuba, 1989), it is very difficult to determine the value of the target wafers since many factors are involved. As shown in Fig. 2, the critical stress of the target wafer is determined using the experimental data.

The paper is organized as follows: Firstly, a thermoelastic wafer model is proposed for predicting the conditions of defect onset during the heat cycle in a furnace. Then, an analysis of thermal stress characteristics simulated by the thermoelastic wafer model formulated for application is given in Section II (Mokuya et al., 1985). This model is formulated for application to the plane

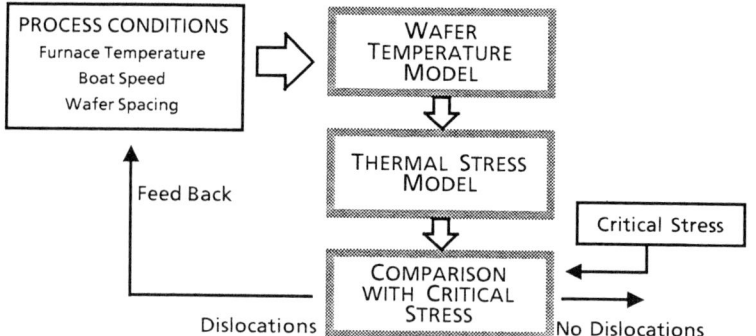

FIG. 2. Schematic diagram of the proposed system.

stress state under thermal loading in which the wafer temperature is calculated by the wafer temperature model. In Section III, using the proposed model, methods are described for determining the critical stress curve and for predicting the conditions of defect onset corresponding to the given heat process conditions. Predictions are executed by comparing the thermal stresses resolved on the slip systems of the silicon crystal under the process conditions. By applying to the Φ 125-mm wafer and Φ 150-mm wafer processes, it is shown that the thermal stress level is reduced to about half by increasing the wafer spacing by a factor of two or three. Accordingly, prediction of defect onset based on this model is shown to be in reasonable accordance with related experiments. In Section IV, an application to the real fabrication process is presented. Finally, conclusions are given in Section V.

II. Thermoelastic Model of Dislocations

2. Wafer Temperature Model

The N_w wafers are all of the same radius R_w and thickness ℓ_w. They are positioned vertically and are equally spaced apart from one another at a distance H in a furnace tube with a radius R_f, as shown in Fig. 1. It is assumed that the wafer thickness is small enough that the temperature distribution can be made uniformly across the thickness of the wafer, and that the thickness of the vitreous slab flat boat is also small enough that the temperature distribution can be made uniformly across the thickness of the boat. Let $T_i(\mathbf{r}, t)$, $T_b(\mathbf{r}, t)$, and $T_f(\mathbf{r}, z, t)$ denote the temperature distribution in the ith wafer, the boat, and the tube wall at time t, respectively, $\mathbf{r} = (x, y)$

8. PROCESSING AND SEMICONDUCTOR THERMOELASTIC BEHAVIOR

being the two-dimensional coordinates perpendicular to the furnace axis (z). With these assumptions, the dynamic behavior of the temperature distribution is described by the following nonlinear partial differential equation by using the two-dimensional Laplacian $\nabla^2 = \partial^2/\partial x^2 + \partial^2/\partial y^2$ (Matsuba et al., 1985):

$$\varepsilon \frac{\partial}{\partial t} T_i(\mathbf{r}, t) = -2T_i^4(\mathbf{r}, t) + 2a_w \int T_i^4(\mathbf{r}', t) F_s(\mathbf{r}, \mathbf{r}') \, d\mathbf{r}'$$

$$+ a_w \sum_{j=i\pm 1} \int T_j^4(\mathbf{r}', t) F_a(\mathbf{r}, \mathbf{r}') \, d\mathbf{r}'$$

$$+ \varepsilon_{fw} a_w \int T_f^4(\mathbf{r}'', z'', t) F_{wf}(\mathbf{r}, \mathbf{r}'', t) \, d\mathbf{r}'' \qquad (1)$$

$$+ \varepsilon_{bw} a_w \int T_b^4(\mathbf{r}''', t) F_{wb}(\mathbf{r}, \mathbf{r}''') \, d\mathbf{r}'''$$

$$+ K_w \nabla^2 T_i(\mathbf{r}, t) \qquad (i = 2, 3, \ldots, N_w - 1),$$

$$\frac{\partial}{\partial t} T_b(\mathbf{r}, t) = -2T_b^4(\mathbf{r}, t) + \varepsilon_{wb} a_b \sum_j \int T_j^4(\mathbf{r}', t) F_{bw}(\mathbf{r}, \mathbf{r}') \, d\mathbf{r}'$$

$$+ \varepsilon_{fb} a_b \int T_f^4(\mathbf{r}'', z'', t) F_{bf}(\mathbf{r}, \mathbf{r}'', t) \, d\mathbf{r}'' + K_b \nabla^2 T_b(\mathbf{r}, t), \qquad (2)$$

$$e \frac{\partial}{\partial t} T_f(\mathbf{r}, z, t) = K_f(\nabla^2 + \partial^2/\partial z^2) T_f(\mathbf{r}, z, t) + Q(\mathbf{r}, z, t), \qquad (3)$$

with appropriate boundary conditions. For the end wafer ($i = 1$),

$$\varepsilon \frac{\partial}{\partial t} T_1(\mathbf{r}, t) = -2T_1^4(\mathbf{r}, t) + a_w \int T_1^4(\mathbf{r}', t) F_s(\mathbf{r}, \mathbf{r}') \, d\mathbf{r}'$$

$$+ a_w \int T_2^4(\mathbf{r}', t) F_a(\mathbf{r}, \mathbf{r}') d\mathbf{r}' + \varepsilon_{fw} a_w \int T_f^4(\mathbf{r}'', z'', t) F_{wf}(\mathbf{r}, \mathbf{r}'', t) \, d\mathbf{r}''$$

$$+ \varepsilon_{bw} a_w \int T_b^4(\mathbf{r}''', t) F_{wb}(\mathbf{r}, \mathbf{r}''') \, d\mathbf{r}''' + K_w \nabla^2 T_1(\mathbf{r}, t).$$

A similar equation is obtained for another end wafer ($i = N_w$). $\partial/\partial t$ denotes the time derivative. In the equations, $\varepsilon = c_w \rho_w \ell_w \varepsilon_b / c_b \rho_b \ell_b \varepsilon_w$, $e = c_f \rho_f / c_b \rho_b$, $\varepsilon_{bw} = \varepsilon_b/\varepsilon_w$, $\varepsilon_{fw} = \varepsilon_f/\varepsilon_w$, etc.; c_w, c_b, and c_f are the specific heat of the wafer, the boat, and the tube wall, respectively; ρ_w, ρ_b, and ρ_f are the density of the wafer, the boat, and the tube wall, respectively; ε_w, ε_b, and ε_f are the emittance of the wafer, the boat, and the tube wall, respectively; a_w, a_b, and a_f are the absorbance of the wafer, the boat, and the tube wall, which are equal to ε_w, ε_b, and ε_f in thermal equilibrium, respectively; σ is the Stefan–Boltzmann constant, $F_s(\mathbf{r}, \mathbf{r}')$ is the geometric factor governing the radiant heat transfer from one element to another element in the same wafer via reflections between neighboring wafers, $F_a(\mathbf{r}, \mathbf{r}')$ is the geometric factor between two elements in neighboring wafers, $F_{wf}(\mathbf{r}, \mathbf{r}'', t)$ is the geometric factor between one element in the wafer and one element in the tube wall, $F_{wb}(\mathbf{r}, \mathbf{r}''')$ is the geometric factor between the wafer and the boat, $d\mathbf{r} = dx\,dy$ is an infinitesimal element on the wafer, $d\mathbf{r}'' = R_f d\theta\,dz''$ is an infinitesimal element on the tube wall, and $d\mathbf{r}''' = dx\,dz$ is an infinitesimal element on the boat surface. Similar notations are used in Eq. (2). In the right-hand side of Eq. (1), the first term is the surface radiant heat loss from the wafer, the second term is the radiant interchange between two elements in the same wafer via reflection, the third term is the radiant interchange between two neighboring wafers, the fourth term is the radiant heat received from the tube wall, the fifth term is the radiant interchange between the wafers and the boat, and the sixth term is the thermal diffusion in the wafers. Similarly, each term in Eqs. (2) and (3) is easily understood. The factor 2 comes from the fact that the radiant heat transfer takes place on both surfaces of the wafers. The transient wafer temperature distributions are mainly determined by the effects of the rate of radiation increase localized in a wafer over time and the radiative exchange between the wafers and the boat. The dimensionless variables $\mathbf{r}$, t, T_i scaled by R_w, $c_b\rho_b\ell_b/\varepsilon_b\sigma T_0^3$, T_0, respectively, and $K_w = k_w\ell_w/\varepsilon_b\sigma T_0^3 R_w^2$, $K_b = k_b\ell_b/\varepsilon_b\sigma T_0^3 R_w^2$ and $K_f = k_f/\varepsilon_b\sigma T_0^3 R_w$, T_0 being a certain reference temperature, where k_w, k_b and k_f are the thermal conductivity of the wafer, the boat and the tube wall, respectively, are used in the above equations. In Eq. (3), Q denotes the dimensionless supplied heat source, and the tube wall temperature is coupled with the wafer and the boat temperature through the following boundary condition at the tube wall surface:

$$-K_f \frac{\partial}{\partial r} T_f^4(\mathbf{r}, z, t)|_{r=R_f} = -T_f^4(\mathbf{r}, z, t) + \varepsilon_{fw} a_f \int_{r''=r} T_f^4(\mathbf{r}'', z'', t) F_{ff}(\mathbf{r}, \mathbf{r}'', t)\, d\mathbf{r}''$$

$$+ \varepsilon_{wf} a_f \sum_j \int T_j^4(\mathbf{r}', t) F_{fw}(\mathbf{r}, \mathbf{r}', t)\, d\mathbf{r}' \qquad (4)$$

$$+ \varepsilon_{bf} a_f \int T_b^4(\mathbf{r}''', t) F_{fb}(\mathbf{r}, \mathbf{r}'', t)\, d\mathbf{r}'',$$

where each term has the same meaning as in Eq. (1), and the notation $r = |\mathbf{r}|$ is used. It should be noted that in the steady state in which the wafers are held stationary, the effect of the boat was proved to be unimportant from the experiments. Therefore, Eq. (2) is not considered to be in a steady state. However, in the transient state, the heat exchange between the boat and the wafers plays an essential role to generate a distribution in temperature within the wafers. When the wafers are inserted in a furnace, the wafer temperature gradually increases from room temperature to a temperature of about 1,000°C. Since the temperature variation over time is not so large that the wafers exhibit a thermal shock phenomenon, the temperature distribution in the wafers for each time is assumed to be in a quasi-steady state.

3. Steady Temperature Profile

In order to investigate the steady wafer temperature, the experiments were performed with 4-inch wafers equally spaced at a distance $H = 4.8$ mm apart in the furnace at 1,200°C. The oxidation time is 900 minutes, which is long enough to obtain a steady state. Making use of the well-known Deal–Grove's formula (Sze, 1981), the wafer temperature is calculated from the dioxide film thickness measured with an Applied Materials Ellipsometer.

The results are summarized in Fig. 3 and Fig. 4 (Matsuba et al., 1984). In Fig. 3, the temperature in the end wafer seems to be transferred toward the central wafers with decreasing distribution but increasing absolute value, and thus the distribution becomes almost uniform in the central wafers (Fig. 4).

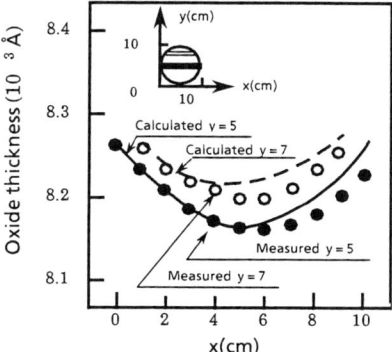

Fig. 3. Comparison of calculated and observed temperature distributions at the end wafer. [After Matsuba et al. (1984).]

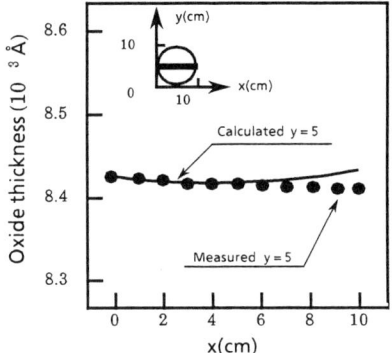

FIG. 4. Comparison of calculated and observed temperature distributions at the central wafer. [After Matsuba et al. (1984).]

For this reason the radiant heat transfer from the tube wall reflects many times between neighboring wafers and diffuses in the wafer; thus, the temperature distribution within wafers becomes uniform. On the other hand, the end wafer loses its heat out of the furnace in an axial direction and receives reflections on only one surface, and thus the distribution becomes highly nonuniform (Fig. 3). This shows that the emittance takes a small value compared with 1.0. It was also found that the temperature distribution was almost symmetric in the axial direction, and thus the dioxide film thickness distribution was determined mainly by the wafer temperature independent of O_2 density distribution in the steady state. These figures show that simulation results were in good agreement with experiments, and that the distribution is almost symmetric in the direction perpendicular to the boat surface. This is consistent with the assumption that the distribution in the steady state is not affected by the boat, as might be expected. It is worth noting that keeping the distance between the furnace edge and the end wafer and increasing the furnace length and thus, of course, the number of wafers, does not change the temperature distribution.

4. Transient Temperature Behavior

The temperature distribution within wafers plays an important role to generate dislocations in the transient state. It is well known that temperature distributions within wafers, which cause thermal stresses, appear during the transient stage of the heat cycle (Morizane and Gleim, 1969). This temperature distribution depends strongly on the heat capacity ratio of wafer to boat

(Matsuba et al., 1985). Before the numerical calculation is performed, the qualitative behavior of the wafer temperature is described on the basis of singular perturbation methods (Matsuba et al., 1985; Nayfeh, 1973). The parameter multiplied by the time derivative in Eq. (1) characterizes the wafer temperature transient behavior. This problem is a typical singular perturbation problem in which the parameter takes a small value. In this case the numerical value of ε is of order $O(10^{-1})$, taking into account the fact that $\varepsilon_b/\varepsilon_w$ is of order $O(1)$ for the silica boat. Note that $e > O(1)$. According to the so-called matched asymptotic expansion method (Nayfeh, 1973), time is divided into two regions, that is, (i) $0 < t < O(\varepsilon)$, the inner region, and (ii) $O(\varepsilon) < t < O(1)$, the outer region. As time goes on and $t = O(1)$ is achieved, the order of the system is reduced by neglecting the temperature dynamics to obtain the dynamic equation for the boat and the tube wall temperature and the steady equation for the wafer temperature. The solution of the reduced-order model is called the outer solution by analogy to the boundary layer theory in fluid mechanics. The reduced-order solution is unable to satisfy the initially conditions imposed in the original problem. This discrepancy is corrected by the inner solution in (i), the order of t being so small that the original equations can be linearized to obtain the linear equation, allowing rapid changes of the variables using the stretched time scale t/ε. The inner solution has to satisfy the violated initial conditions and match the outer solution. An additive composite of the outer solution in (ii) and inner solution in (i) by the so-called matching condition presents a uniformly valid approximation of the original problem.

In the inner region (i), introducing the scaled time $\tau = t/\varepsilon$ into the basic equations, the limit $\varepsilon \to 0$ is taken with τ fixed. Expanding T_i, T_b, and T_f in powers of ε around initial values, $\partial T_b/\partial \tau = 0$, $\partial T_f/\partial \tau = 0$ is obtained, and a linear equation for T_i up to the first order of ε. Therefore, T_i grows exponentially, while T_b and T_f remain at the initial values in the vicinity of the initial time. The wafer temperature reaches the order $O(1)$ at $t = O(\varepsilon)$ or $\tau = O(1)$, that is, the boundary between the inner and outer region. As time goes on further and the system goes into the outer region (iii), the original system can be approximated by expanding Eqs. (1)–(3) in powers of ε and retaining the first-order term, and then a nonlinear equation for T_i is obtained. Since T_b remains at the initial value in the inner region, it gradually grows in the vicinity of the boundary in the outer region, and thus it is found that the equation of the boat can be linearized. By applying the singular perturbation technique for large e, it turns out that T_f still remains at its initial value T_{f0}, and that if we assume that the temperature of the lower part of the wafer coincides with that of the boat at the contact place, the temperature difference within the wafers becomes of the order $T_{f0}(1 - A\exp(B\varepsilon))$, where A and B are positive constants. This conclusion indicates that the furnace

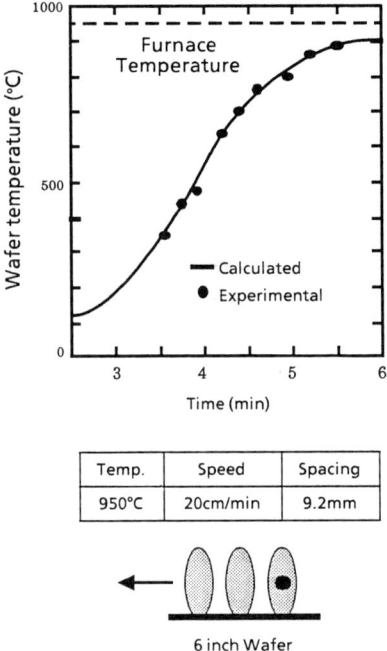

FIG. 5. Comparison of calculated and observed wafer temperature distributions at the center of wafer as a function of time. [After Mokuya et al. (1987).]

designer must reduce the heat capacity of the boat as much as possible in order to obtain wafers of good quality.

One comparison of simulated and observed temperature distribution at the center of the wafer as a function of time t is shown in Fig. 5. Simulation is performed with a spacing of $H = 9.2$ mm, a furnace temperature of $T_f = 950°C$, and a loading speed of $V = 20$ cm/min. The simulated results are in good agreement with experiments. Figure 6 shows the simulated and observed temperature distribution along the vertical direction of the wafer as a function of time t with the same process conditions as in Fig. 5. From this figure, the dislocation is likely to occur near the top of the wafer, since the thermal stress is nearly proportional to the temperature gradient.

5. THERMAL STRESS MODEL

It is assumed that the wafer is an isotropic and homogeneous elastic body. Thus, the thermoelastic wafer model is formulated as a plane stress problem based only on thermal loading. The plane stress components, $\{\sigma\} = \{\sigma_{xx}, \sigma_{yy}, \sigma_{xy}\}^T$ (the suffix T is the transform notation), are calculated by the finite-

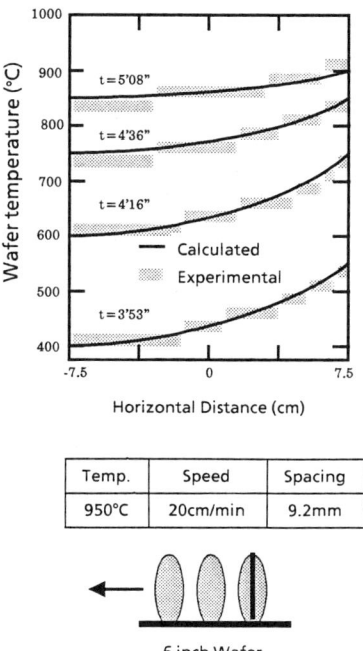

FIG. 6. Comparison of calculated and observed wafer temperature distributions at $x = 0$ and at different times. [After Matsuba et al. (1986).]

element method (Zienkiewicz, 1970) using triangular constant strain elements. The example of finite-element division for the half-plane of the wafer is shown in Fig. 7. Mesh points show the temperature definition point in the finite-difference method in the wafer temperature model (Matsuba et al., 1985, Mokuya et al., 1985). However, points not at mesh points are extrapolated from the mesh-point values. The model simulates only half of the area of one face of the wafer because of the y-axis symmetry of the temperature distribution. It is possible arbitrarily to choose a triangular element, assuming the temperatures given by the three nodes (i, j, k) calculated by the temperature model, such as T_{wi}, T_{wj}, and T_{wk}. The representative value of the element is given by $\frac{1}{3}(T_{wi} + T_{wj} + T_{wk})$. This temperature deviation ΔT from the arbitrary standard temperature value gives the initial strain, which is described by

$$[\varepsilon_0] = \left\{ \begin{array}{c} \varepsilon_{x0} \\ \varepsilon_{y0} \\ \varepsilon_{xy0} \end{array} \right\} = \left\{ \begin{array}{c} \alpha \Delta T \\ \alpha \Delta T \\ 0 \end{array} \right\}, \qquad (5)$$

where α is the thermal expansion coefficient.

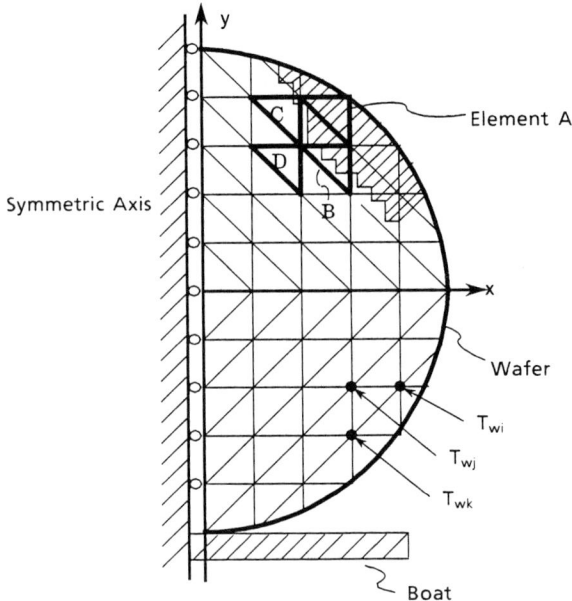

FIG. 7. Finite-element division for a part of the wafer. [After Mokuya and Matsuba (1989).]

Under this initial strain condition, every element must satisfy the balance-of-force condition. Of course, the external body force does not exist in the problem considered here. Furthermore, the boundary conditions assume that the x-direction displacement u on the y-axis is zero because of the y-axis symmetry, and that point where the wafer and the boat are attached is a fixed point. That is, the y-direction displacement v is also zero, if one takes into account that the rigid displacement of the wafer (such as rotation) is prevented. Under the boundary conditions, the strain $\{\varepsilon\} = \{\varepsilon_{xx}, \varepsilon_{yy}, \varepsilon_{xy}\}^T$ can be resolved by the finite-element method (Zienkiewicz, 1970). After that, the lane stress $\{\sigma\}$ for each element is calculated by the linear stresses and strains relation under an initial strain loading as follows:

$$\begin{Bmatrix} \sigma_{xx} \\ \sigma_{xx} \\ \sigma_{xy} \end{Bmatrix} = [D] \begin{Bmatrix} \varepsilon_{xx} - \varepsilon_{x0} \\ \varepsilon_{xx} - \varepsilon_{y0} \\ \varepsilon_{xy} - \varepsilon_{xy0} \end{Bmatrix}, \qquad (6)$$

where $[D]$ is the elastic stress–strain matrix containing Young's modulus E and Poisson's ratio v. In the plane stress condition considered here, $[D]$ is given by

$$[D] = E/(1-v^2) \begin{Bmatrix} 1 & v & 0 \\ v & 1 & 0 \\ 0 & 0 & (1-v)/2 \end{Bmatrix} \quad (7)$$

For the symmetric temperature distribution,

$$\sigma_{rr}(r) = \alpha E [R_w^{-2} \int_0^{R_w} T(r) r \, dr - r^{-2} \int_0^r T(r) r \, dr],$$

$$\sigma_{\theta\theta}(r) = \alpha E [R_w^{-2} \int_0^{R_w} T(r) r \, dr - r^{-2} \int_0^r T(r) r \, dr - T(r)], \quad (8)$$

$$\sigma_{r\theta} = 0$$

are easily obtained (Boley and Weiner, 1960). Here, R_w is wafer radius, r is distance from the wafer center, and $T(r)$ is the radius temperature distribution. The results, $\{\sigma_{rr}, \sigma_{\theta\theta}, \sigma_{r\theta}\}^T$, of (6) in cylindrical coordinates are transformed in $\{\sigma_{xx}, \sigma_{yy}, \sigma_{xy}\}^T$ by the following equation:

$$\sigma_{xx} = \tfrac{1}{2}(\sigma_{rr} + \sigma_{\theta\theta}) + \tfrac{1}{2}(\sigma_{rr} - \sigma_{\theta\theta})\cos(2\theta),$$

$$\sigma_{yy} = \tfrac{1}{2}(\sigma_{rr} + \sigma_{\theta\theta}) - \tfrac{1}{2}(\sigma_{rr} - \sigma_{\theta\theta})\cos(2\theta), \quad (9)$$

$$\sigma_{xy} = \tfrac{1}{2}(\sigma_{rr} - \sigma_{\theta\theta})\sin(2\theta).$$

For a parabolic radial temperature distribution indicated, the normal stress components σ_{xx} and σ_{yy} on the x axis show the maximum tensile stress at the center of the wafer. The x component σ_{xx} decreases to zero, and the y component σ_{yy} shows the maximum compressive stress at the periphery. This compressive stress is much larger than the maximum tensile stress at the wafer center. On the other hand, the shear stress component σ_{xy} in the $\theta = 45°$ direction is zero at the wafer center and shows the maximum stress at the periphery. Moreover, the results obtained by refining the mesh divisions by four are similar to the previous ones.

By using the plane stress components $\{\sigma\} = \{\sigma_{xx}, \sigma_{yy}, \sigma_{xy}\}^T$ obtained here, the resolved shear stresses corresponding to the slip systems of the silicon crystal are calculated as follows: The absolute values of these shear stresses should be compared to the critical stress. Considering this point, a total of 12 possible slip systems should be reduced to five independent slip systems (Bentini et al., 1984). Those are the A − B, B − C, and A − C slip directions on the slip plane ΔABC, and A′ − C and A − C slip directions on the slip

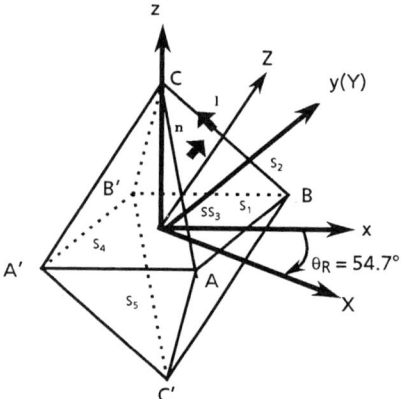

FIG. 8. Elemental octahedron showing slip directions in single-crystal silicon. Resolved shear stresses are calculated by using the unit normal vector **n** on the slip planes {111} and the unit vector **l** in the slip directions [110] in (x, y, z) coordinates for the (100)-oriented wafer. Rotational coordinates (X, Y, Z) are used for the (111)-oriented wafer. [After Mokuya and Matsuba (1989).]

plane $\triangle ACA'$ in the $(x-y-z)$ Cartesian coordinates as shown in Fig. 8. As the wafers vertically stand in a row so that the "primary flat" locates in the top for each wafer, the y axis coincides with the "primary flat" in the analysis of the wafer temperature and stress. Therefore, the x and y axis are along the $\langle 110 \rangle$ and $\langle 110 \rangle$ slip directions, respectively. Introducing the unit normal vector **n** (n_x, n_y, n_z) at each slip plane and the unit vector **l** (l_x, l_y, l_z) in each slip direction, the resolved shear stresses S_m are derived as (Dieter, 1970)

$$S_m = \Sigma F_{mi} \cdot l_i \quad (i = x, y, z),$$
$$F_{mi} = \Sigma \sigma_{ij} \cdot n_j \quad (j = x, y, z), \tag{10}$$

where F_{mi} is the coordinate component of the force that is exerted on the plane with the unit normal vector **n** due to the plane-stress components, $\{\sigma\}$. By using (8), the absolute values of the resolved shear stresses (Bentini et al., 1984) are given by

$$S_1 = \sqrt{2/3}|\sigma_{xy}|, \quad S_2 = \sqrt{1/6}|\sigma_{xx} + \sigma_{xy}|, \quad S_3 = \sqrt{1/6}|\sigma_{xx} - \sigma_{xy}|,$$
$$S_4 = \sqrt{1/6}|\sigma_{yy} + \sigma_{xy}|, \quad S_5 = \sqrt{1/6}|\sigma_{yy} + \sigma_{xy}|. \tag{11}$$

The azimuthal distribution of the stresses resolved on the five independent slip systems of the silicon crystal is shown in Fig. 9. The calculated results correspond to the hatched elements indicated in Fig. 7, and the thermal-

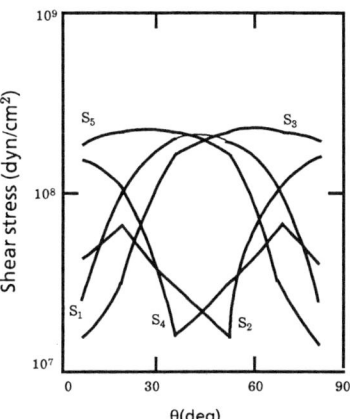

FIG. 9. Azimuthal distribution of the stresses resolved on the five independent slip systems for the (100)-oriented wafer. Calculated results correspond to the hatched elements indicated in Fig. 7. [After Mokuya and Matsuba (1989).]

loading condition is the same as above. In the first quadrant region $\theta = 0° - 90°$, the slip systems S_1, S_3, and S_5 are dominant, and their maximum values are shown at $\theta = 45°$, $\theta = 65°$, and $\theta = 25°$, respectively. These results calculated by FEM are in good agreement with the analytical results $\theta = 45°$, $\theta = 67.5°$, and $\theta = 22.5°$. Furthermore, as shown in Fig. 9, the slip systems S_3 and S_5 intersect at $\theta = 45°$ where the slip system S_1 indicates the maximum stress. Therefore, multiple cross slip occurs easily at $\theta = 45°$, so that the slip lines should primarily be produced in this region. These results agree with the experiments on the (100)-oriented wafer. In later defect onset evaluations, the resolved shear stress S_1 at the element R indicated in Fig. 7 is adopted as the representative value that should be compared with the critical stress of the target wafer.

Next, in a (111)-oriented wafer the resolved shear stresses are derived in the same way as for the (100)-oriented one by introducing the rotated coordinates $(X\text{-}Y\text{-}Z)$ as shown in Fig. 10. The normal unit vector n and the unit vector l are transformed using the following rotational transform matrices:

$$\begin{pmatrix} \sqrt{1/3} & 0 & -\sqrt{2/3} \\ 0 & 1 & 0 \\ \sqrt{2/3} & 0 & \sqrt{1/3} \end{pmatrix}. \qquad (12)$$

Here, the matrix components are calculated by the rotational angle $\theta_R = 54.7°$ around the y axis. There are six independent slip systems for $0° < \theta < 90°$ in the (111)-oriented wafer as shown in Fig. 10. Each resolved shear stress

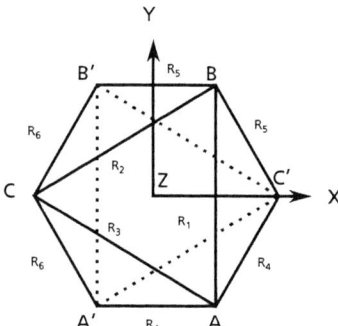

FIG. 10. Six independent slip systems, R_1–R_6, for the (111)-oriented wafer projected on the (111) plane. [After Mokuya and Matsuba. (1989).]

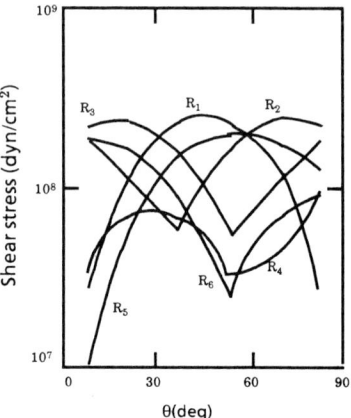

FIG. 11. Azimuthal distribution of the stresses on the six independent slip systems for the (111)-oriented wafer. [After Mokuya and Matsuba (1989).]

is given by

$$R_1 = 2\sqrt{2/3}|\sigma_{xy}|, \qquad R_2 = \sqrt{1/6}|\sigma_{xx} - \sigma_{yy} - 2/\sqrt{3}\sigma_{xy}|.$$
$$R_3 = \sqrt{1/6}|\sigma_{xx} - \sigma_{yy} + 2/\sqrt{3}\sigma_{xy}|, \qquad R_4 = \sqrt{1/6}|2/3\sigma_{xx} + 2/\sqrt{3}\sigma_{xy}|,$$
$$R_5 = \sqrt{1/6}|2/3\sigma_{xx} - 2/\sqrt{3}\sigma_{xy}|, \qquad R_6 = \sqrt{1/6}|1/3\sigma_{xx} - \sigma_{yy}|. \qquad (13)$$

Note that $R_i(\theta) = R_i(-\theta)$, $i = 1, 2, \ldots, 5$, for $-90° < \theta < 0°$. Similarly to the (100)-oriented wafer, the azimuthal distributions of the stresses resolved on the six independent slip systems of the silicon crystal are shown in Fig. 11. The maximum values of the slip systems that produce the slip on the wafer surface occur at $\theta = -60°$, $\theta = 60°$, and $\theta = 0°$, respectively. Furthermore, R_4, R_5, and R_6 intersect the other slip systems, (R_1 and R_3), (R_1 and R_2), and (R_2 and R_3), respectively. The calculated results by FEM agree with the experiments that give a slip pattern resembling a hexadentate star in the case of the (111)-oriented wafer.

III. Prediction of Defect Onset

6. Estimation of Critical Stress

Based on the proposed thermoelastic model, the transient thermal stresses corresponding to the insertion of a wafer into a furnace are calculated and plotted on the semilogarithm diagram with wafer temperature on the abscissa and the resolved shear stress on the ordinate. Since the critical stress is described by the experimental formula (Siethoff and Haasen, 1968)

$$\sigma_E(T) = C_\varepsilon^{1/n} \exp(U/nkT), \tag{14}$$

prediction of the defect onset is performed simply by comparing this value with the calculated stress. In (14), T is the representative temperature, U is the activation energy for the slide movement, k is the Boltzmann constant, ε is the strain rate, and C and n are experimental constants depending on the material considered. It is sufficient simply to determine whether the transient thermal stress curve corresponding to the given processes conditions is beyond the critical stress curve. In practical fabrication processes, the critical stress drifts because of the characteristics of the objective wafer strength and the hysteresis of heat processing (Mokuya and Matsuba, 1989; Patel and Chaudhuri, 1963). Therefore, it is important to evalute the defect onset conditions using the critical stress estimated from the experiments corresponding to each wafer. Two equivalent methods for the prediction of the defect onset are described below. In our applications, the wafer is (100)-oriented so that the simulated results of the transient stress curves are related to the element R as mentioned before. The evaluation process consists of two parts, the determination of the critical stress and the prediction of the defect onset. The former is significant in maintaining prediction accuracy.

The critical stress σ_E as a function of wafer temperature is determined based on experiments performed with wafers vertically and equally spaced

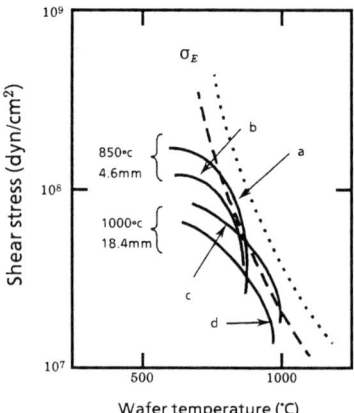

Process	T_f (°C)	V (cm/min)	H (mm)	Defect
a	850	30	4.6	Onset
b	850	20	4.6	Free
c	1000	30	18.4	Onset
d	1000	20	18.4	Free

FIG. 12. Estimation of the critical stress σ_E as a function of wafer temperature T. [After Matsuba et al. (1986).]

apart at a distance of H, with a loading speed of V and a furnace temperature of T_f. Figure 12 shows four transient stress curves a–d corresponding to the experiments with different process conditions a–d for the MOS device fabricating process using Φ150-mm wafers. While cases a and c correspond to defect onset conditions, cases b and d correspond to defect-free conditions. Since the critical stress is the threshold of the defect onset, it must lie between the curves a and b, and also between the curves c and d. Since the critical curve is of the activation energy type $\exp(U/nkT_w)$, it is thus approximately determined by the broken curve plotted in Fig. 12. If more experiments are taken into account, a more accurate curve can be obtained. For comparison, σ_E for wafers without processing is plotted by the dotted curve in the same figure. As expected, it is much greater than σ_E for the processed wafers. This means that as the processing steps proceed, σ_E decreases, which indicates that the wafers more easily generate dislocations.

The estimated critical stress is 5.3×10^7 dyn/cm² at a wafer temperature of 800°C. This value is smaller than 7.6×10^7 dyn/cm² obtained by the rapid flash annealing with a strain rate of $\varepsilon = 4.8 \times 10^{-5}$/s (Bentini et al., 1984). The reason is that the strain rate when wafers are brought into an ordinary electric furnace with a loading speed of between 20 and 30 cm/min is smaller than that of the rapid annealing. Next, using this critical stress σ_E we will predict defect onset under two process conditions. One case is when the furnace temperature is $T_f = 950°C$, the loading speed $V = 20$ cm/min, and the wafer spacing $H = 9.2$ mm (Case 1). The other case employs the same conditions as Case 1, except that $H = 4.6$ mm (Case 2). According to the conditions for Cases 1 and 2, the calculated transient thermal stresses are shown in Fig. 13. Curves A and B represent the values at the element R of the center wafer and of the end wafer in the series for Case 1, respectively. Curves C and D are the values for Case 2. Furthermore, the broken line is the critical stress curve of the wafer determined in the previous experiments shown in Fig. 12.

In Fig. 13, curves C and D for Case 2 have higher stress values than those for Case 1, and the thermal stress level becomes higher. This is because the narrower the wafer spacing is, the steeper the wafer temperature gradient becomes. On the other hand, comparing the central and the end wafers, the former has a higher stress level. This result corresponds to the fact that the characteristics of the temperature distribution in the center wafer are different than those of the last one (Mokuya et al., 1987). Comparing the transient stress curves A, B, C, and D with σ_E, A and C both exceed σ_E. However,

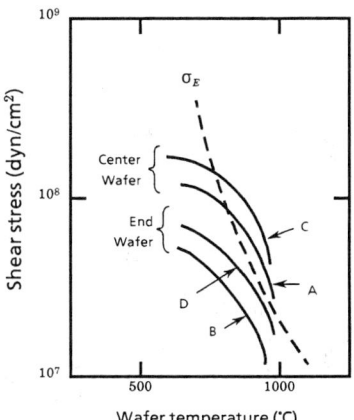

Wafer	T_f (°C)	V (cm/min)	H (mm)	Wafer Position
A	950	20	9.2	Center
B	950	20	9.2	End
C	950	20	4.6	Center
D	950	20	4.6	End

FIG. 13. Calculated thermal stress curves. [After Matsuba et al. (1986).]

curves B and D do not exceed σ_E, and thus dislocations are expected to be induced into the wafers at the central position in the row under the process conditions for Cases 1 and 2, and wafers in the end position will not show any dislocations.

7. Comparison with Experiments

The predictions based on the thermoelastic wafer model are in good agreement with the experiments, as shown in Fig. 14. The wafers were Wright-etched to allow for easier observation of the slip lines. However, dislocations are not introduced in the first quadrant in the wafer. This result might be explained by considering the warpage of the wafer (Leroy and Plougonven, 1980). It has been reported that the wafer buckles in the saddle shape (Otsuka et al., 1979). The first and second quadrant surfaces of the wafer have the convext and concave curvatures, respectively. These bending stresses correspond to tensile and compressive strengths. On the other hand, the membrane stress is mainly compressive in the periphery of the wafer. Assuming that the total of the membrane and bending stresses decreases in the first quadrant, it increases in the second quadrant.

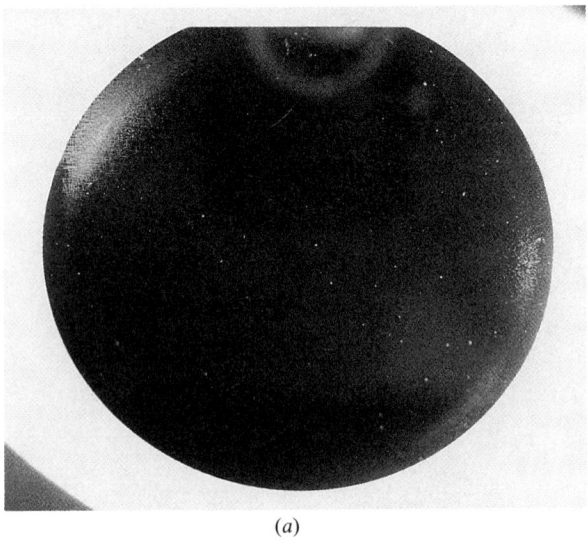

(a)

Fig. 14. Photographs of a Φ150 mm wafer that was Wright-etched after heat processing under the condition of Case 2 in Fig. 13. (a) Center wafer corresponding to the thermal stress C in Fig. 13, (b) End wafer D. [After Mokuya and Matsuba (1989).]

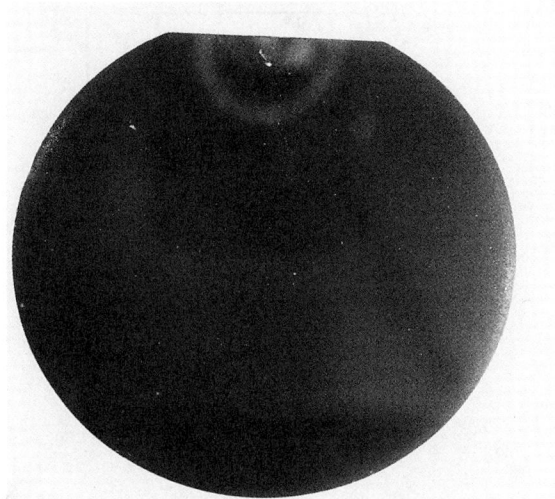

FIG. 14(b).

Hereafter, another method for determing the critical stress is proposed by using one wafer. In this method the critical stress is estimated by means of the experimental data related to the dislocated area indicated as the width of the hatched bar in Fig. 7. The dislocated area was obtained in a Φ100-mm wafer process for fabricating bipolar devices. Transient thermal stress curves are calculated at the elements "A", "B", "C", and "D" by using the thermoelastic model. The more central the element is, and the further it is from the $\theta = 45°$ direction in a wafer, the lower the thermal stress level becomes. These qualitative characteristics agree with the degree of defect onset in each target wafer element. Now, assuming $\varepsilon = 4 \times 10^{-6}/s$ and $n = 2.6$ in Eq. 14, the variation of the parameter C corresponds to a parallel displacement of the critical stress curve in the stress figure like Fig. 12. Comparing the critical stress curves with the calculated stress curves for each element, $C = 3.0 \times 10^5$ dyn/cm^2 was selected for this heat processing. Of course, this method is equivalent to the method described in the previous section. The details of this method are described and used in Section IV.

8. DEPENDENCE OF CRITICAL STRESS ON STRAIN RATE.

The critical stress σ_E depends on the strain rate ε as shown in Fig. 15. The larger the strain rate is, the larger the critical stress values become. For a larger strain rate, for example, when the loading speed is very large, it is assumed that the critical stress values become larger than those for the

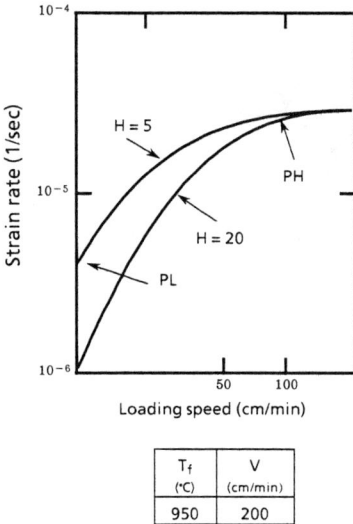

FIG. 15. Strain rate as a function of loading speed V. [After Mokuya and Matsuba (1989).]

process conditions corresponding to an ordinary strain rate. Now, if the strain rate ε is given as the time derivatives of the initial strain ε_0 in a wafer under the thermal loading, it is defined by

$$\varepsilon = (\varepsilon_0(t_2) - \varepsilon_0(t_1))/(t_2 - t_1), \tag{15}$$

where $\varepsilon_0(t_1)$ and $\varepsilon_0(t_2)$ are the initial strains at the times t_1 and t_2 when the wafer temperatures are around 800°C and 900°C, respectively. The initial strains in these temperature regions are used because the transient thermal stress curve values are usually the closest to the critical stress curve values in these regions. The relationship between the strain rate ε and the loading speed V is depicted in Fig. 15. The strain rate increases as the loading speed becomes larger until $V = 200$ cm/min. Then, it saturates at $\varepsilon = 1.8 \times 10^{-5}$/s.

The effect on the critical stress is displayed in Fig. 16. Curves L and H show the transient thermal stress corresponding to $\varepsilon = 0.39 \times 10^{-5}$/s and $\varepsilon = 1.7 \times 10^{-5}$/s, respectively. If the defect onset is predicted using the critical stress σ_{EL} determined from the experimental results with $\varepsilon = 0.39 \times 10^{-5}$/s, then the dislocated area in the wafer for Case PH should be larger than that for Case PL because the thermal transient stress curve H is much higher than curve L. However, a wafer processing utilizing Case L showed a dislocated area, while a wafer processing utilizing Case H was defect-free. This discrepancy is explained by considering the effect of the strain rate on the critical

8. PROCESSING AND SEMICONDUCTOR THERMOELASTIC BEHAVIOR 361

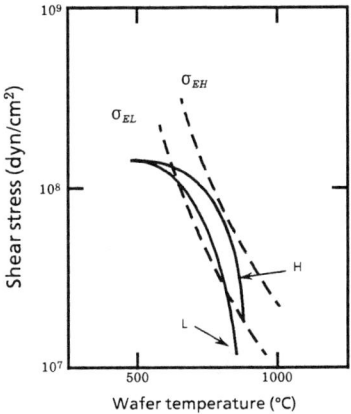

Wafer	T_f (°C)	V (cm/min)	H (mm)	Defect
L	950	10	5.0	Onset
H	950	100	10.0	Free

FIG. 16. Stress curves by taking into account the strain rate. [After Mokuya and Matsuba (1989).]

stress. As shown in Fig. 16, for the prediction of defect onset for Case H alternating the critical stress σ_{EH} corresponding to $\varepsilon = 1.7 \times 10^{-5}$/s instead of σ_{EL}, the transient thermal stress curve H is not beyond the σ_{EH} values, where it is assumed that the experimental constants n and C are 2.6 and 5.5×10^5 dyn/cm^2, respectively.

IV. Application of Thermoelastic Model

9. Optimization of Heat Processing

A systematic approach for the optimization of heat processing to prevent dislocations in the manufacturing of VLSIs, involving the use of a thermoelastic model that describes temperature and thermal stress distributions induced in wafers during transient periods of the heat cycle, is proposed. This approach is found to be effective by illustrating examples of its application to typical processing problems as shown in Fig. 17.

The wafer spacing results in defect-free conditions under the parameters $T_f = 950°C$ and $V = 15$ cm/min in a Φ125-mm wafer process. This process is similar to the Φ100-mm wafer process except that the wafer radius is larger.

No.	Process	T_f (°C)	V (cm/min)	H (mm)
5	Annealing	940	13	5
10	Deposition	950	15	10
11	Diffusion	950	15	5
12	Annealing	940	13	5

FIG. 17. Semiconductor fabrication process flow.

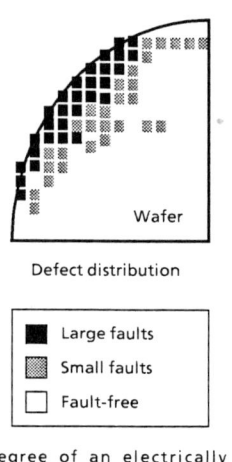

Defect distribution

■ Large faults
▨ Small faults
☐ Fault-free

Degree of an electrically degraded chip

FIG. 18. Observed yield at each chip location ($T_f = 960°C$, $V = 20$ cm/min, $H = 9.2$ mm).

In this situation, the critical stress from the Φ100-wafer process obtained above is found to be applicable to the Φ125-mm wafer process. It is obvious that the thermal stress with $H = 15$ mm does not exceed σ_E. On the other hand, the thermal stress with $H = 5$ mm is much larger than σ_E. Therefore, it is expected to introduce a significant defect over a wafer. The yield map obtained under these process conditions is shown in Fig. 18. The dark parts indicate the electrically degraded chips. As predicted based on the model, the fraction of the degraded chips is largely reduced by the reduction of the thermal stresses. The device yield is clearly shown to reveal a one-to-one correlation between refresh failures and dislocations, as shown in Figs. 18 and 19. Under the condition $H = 15$ mm, some degraded chips were observed in the periphery. It is considered that the critical stress values of the target wafer here are possibly only slightly in disagreement with σ_E determined above in the Φ100-mm wafer process. Furthermore, it is assumed that a few chips' inner regions are electrically degraded by other factors than thermal stress.

The critical stress is determined based on experiments performed with 125 mm wafers equally spaced at a distance of H, with a loading speed of V, and a furnace temperature of T_f. Curve A in Fig. 20 corresponds to the stress in damaged chips, while curve C corresponds to the stress in chips with no dislocations. Curve B is just on the boundary between areas of dislocations and that of no dislocations. Thus, σ_E must be tangential to curve B. Since the critical stress is of the activation energy type, σ_E as a function of wafer temperature can be determined approximately by the broken curve in this figure.

By use of this critical stress σ_E, we optimized the practical processing (Fig. 17), which consists of various types of heat cycles. We first examined in which process the largest stress was induced. It turned out that process No. 11 was the most likely candidate to induce dislocations. We can now determine the

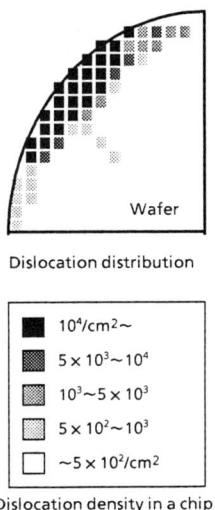

FIG. 19. Observed dislocations at each chip location.

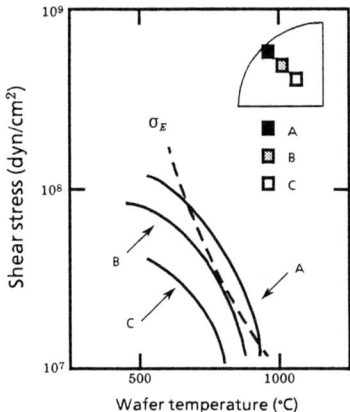

FIG. 20. Determination of the yield stress σ_E as a function of wafer temperature. [After Mokuya and Matsuba (1989).]

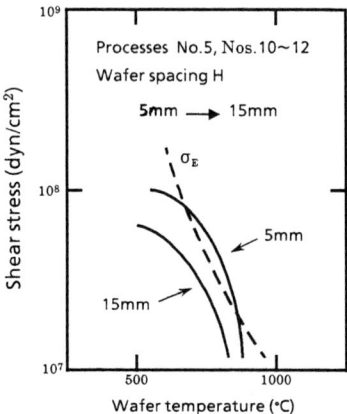

FIG. 21. Thermal stress induced in the processes No. 5 and Nos. 10-12. [After Mokuya and Matsuba (1989).]

conditions under which the dislocations do not appear. In order to reduce the stress, the wafer spacing was changed from 5 mm to 15 mm for processes No. 5 and Nos. 10-12. The calculated stress was found to be less than σ_E, and thus the dislocations were not expected to be observed (Fig. 21). The experimental results in Fig. 22 show that damaged chips prior to optimization were reduced considerably.

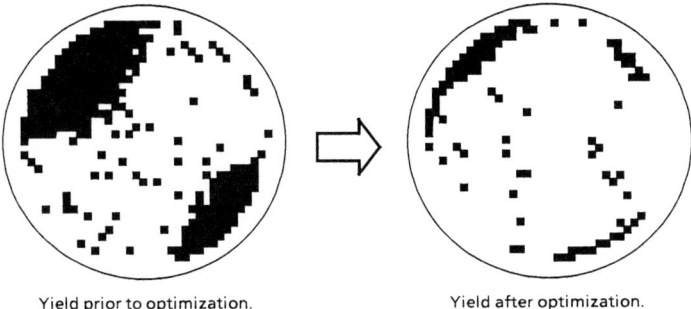

Yield prior to optimization.   Yield after optimization.

FIG. 22. Experimental results on a Φ-125 mm wafer. Chips that pass (fail) test are indicated by □(■), [After Mokuya and Matsuba. (1989).]

V. Conclusions

We have developed the thermoelastic model for the radiant heat transfer between the wafers, the boat, and the furnace tube wall to estimate the temperature of the wafers. Since the temperature distribution within a wafer is enhanced up to more than 150°C over time, according to the heat capacity ratio between the wafer and the boat, it was found that the boat plays an essential role in the study of thermal stress. Thermal stress calculated by applying the finite element method is determined from the temperature distributions in wafers as they are inserted into a diffusion furnace. The process conditions necessary to prevent dislocations during processing have been established by comparing the critical stress with the calculated stress based on the temperature distribution that has been modeled. An effective way of determining the critical stress has been proposed by comparing the experiments and simulations. Once the critical stress is known, it is easy to determine the conditions under which the dislocations do not occur during heat cycles.

A thermoelastic wafer model simulates the thermal stresses in a row of wafers during the heat cycling periods in a furnace. One can predict the defect onset under the given process conditions based on this model. The plane-stress components $\{\sigma_{xx}, \sigma_{yy}, \sigma_{xy}\}$ calculated by the finite-element method agreed with the analytical ones under the thermal-loading conditions of parabolic temperature distribution in a wafer. It was shown that the resolved shear stresses in each slip system of the silicon crystal can explain the ordinary experimental results with respect to slip patterns in (100)-oriented and (111)-oriented wafers. As a first step in the prediction of defect onset, the critical stress was estimated on the semi-logarithm diagram with the wafer temperature on the abscissa and the resolved shear stress on the ordinate. The transient thermal stress curves corresponding to the process conditions in previous experiments were used. After that, it was sufficient to determine whether the transient thermal stress curve was beyond the critical stress determined in the previous experiments to predict the defect onset. The results of experiments on the Φ150-mm wafer MOS process and the Φ125-mm wafer bipolar process were in good agreement with the predicted results. It is expected that optimization of the annealing process conditions can be executed quickly by using the thermoelastic wafer model proposed here. However, for a more detailed analysis of the defect, it will be necessary to take into account the wafer warpage effects.

Acknowledgments

The authors are grateful to H. Nagatomo, Dr. A. Yoshinaka and T. Aoshima of the Musashi works, and Y. Sakamoto and Y. Ohyama of the Takasaki works for their advice and support during the course of this research. We also wish to thank Y. Dohmen, the head, Dr. J. Kawasaki, the former head of the Systems Development Laboratory, and Dr. K. Haruna, F. Funabashi, and Dr. K. Matsumoto, also of the Systems Development Laboratory, for their daily guidance and encouragement.

References

Bentini, G., Correra, L., and Donolato, C. (1984). "Defects introduced in silicon wafers during rapid isothermal annealing: Thermoelastic and thermoplastic effects,"*J. Appl. Phys.* **56**, 2922.
Boley, B. A., and Weiner, J. H. (1960). "Theory of Thermal Stresses," p. 290. Wiley, New York.
Dieter, G. E. (1970). "Mechanical Metallurgy," p. 28. McGraw-Hill, New York.
Hu, S. M. (1969). "Temperature distribution and stresses in circular wafers in a row during radiative cooling," *J. Appl. Phys.* **40**, 4413.
Leroy, B., and Plougonven, C. (1980). "Warpage of silicon wafers," *J. Electrochem. Soc.* **27**, 961.
Matsuba, I., Matsumoto, K., and Yoshinaka, A. (1984). "Temperature distribution in a furnace for semiconductor fabrication processes," *Trans. Inst. Electron. & Commun. Eng. Japan* J67-C, 332. In Japanese.
Matsuba, I., Mokuya, K., Matsumoto, K., and Yoshinaka, A. (1985). "Mathematical model of temperature distribution in wafers in a furnace for semiconductor fabrication processes," *Proc. NASECODE IV*, pp. 405-410.
Matsuba, I., Mokuya, K., and Aoshima, T. (1986). "Thermoelastic model of dislocations in wafers," *Int. Electron Devices Meeting (IEDM) Tech. Dig., Los Angeles*, p. 530.
Mokuya, K., and Matsuba, I. (1989). "Prediction of defect onset condition in heat cycling based on a thermoelastic wafer model," *IEEE Trans. Electron Devices* **ED-36**, 319.

Mokuya, K., Matsuba, I., Matsumoto, K., and Yoshinaka, A. (1985). "Transient model of temperature in a furnace for semiconductor fabrication processes," *Trans. Inst. Electron. & Commun. Eng. Japan* **J68–C**, 425. In Japanese.

Mokuya, K., Matsuba, I., and Aoshima, T. (1987). "Characteristics of the transient wafer temperature distribution in a furnace for semiconductor fabrication processes," *Trans. IECE Japan* **J70–C**, 324. In Japanese.

Morizane, K., and Gleim, P. S. (1969). "Thermal stress and plastic deformation of thin silicon slices," *J. Appl. Phys.* **40**, 4103.

Nayfeh, A. H. (1973). "Perturbation Methods." John Wiley & Sons, New York.

Otsuka, H., *et al.* (1979). "Effects of backside damage on thermal warpage of silicon wafer," *Electrochem. Soc. Extended Abstracts* **79–2**, 1366.

Patel, J. R., and Chaudhuri, A. R. (1963). "Macroscopic plastic properties of dislocation-free germanium and other semiconductor crystals. I. Yield behavior," *J. Appl. Phys.* **34**, 2788.

Siethoff, H., and Haasen, P. (1968). *In* "Lattice Defects in Semiconductors" (Hasiguti, R. R., ed.), p. 491. University of Tokyo, Tokyo.

Sze, S. M. (1981). "Physics of Semiconductor Devices," 2nd edition. Wiley, New York.

Zienkiewicz, O. C. (1970). "The Finite Element Method in Engineering Science," p. 48. McGraw-Hill, New York.

List of Symbols

a_b	Absorbance of boat
a_f	Absorbance of furnace tube wall
a_w	Absorbance of wafer
c_b	Specific heat of boat
c_f	Specific heat of furnace tube wall
c_w	Specific heat of wafer
$[D]$	Elastic stress–strain matrix
E	Young's modulus
F_a	Geometric factor between two elements in neighboring wafers
F_s	Geometric factor between two elements in the same wafer
F_{wb}	Geometric factor between wafer and boat
F_{wf}	Geometric factor between wafer and furnace tube wall
F_{mi}	Component of force exerted on stress plane
H	Wafer spacing in a row
k	Boltzmann constant
k_b	Thermal conductivity of boat
k_f	Thermal conductivity of furnace tube wall
k_w	Thermal conductivity of wafer
K_b	Dimensionless thermal conductivity of boat
K_f	Dimensionless thermal conductivity of furnace tube wall
K_w	Dimensionless thermal conductivity of wafer

l	Unit vector in slip direction
ℓ_w	Thickness of wafer
n	Unit normal vector on slip plane
N_w	Number of wafers on the boat
Q	Dimensionless supplied heat source
$\mathbf{r} = (x, y)$	Two-dimensional coordinates
R_m	Resolved shear stress
R_f	Radius of furnace tube
R_w	Radius of wafer
t	Time
S_m	Resolved shear stress
T_0	Reference temperature
T_b	Boat temperature
T_f	Furnace temperature
T_{f0}	Initial furnace temperature
T_i	Temperature distribution in the ith wafer
U	Activation energy
V	Loading speed into the furnace
z	Coordinate along the furance axis

Greek Letters

α	Thermal expansion coefficient
ε	Strain rate
$\{\varepsilon\}$	Strain vector
ε_0	Initial strain
$\{\varepsilon_0\}$	Initial strain vector
ε_b	Emittance of boat
ε_f	Emittance of furnace tube wall
ε_w	Emittance of wafer
ΔT	Temperature change from an arbitrary temperature
σ	Stefan–Boltzmann constant
$\{\sigma\}$	Plane stress vector
$\sigma_E, \sigma_{EH}, \sigma_{EL}$	Critical stress
σ_b	Density of boat
ρ_f	Density of furnace tube wall
ρ_w	Density of wafer
ν	Poisson's ratio

INDEX

A

Ab-initio theory
  density function theory, 25, 72
  exchange correlation potential, 27
Activation energy, of
  climb, 166f, 175, 183
  cross-slip, 170f, 172
  dislocation mobility, 154, 157f, 163, 175, 181, 182
Alloys
  bond length, 57
  bulk modulus, 56, 60
  cluster population, 59
  disordered, 56
    binary, 56
    pseudobinary, 56, 57
  ordered, 51
Antisite defects
  in GaAs, 215, 219

B

Barenblatt model, 92
Binding energy of doping atom and dislocation, 176, 177
Bond order model (BOM)
  analytical expressions, 39–41
  results, 45, 49–51
  sp^3 hybrid, 36
  universal parameters, 38
Bonding
  anodic, 318
  compound semiconductors, 330–332
  direct, 321–330
  electrostatic, 318
  field-assisted, 318–320
  fusion, 321
  glass-to-metal seal, 318
  solid phase, 333
  thermal, 321
  van der Waals, 332
Born–Oppenheimer adiabatic approximation, 25
Brittle-to-ductile transition, 103, 116–133
  doping effects, 120
  models for, 129
  strain rate dependence, 118
  temperature, 192
  TEM studies, 125
  x-ray topography, 123
Burgers vector, influence on dynamical recovery, 146, 170

C

CdTe
  effect of illumination on fracture, 115
Cleavage planes in semiconductors, 106
Climb, of edge dislocations, 165f
Coherency strains, 267
Complex formation, influence on yield point, 163, 181, 182
Core structure, 270
Corrections to pulse measurements, 18
  bond thickness, 18
  diffraction corrections, 20
Crack advance mechanisms, 98, 99
Crack shielding, 105

Crack tip models, 91–98
Creep
  curve, 145, 153f
  inflection point, 153, 154, 157f, 181f
  natural, 184
  power law, 193
    of GaAs, 207
    of Si, 199
  steady state, 153, 165, 170, 171f, 174
Critical resolved shear stress
  of CdTe, 217
  of GaAs, 207
    B-doped, 214
    In-doped, 207, 214
    (In, Si)-doped, 214
    P-doped, 214
    Sb-doped, 214
  of MnCdTe, 217
  of ZnCdTe, 217
  strain rate dependence of GaAs, 213
Cross-slip, 146, 153, 168, 169f, 174, 184
Crystal growth
  of III–V compounds, 222ff
  nonstoichiometry during, 223
  role of dopants in, 222ff
Crystal structure
  of Ge, 190
  of Si, 106, 107, 190
  of III–V compounds, 191

D

Defect onset, 355
  activation energy for, 355
  critical stress for, 355
    strain rate dependence, 359
  in MOS devices, 356
  in Wright-etched wafers, 358
  slip line to detect, 358
  wafer strength, effect on, 355
  warpage to detect, 358
Deformation
  activation energy
    for Ge, 204
    for Si, 204
  high temperature, 250
  intermediate temperature, 245
  low temperatures and high stresses, 237

  map for Si, 199, 204
  temperature dependence of, 192
  temperature-independent region of, 193
Direct silicon bonding, 321–330
  applications, 328–330
  bond strength, 325
  contamination, 326
  electrical properties of, 327
  mechanisms of, 322–325
  voids in, 326
Dislocation, 233
  glide set, 235
  interface, 269
  interaction with doping atoms
    elastic and electrostatic, 177, 184
    local pinning, 184
    non-local, 181f
  measurement
    by transmission electron microscopy, 209
    by x-ray topography, 209
  misfit, 267, 280
  mobility, 70, 145, 155f, 168, 175, 181f, 182
    stress exponent of 154, 157f, 163, 181f, 182
  multiplication, 289
  nucleation, 284
  pile-up ahead of cracks, 101, 104, 124, 130
  shuffle set, 235
  structures
    in Group IV elements, 190–191
    in III–V compounds
      α, 191
      β, 191
    in II–VI compounds, 215
  velocity, 193
    in GaAs
      In-doped, 213, 219
      Te-doped, 218
      Zn-doped, 218
    in Ge
      As-doped, 218
Dry etching
  anisotropy, 312
  ashing, 312
  ion milling, 311
  isotropy, 312

plasma, 311
reactive ion etching, 311
reactive ion beam etching, 312
selectivity, 312–313
Dynamical recovery, 164*ff*
  first stage, 147ff, 164*f*, 183
  intersection of mechanisms, 173*f*
  orientation, effect of, 145, 153, 174
  second stage, 147*f*, 169*f*, 184
  stacking fault energy, effect of, 146, 170, 183

E

Elastic constant
  analytical expression from bond orbital model, 39–43
  definition, 2
    bulk modulus, 4
    $C_{11} - C_{12}$, 4
    C_{44}, 5
  from valence force-field model, 29–33
  identity relation, 30, 31, 32, 41, 42
    from bond-orbital model, 41
    Keating identity, 30
    Martin identity, 32
  independent components in crystals, 9
  of semiconductor alloys, 63
  quantitative parametrization, 46–51
  relation to sound velocity, 10
Effective stress, 155, 166, 172
Epitaxial lift off, 331
Equation of state (mechanical), 153, 154

F

Field-assisted bonding, 318–320
  glass, role of, 319
  stress, 319
  temperature, role of, 320
Fracture
  effect of illumination, 113
  effect of environment, 133, 134
Fracture resistance, 86
  dislocation models, 103–106
  environmental effects, 133
  measurement by indentation, 140
  R-curve, 90
  strain rate dependence, 113
  temperature dependence, 117–119, 120, 122
Fracture surfaces formed by cleavage, 108
Fracture surface energies
  calculated, 110
  measured, 111
Furnace techniques for oxide growth, 339
  annealing, 339
  boat, 340
  loading speed of, 341
  growing of SiO_2 in, 339
  wafer configurations, 340, 341

G

GaAs
  brittle-to-ductile transition, 117, 118, 126
  dislocation mobility under illumination, 114
  electron microscopy of cracks, 126–128
  fracture resistance with In-doping, 136
  fracture surface, 108, 109
  fracture surface energies, 111
  indentation cracking, 82, 115
  strain rate dependence of fracture, 113
  surface energy
    calculated, 110
  temperature dependence of fracture, 117
Ge
  strength of whiskers, 81
Glide system, 146, 150, 160
Griffith theory of fracture, 85, 86
Griffith equation, 85

H

Hardness, 63–64, 196
  cracking at indentations, 197
  nanoindenter measurement, 64

of CdHgTe, 216, 217
of GaAs, 197
  B-doped, 199
  In-doped, 197
of Ge, 199
of Si, 199
temperature dependence, 70, 71
Vickers' measurement, 64

I

Indentation measurement of fracture resistance, 140
Integrated circuits, three-dimensional, 307–311
  MU-STRATE, 308, 310–311
Internal displacement, 5, 30, 41, 46, 76
Irreversible crack extension, 88

J

Jerky (serrated) flow, 162, 178, 180, 182

K

Kelly, Tyson, Cottrell model, 100
Kinetic theory of crack advance, 98
Kink mechanism, 155, 156, 175, 181
Lagrangian strain, 7

L

Lamé constant, 10, 11
Light-emitting diodes
  degradation resistance, 257
    atomic ordering effects, 257
    surface phase separation effects, 257
Line tension, 284
Liquid encapsulation, during deformation, 156, 163, 165, 200, 201
Lüders band, 179f

M

Membranes, 328
Micromachining, 293
Micromirror
  additives, role of, 306
  compound semiconductors, 307
  silicon, 303–306
Microyield, 202, 223
  in GaAs, 209
Misfit strain
  calculated for dopants
    in III–V systems, 196
    in II–VI, 196
  due to isovalent solute atom, 194

O

Optical contacting, 317, 318, 322
Optical techniques, 21–24
  Brillouin scattering, 25
  diffraction
    Bragg, 23
    Fraunhofer, 21
    Freshnel, 23
    Raman–Nath, 22, 23
Optimization of heat processing, 361
  defect free condition, 361
  degraded chip, 363
  yield map, 362
Orowan theory of strength, 83

P

Patel effect, 121
Peierls barrier, 65–67, 190, 192
Perturbation theory, 35, 38
Photoplasticity in compound semiconductors, 252
Plasma etching
  anisotropy, 314, 315
  chemical plasma, 315
  dry etching, 311
  surface damage due to, 316
Point defects
  in GaAs, 219, 221
  influence on plasticity, 147, 160, 165, 175, 183

precipitation, 160, 163, 180
Portevin-LeChatelier effect, 162, 180

R

Reactive ion etching
  anisotropy, 314, 315
  Al, 313
  dry etching, 312
  metal, 313-314
  polyimide, 313
  polysilicon, 313
  Si, 313
  SiO_2, 313
Recovery
  first stage
    effect of doping on, 180, 184
    effect of self-diffusion on, 146, 166f, 175, 183
    stress exponent, 166f, 183f
Resistance curve or R-curve, 90
Rice-Thompson model, 100

S

Shear modulus, effect on plasticity, 146
Si
  brittle-to-ductile transition, 118-127
  computed atomic structure of crack tip, 91, 93
  crystallography aspects, 106, 107, 190
  doping effects, 120-123, 134
  electron microscopy of cracks, 95, 97, 101, 125
  flaw size dependence on strength, 86
  hydrogen effects, 134
  on insulator, 328
  porous, 302-303
  strengths, reported, 84
  surface of
    hydrophilic, 321
    hydrophobic, 322
    native oxide, 321
    (111) orientation, 296
  topographic studies, 123
Slip system, 65
Sonic wave, 6
  in cubic crystal, 9
  in isotropic solid, 10
  longitudinal, 10, 11
  transverse, 10, 11
  velocity, 11
Solid solution strengthening (or hardening), 184, 194ff, 206ff
Solute
  atmosphere breaking away, 177f
  drag (Cottrell atmosphere), 176f, 194, 196
  hardening, 192
  softening, 192
Stacking fault
  deformation-induced, 233
  energy
    GaAs, In-doped, 211
    influence on dynamical recovery, 146, 170, 183
Steady state crack extension, 89
Strain energy release rate, 86-89
  dependence on energy gap, 112
Strain tensor, 3
Strained multilayer structures, 268
Strength
  compressive, 200
  of Si, 84
  tensile, 200
  non-isovalent dopants, 218ff
Stress
  at transition from stage I to stage II, 152
  plateau, 153, 163, 177
Stress intensity factor, 87
Stress relaxation
  of GaAs, 211, 212
  In-doped, 211, 212
Stress-strain behavior, 145, 147f, 153f, 164
  of GaAs, 206ff
    Al-doped, 215
    In-doped, 206ff
    P-doped, 208, 214
    Sb-doped, 208, 214
    Te-doped, 219
    Zn-doped, 219
  of GaP, 205, 206
  of Ge
    function of dislocation density, 201
  of InP, 205

of Si
   function of dislocation density, 201
   multiple slip orientation, 202, 203
   single slip orientation, 203
Surface energies, 110

T

Theoretical strength, 83
Thermal stress model, 348
   finite-element method, 348
   (111)-oriented wafer, 353
   (100)-oriented wafer, 353
   plane stress, 348
   resolved shear stress, 352
   stress-strain matrix, 350
Thermoelastic model, 342
III-V compounds
   brittle-to-ductile transition, 118
   dependence on energy gap, 112
   surface energies
      calculated, 110
      measured, 111
Tight binding (TB) theory, 33, 34, 36, 43
   calculations, 53
   Hamiltonian, 33, 34, 44
   parameters, 44–48
Transmission electron microscopy, 94, 125
Transverse optical phonon, 5, 46
Twin boundary, 269
Twinning, at lower yield point, 184

U

Ultrasonic measurement, 15
   accuracy of measurement, 15
   non-contacting transducer, 17
   piezoelectric transducer, 16

V

Valence force field (VFF) model, 29, 34, 43, 54, 55
   diamond structure, 29
   zincblende structure, 30

W

Wafer temperature model, 342
   Deal-Grove's formula, 345
   dioxide film, 345
   radiant heat transfer, 344
   singular perturbation method, 347
   steady state temperature, 345
   transient temperature, 346
Wet chemical etching of single crystal silicon, 295–303
   anisotropy, 295, 296, 298–299
   EDA etching, 295, 298
   etch rate ratio (etch ratio), 295, 296, 298–299
   KOH etching, 295, 296–297
   etch stopping, 301
   models, 296–297
      electrochemical, 300–301
   (111) surface, 296
   stopping, 301, 330
   stress, effect of etch ratio, 298
   undercutting, 298
Whiskers, 80, 84
   Si, 81
   Ge, 81
   ZnO, 81
Work (strain) hardening
   coefficient, 148f, 150f, 168
   stage, 148, 150f, 169

X

X-ray topography, 123

Y

Yield point, 147, 153, 154ff, 175ff, 182f
   environment, effect of, 161f
   Fermi level, effect of, 181
   hydrostatic pressure, effect of, 158
   lower, 201
      effect of doping on, 175ff
      twinning at, 184
   oxygen, effect of, 180
   pre-deformation, effect of, 155, 156, 163

upper, 201
Yield stress
    of CdHgTe, 216
    of GaAs
        (In, Si)-doped, 220
    of InP
        S-doped, 219
        Te-doped, 219

Zn-doped, 219
see also Critical resolved shear stress

Z

ZnO
    strength of whiskers, 81

Contents of Volumes in this Series

Volume 1   Physics of III–V Compounds

C. Hilsum, Some Key Features of III–V Compounds
Franco Bassani, Methods of Band Calculations Applicable to III–V Compounds
E. O. Kane, The $k \cdot p$ Method
V. L. Bonch-Bruevich, Effect of Heavy Doping on the Semiconductor Band Structure
Donald Long, Energy Band Structures of Mixed Crystals of III–V Compounds
Laura M. Roth and Petros N. Argyres, Magnetic Quantum Effects
S. M. Puri and T. H. Geballe, Thermomagnetic Effects in the Quantum Region
W. M. Becker, Band Characteristics near Principal Minima from Magnetoresistance
E. H. Putley, Freeze-Out Effects, Hot Electron Effects, and Submillimeter Photoconductivity in InSb
H. Weiss, Magnetoresistance
Betsy Ancker-Johnson, Plasmas in Semiconductors and Semimetals

Volume 2   Physics of III–V Compounds

M. G. Holland, Thermal Conductivity
S. I. Novkova, Thermal Expansion
U. Piesbergen, Heat Capacity and Debye Temperatures
G. Giesecke, Lattice Constants
J. R. Drabble, Elastic Properties
A. U. Mac Rae and G. W. Gobeli, Low Energy Electron Diffraction Studies
Robert Lee Mieher, Nuclear Magnetic Resonance
Bernard Goldstein, Electron Paramagnetic Resonance
T. S. Moss, Photoconduction in III–V Compounds
E. Antončik and J. Tauc, Quantum Efficiency of the Internal Photoelectric Effect in InSb
G. W. Gobeli and F. G. Allen, Photoelectric Threshold and Work Function
P. S. Pershan, Nonlinear Optics in III–V Compounds
M. Gershenzon, Radiative Recombination in the III–V Compounds
Frank Stern, Stimulated Emission in Semiconductors

Volume 3   Optical of Properties III–V Compounds

Marvin Hass, Lattice Reflection
William G. Spitzer, Multiphonon Lattice Absorption
D. L. Stierwalt and R. F. Potter, Emittance Studies
H. R. Philipp and H. Ehrenreich, Ultraviolet Optical Properties
Manuel Cardona, Optical Absorption above the Fundamental Edge
Earnest J. Johnson, Absorption near the Fundamental Edge
John. O. Dimmock, Introduction to the Theory of Exciton States in Semiconductors
B. Lax and J. G. Mavroides, Interband Magnetooptical Effects

H. Y. Fan, Effects of Free Carries on Optical Properties
Edward D. Palik and George B. Wright, Free-Carrier Magnetooptical Effects
Richard H. Bube, Photoelectronic Analysis
B. O. Seraphin and H. E. Bennett, Optical Constants

Volume 4   Physics of III–V Compounds

N. A. Goryunova, A. S. Borschevskii, and D. N. Tretiakov, Hardness
N. N. Sirota, Heats of Formation and Temperatures and Heats of Fusion of Compounds $A^{III}B^V$
Don L. Kendall, Diffusion
A. G. Chynoweth, Charge Multiplication Phenomena
Robert W. Keyes, The Effects of Hydrostatic Pressure on the Properties of III–V Semiconductors
L. W. Aukerman, Radiation Effects
N. A. Goryunova, F. P. Kesamanly, and D. N. Nasledov, Phenomena in Solid Solutions
R. T. Bate, Electrical Properties of Nonuniform Crystals

Volume 5   Infrared Detectors

Henry Levinstein, Characterization of Infrared Detectors
Paul W. Kruse, Indium Antimonide Photoconductive and Photoelectromagnetic Detectors
M. B. Prince, Narrowband Self-Filtering Detectors
Ivars Melngailis and T. C. Harman, Single-Crystal Lead-Tin Chalcogenides
Donald Long and Joseph L. Schmit, Mercury-Cadmium Telluride and Closely Related Alloys
E. H. Putley, The Pyroelectric Detector
Norman B. Stevens, Radiation Thermopiles
R. J. Keyes and T. M. Quist, Low Level Coherent and Incoherent Detection in the Infrared
M. C. Teich, Coherent Detection in the Infrared
F. R. Arams, E. W. Sard, B. J. Peyton, and F. P. Pace, Infrared Heterodyne Detection with Gigahertz IF Response
H. S. Sommers, Jr., Macrowave-Based Photoconductive Detector
Robert Sehr and Rainer Zuleeg, Imaging and Display

Volume 6   Injection Phenomena

Murray A. Lampert and Ronald B. Schilling, Current Injection in Solids: The Regional Approximation Method
Richard Williams, Injection by Internal Photoemission
Allen M. Barnett, Current Filament Formation
R. Baron and J. W. Mayer, Double Injection in Semiconductors
W. Ruppel, The Photoconductor-Metal Contact

Volume 7   Application and Devices
PART A

John A. Copeland and Stephen Knight, Applications Utilizing Bulk Negative Resistance
F. A. Padovani, The Voltage-Current Characteristics of Metal-Semiconductor Contacts
P. L. Hower, W. W. Hooper, B. R. Cairns, R. D. Fairman, and D. A. Tremere, The GaAs Field-Effect Transistor
Marvin H. White, MOS Transistors
G. R. Antell, Gallium Arsenide Transistors
T. L. Tansley, Heterojunction Properties

PART B

T. Misawa, IMPATT Diodes
H. C. Okean, Tunnel Diodes
Robert B. Campbell and Hung-Chi Chang, Silicon Carbide Junction Devices
R. E. Enstrom, H. Kressel, and L. Krassner, High-Temperature Power Rectifiers of $GaAs_{1-x}P_x$

Volume 8 Transport and Optical Phenomena

Richard J. Stirn, Band Structure and Galvanomagnetic Effects in III–V Compounds with Indirect Band Gaps
Roland W. Ure, Jr., Thermoelectric Effects in III–V Compounds
Herbert Piller, Faraday Rotation
H. Barry Bebb and E. W. Williams, Photoluminescence 1: Theory
E. W. Williams and H. Barry Bebb, Photoluminescence II: Gallium Arsenide

Volume 9 Modulation Techniques

B. O. Seraphin, Electroreflectance
R. L. Aggarwal, Modulated Interband Magnetooptics
Daniel F. Blossey and Paul Handler, Electroabsorption
Bruno Batz, Thermal and Wavelength Modulation Spectroscopy
Ivar Balslev, Piezooptical Effects
D. E. Aspnes and N. Bottka, Electric-Field Effects on the Dielectric Function of Semiconductors and Insulators

Volume 10 Transport Phenomena

R. L. Rode, Low-Field Electron Transport
J. D. Wiley, Mobility of Holes in III–V Compounds
C. M. Wolfe and G. E. Stillman, Apparent Mobility Enhancement in Inhomogeneous Crystals
Robert L. Peterson, The Magnetophonon Effect

Volume 11 Solar Cells

Harold J. Hovel, Introduction; Carrier Collection, Spectral Response, and Photocurrent; Solar Cell Electrical Characteristics; Efficiency; Thickness; Other Solar Cell Devices; Radiation Effects; Temperature and Intensity; Solar Cell Technology

Volume 12 Infrared Detectors (II)

W. L. Eiseman, J. D. Merriam, and R. F. Potter, Operational Characteristics of Infrared Photodetectors
Peter R. Bratt, Impurity Germanium and Silicon Infrared Detectors
E. H. Putley, InSb Submillimeter Photoconductive Detectors
G. E. Stillman, C. M. Wolfe, and J. O. Dimmock, Far-Infrared Photoconductivity in High Purity GaAs
G. E. Stillman and C. M. Wolfe, Avalanche Photodiodes
P. L. Richards, The Josephson Junction as a Detector of Microwave and Far-Infrared Radiation
E. H. Putley, The Pyroelectric Detector–An Update

Volume 13 Cadmium Telluride

Kenneth Zanio, Materials Preparation; Physics; Defects; Applications

CONTENTS OF VOLUMES IN THIS SERIES

Volume 14 Lasers, Junctions, Transport

N. Holonyak, Jr. and M. H. Lee, Photopumped III–V Semiconductor Lasers
Henry Kressel and Jerome K. Butler, Heterojunction Laser Diodes
A. Van der Ziel, Space-Charge-Limited Solid-State Diodes
Peter J. Price, Monte Carlo Calculation of Electron Transport in Solids

Volume 15 Contacts, Junctions, Emitters

B. L. Sharma, Ohmic Contacts to III–V Compound Semiconductors
Allen Nussbaum, The Theory of Semiconducting Junctions
John S. Escher, NEA Semiconductor Photoemitters

Volume 16 Defects, (HgCd)Se, (HgCd)Te

Henry Kressel, The Effect of Crystal Defects on Optoelectronic Devices
C. R. Whitsett, J. G. Broerman, and C. J. Summers, Crystal Growth and Properties of $Hg_{1-x}Cd_xSe$ Alloys
M. H. Weiler, Magnetooptical Properties of $Hg_{1-x}Cd_xTe$ Alloys
Paul W. Kruse and John G. Ready, Nonlinear Optical Effects in $Hg_{1-x}Cd_xTe$

Volume 17 CW Processing of Silicon and Other Semiconductors

James F. Gibbons, Beam Processing of Silicon
Arto Lietoila, Richard B. Gold, James F. Gibbons, and Lee A. Christel, Temperature Distributions and Solid Phase Reaction Rates Produced by Scanning CW Beams
Arto Lietoila and James F. Gibbons, Applications of CW Beam Processing to Ion Implanted Crystalline Silicon
N. M. Johnson, Electronic Defects in CW Transient Thermal Processed Silicon
K. F. Lee, T. J. Stultz, and James F. Gibbons, Beam Recrystallized Polycrystalline Silicon: Properties, Applications, and Techniques
T. Shibata, A. Wakita, T. W. Sigmon, and James F. Gibbons, Metal-Silicon Reactions and Silicide
Yves I. Nissim and James F. Gibbons, CW Beam Processing of Gallium Arsenide

Volume 18 Mercury Cadmium Telluride

Paul W. Kruse, The Emergence of $(Hg_{1-x}Cd_x)Te$ as a Modern Infrared Sensitive Material
H. E. Hirsch, S. C. Liang, and A. G. White, Preparation of High-Purity Cadmium, Mercury, and Tellurium
W. F. H. Micklethwaite, The Crystal Growth of Cadmium Mercury Telluride
Paul E. Petersen, Auger Recombination in Mercury Cadmium Telluride
R. M. Broudy and V. J. Mazurczyck, (HgCd)Te Photoconductive Detectors
M. B. Reine, A. K. Sood, and T. J. Tredwell, Photovoltaic Infrared Detectors
M. A. Kinch, Metal-Insulator-Semiconductor Infrared Detectors

Volume 19 Deep Levels, GaAs, Alloys, Photochemistry

G. F. Neumark and K. Kosai, Deep Levels in Wide Band-Gap III–V Semiconductors
David C. Look, The Electrical and Photoelectronic Properties of Semi-Insulating GaAs
R. F. Brebrick, Ching-Hua Su, and Pok-Kai Liao, Associated Solution Model for Ga-In-Sb and Hg-Cd-Te
Yu. Ya. Gurevich and Yu. V. Pleskov, Photoelectrochemistry of Semiconductors

CONTENTS OF VOLUMES IN THIS SERIES

Volume 20   Semi-Insulating GaAs

R. N. Thomas, H. M. Hobgood, G. W. Eldridge, D. L. Barrett, T. T. Braggins, L. B. Ta, and S. K. Wang, High-Purity LEC Growth and Direct Implantation of GaAs for Monolithic Microwave Circuits
C. A. Stolte, Ion Implantation and Materials for GaAs Integrated Circuits
C. G. Kirkpatrick, R. T. Chen, D. E. Holmes, P. M. Asbeck, K. R. Elliott, R. D. Fairman, and J. R. Oliver, LEC GaAs for Integrated Circuit Applications
J. S. Blakemore and S. Rahimi, Models for Mid-Gap Centers in Gallium Arsenide

Volume 21   Hydrogenated Amorphous Silicon
Part A

Jacques I. Pankove Introduction
Masataka Hirose, Glow Discharge; Chemical Vapor Deposition
Yoshiyuki Uchida, dc Glow Discharge
T. D. Moustakas, Sputtering
Isao Yamada, Ionized-Cluster Beam Deposition
Bruce A. Scott, Homogeneous Chemical Vapor Deposition
Frank J. Kampas, Chemical Reactions in Plasma Deposition
Paul A. Longeway, Plasma Kinetics
Herbert A. Weakliem, Diagnostics of Silane Glow Discharges Using Probes and Mass Spectroscopy
Lester Guttman, Relation between the Atomic and the Electronic Structures
A. Chenevas-Paule, Experiment Determination of Structure
S. Minomura, Pressure Effects on the Local Atomic Structure
David Adler, Defects and Density of Localized States

Part B

Jacques I. Pankove, Introduction
G. D. Cody, The Optical Absorption Edge of a-Si:H
Nabil M. Amer and Warren B. Jackson, Optical Properties of Defect States in a-Si:H
P. J. Zanzucchi, The Vibrational Spectra of a-Si:H
Yoshihiro Hamakawa, Electroreflectance and Electroabsorption
Jeffrey S. Lannin, Raman Scattering of Amorphous Si, Ge, and Their Alloys
R. A. Street, Luminescence in a-Si:H
Richard S. Crandall, Photoconductivity
J. Tauc, Time-Resolved Spectroscopy of Electronic Relaxation Processes
P. E. Vanier, IR-Induced Quenching and Enhancement of Photoconductivity and Photoluminescence
H. Schade, Irradiation-Induced Metastable Effects
L. Ley, Photoelectron Emission Studies

Part C

Jacques I. Pankove, Introduction
J. David Cohen, Density of States from Junction Measurements in Hydrogenated Amorphous Silicon
P. C. Taylor, Magnetic Resonance Measurements in a-Si:H
K. Morigaki, Optically Detected Magnetic Resonance
J. Dresner, Carrier Mobility in a-Si:H

T. Tiedje, Information about Band-Tail States from Time-of-Flight Experiments
Arnold R. Moore, Diffusion Length in Undoped a-Si:H
W. Beyer and J. Overhof, Doping Effects in a-Si:H
H. Fritzche, Electronic Properties of Surfaces in a-Si:H
C. R. Wronski, The Staebler-Wronski Effect
R. J. Nemanich, Schottky Barriers on a-Si:H
B. Abeles and T. Tiedje, Amorphous Semiconductor Superlattices

Part D

Jacques I. Pankove, Introduction
D. E. Carlson, Solar Cells
G. A. Swartz, Closed-Form Solution of I–V Characteristic for a-Si:H Solar Cells
Isamu Shimizu, Electrophotography
Sachio Ishioka, Image Pickup Tubes
P. G. LeComber and W. E. Spear, The Development of the a-Si:H Field-Effect Transitor and Its Possible Applications
D. G. Ast, a-Si:H FET-Addressed LCD Panel
S. Kaneko, Solid-State Image Sensor
Masakiyo Matsumura, Charge-Coupled Devices
M. A. Bosch, Optical Recording
A. D'Amico and G. Fortunato, Ambient Sensors
Hiroshi Kukimoto, Amorphous Light-Emitting Devices
Robert J. Phelan, Jr., Fast Detectors and Modulators
Jacques I. Pankove, Hybrid Structures
P. G. LeComber, A. E. Owen, W. E. Spear, J. Hajto, and W. K. Choi, Electronic Switching in Amorphous Silicon Junction Devices

Volume 22 Lightwave Communications Technology
Part A

Kazuo Nakajima, The Liquid-Phase Epitaxial Growth of InGaAsP
W. T. Tsang, Molecular Beam Epitaxy for III–V Compound Semiconductors
G. B. Stringfellow, Organometallic Vapor-Phase Epitaxial Growth of III–V Semiconductors
G. Beuchet, Halide and Chloride Transport Vapor-Phase Deposition of InGaAsP and GaAs
Manijeh Razeghi, Low-Pressure Metallo-Organic Chemical Vapor Deposition of $Ga_xIn_{1-x}As_yP_{1-y}$ Alloys
P. M. Petroff, Defects in III–V Compound Semiconductors

Part B

J. P. van der Ziel, Mode Locking of Semiconductor Lasers
Kam Y. Lau and Amnon Yariv, High-Frequency Current Modulation of Semiconductor Injection Lasers
Charles H. Henry, Spectral Properties of Semiconductor Lasers
Yasuharu Suematsu, Katsumi Kishino, Shigehisa Arai, and Fumio Koyama, Dynamic Single-Mode Semiconductor Lasers with a Distributed Reflector
W. T. Tsang, The Cleaved-Coupled-Cavity (C^3) Laser

Part C

R. J. Nelson and N. K. Dutta, Review of InGaAsP/InP Laser Structures and Comparison of Their Performance
N. Chinone and M. Nakamura, Mode-Stabilized Semiconductor Lasers for 0.7–0.8- and 1.1–1.6-μm Regions
Yoshiji Horikoshi, Semiconductor Lasers with Wavelengths Exceeding 2 μm
B. A. Dean and M. Dixon, The Functional Reliability of Semiconductor Lasers as Optical Transmitters
R. H. Saul, T. P. Lee, and C. A. Burus, Light-Emitting Device Design
C. L. Zipfel, Light-Emitting Diode Reliability
Tien Pei Lee and Tingye Li, LED-Based Multimode Lightwave Systems
Kinichiro Ogawa, Semiconductor Noise-Mode Partition Noise

Part D

Federico Capasso, The Physics of Avalanche Photodiodes
T. P. Pearsall and M. A. Pollack, Compound Semiconductor Photodiodes
Takao Kaneda, Silicon and Germanium Avalanche Photodiodes
S. R. Forrest, Sensitivity of Avalanche Photodetector Receivers for High-Bit-Rate Long-Wavelength Optical Communication Systems
J. C. Campbell, Phototransistors for Lightwave Communications

Part E

Shyh Wang, Principles and Characteristics of Integratable Active and Passive Optical Devices
Shlomo Margalit and Amnon Yariv, Integrated Electronic and Photonic Devices
Takaaki Mukai, Yoshihisa Yamamoto, and Tatsuya Kimura, Optical Amplification by Semiconductor Lasers

Volume 23 Pulsed Laser Processing of Semiconductors

R. F. Wood, C. W. White, and R. T. Young, Laser Processing of Semiconductors: An Overview
C. W. White, Segregation, Solute Trapping, and Supersaturated Alloys
G. E. Jellison, Jr., Optical and Electrical Properties of Pulsed Laser-Annealed Silicon
R. F. Wood and G. E. Jellison, Jr., Melting Model of Pulsed Laser Processing
R. F. Wood and F. W. Young, Jr., Nonequilibrium Solidification Following Pulsed Laser Melting
D. H. Lowndes and G. E. Jellison, Jr., Time-Resolved Measurements During Pulsed Laser Irradiation of Silicon
D. M. Zehner, Surface Studies of Pulsed Laser Irradiated Semiconductors
D. H. Lowndes, Pulsed Beam Processing of Gallium Arsenide
R. B. James, Pulsed CO_2 Laser Annealing of Semiconductors
R. T. Young and R. F. Wood, Applications of Pulsed Laser Processing

Volume 24 Applications of Multiquantum Wells, Selective Doping, and Superlattices

C. Weisbuch, Fundamental Properties of III–V Semiconductor Two-Dimensional Quantized Structures: The Basis for Optical and Electronic Device Applications
H. Morkoc and H. Unlu, Factors Affecting the Performance of (Al, Ga)As/GaAs and

(Al, Ga)As/InGaAs Modulation-Doped Field-Effect Transistors: Microwave and Digital Applications
N. T. Linh, Two-Dimensional Electron Gas FETs: Microwave Applications
M. Abe et al., Ultra-High-Speed HEMT Integrated Circuits
D. S. Chemla, D. A. B. Miller, and P. W. Smith, Nonlinear Optical Properties of Multiple Quantum Well Structures for Optical Signal Processing
F. Capasso, Graded-Gap and Superlattice Devices by Band-gap Engineering
W. T. Tsang, Quantum Confinement Heterostructure Semiconductor Lasers
G. C. Osbourn et al., Principles and Applications of Semiconductor Strained-Layer Superlattices

Volume 25   Diluted Magnetic Semiconductors

W. Giriat and J. K. Furdyna, Crystal Structure, Composition, and Materials Preparation of Diluted Magnetic Semiconductors
W. M. Becker, Band Structure and Optical Properties of Wide-Gap $A_{1-x}^{II}Mn_xB^{VI}$ Alloys at Zero Magnetic Field
Saul Oseroff and Pieter H. Keesom, Magnetic Properties: Macroscopic Studies
T. Giebultowicz and T. M. Holden, Neutron Scattering Studies of the Magnetic Structure and Dynamics of Diluted Magnetic Semiconductors
J. Kossut, Band Structure and Quantum Transport Phenomena in Narrow-Gap Diluted Magnetic Semiconductors
C. Riqaux, Magnetooptics in Narrow Gap Diluted Magnetic Semiconductors
J. A. Gaj, Magnetooptical Properties of Large-Gap Diluted Magnetic Semiconductors
J. Mycielski, Shallow Acceptors in Diluted Magnetic Semiconductors: Splitting, Boil-off, Giant Negative Magnetoresistance
A. K. Ramdas and S. Rodriquez, Raman Scattering in Diluted Magnetic Semiconductors
P. A. Wolff, Theory of Bound Magnetic Polarons in Semimagnetic Semiconductors

Volume 26   III-V Compound Semiconductors and Semiconductor Properties of Superionic Materials

Zou Yuanxi, III-V Compounds
H. V. Winston, A. T. Hunter, H. Kimura, and R. E. Lee, InAs-Alloyed GaAs Substrates for Direct Implantation
P. K. Bhattacharya and S. Dhar, Deep Levels in III-V Compound Semiconductors Grown by MBE
Yu. Ya. Gurevich and A. K. Ivanov-Shits, Semiconductor Properties of Superionic Materials

Volume 27   High Conducting Quasi-One-Dimensional Organic Crystals

E. M. Conwell, Introduction to Highly Conducting Quasi-One-Dimensional Organic Crystals
I. A. Howard, A Reference Guide to the Conducting Quasi-One-Dimensional Organic Molecular Crystals
J. P. Pouget, Structural Instabilities
E. M. Conwell, Transport Properties
C. S. Jacobsen, Optical Properties
J. C. Scott, Magnetic Properties
L. Zuppiroli, Irradiation Effects: Perfect Crystals and Real Crystals

Volume 28 Measurement of High-Speed Signals in Solid State Devices

J. Frey and D. Ioannou, Materials and Devices for High-Speed and Optoelectronic Applications
H. Schumacher and E. Strid, Electronic Wafer Probing Techniques
D. H. Auston, Picosecond Photoconductivity: High-Speed Measurements of Devices and Materials
J. A. Valdmanis, Electro-Optic Measurement Techniques for Picosecond Materials, Devices, and Integrated Circuits
J. M. Wiesenfeld and R. K. Jain, Direct Optical Probing of Integrated Circuits and High-Speed Devices
G. Plows, Electron-Beam Probing
A. M. Weiner and R. B. Marcus, Photoemissive Probing

Volume 29 Very High Speed Integrated Circuits: Gallium Arsenide LSI

M. Kuzuhara and T. Nozaki, Active Layer Formation by Ion Implantation
H. Hashimoto, Focused Ion Beam Implantation Technology
T. Nozaki and A. Higashisaka, Device Fabrication Process Technology
M. Ino and T. Takada, GaAs LSI Circuit Design
M. Hirayama, M. Ohmori, and K. Yamasaki, GaAs LSI Fabrication and Performance

Volume 30 Very High Speed Integrated Circuits: Heterostructure

H. Watanabe, T. Mizutani, and A. Usui, Fundamentals of Epitaxial Growth and Atomic Layer Epitaxy
S. Hiyamizu, Characteristics of Two-Dimensional Electron Gas in III–V Compound Heterostructures Grown by MBE
T. Nakanisi, Metalorganic Vapor Phase Epitaxy for High-Quality Active Layers
T. Mimura, High Electron Mobility Transistor and LSI Applications
T. Sugeta and T. Ishibashi, Hetero-Bipolar Transistor and Its LSI Application
H. Matsueda, T. Tanaka, and M. Nakamura, Optoelectronic Integrated Circuits

Volume 31 Indium Phosphide: Crystal Growth and Characterization

J. P. Farges, Growth of Discoloration-free InP
M. J. McCollum and G. E. Stillman, High Purity InP Grown by Hydride Vapor Phase Epitaxy
T. Inada and T. Fukuda, Direct Synthesis and Growth of Indium Phosphide by the Liquid Phosphorous Encapsulated Czochralski Method
O. Oda, K. Katagiri, K. Shinohara, S. Katsura, Y. Takahashi, K. Kainosho, K. Kohiro, and R. Hirano, InP Crystal Growth, Substrate Preparation and Evaluation
K. Tada, M. Tatsumi, M. Morioka, T. Araki, and T. Kawase, InP Substrates: Production and Quality Control
M. Razeghi, LP-MOCVD Growth, Characterization, and Application of InP Material
T. A. Kennedy and P. J. Lin-Chung, Stoichiometric Defects in InP

Volume 32 Strained-Layer Superlattices: Physics

T. P. Pearsall, Strained-Layer Superlattices
Fred H. Pollack, Effects of Homogeneous Strain on the Electronic and Vibrational Levels in Semiconductors

J. Y. Marzin, J. M. Gerárd, P. Voisin, and J. A. Brum, Optical Studies of Strained III–V Heterolayers
R. People and S. A. Jackson, Structurally Induced States from Strain and Confinement
M. Jaros, Microscopic Phenomena in Ordered Superlattices

Volume 33 Strained-Layer Superlattices: Materials Science and Technology

R. Hull and J. C. Bean, Principles and Concepts of Strained-Layer Epitaxy
William J. Schaff, Paul J. Tasker, Mark C. Foisy, and Lester F. Eastman, Device Applications of Strained-Layer Epitaxy
S. T. Picraux, B. L. Doyle, and J. Y. Tsao, Structure and Characterization of Strained-Layer Superlattices
E. Kasper and F. Schaffler, Group IV Compounds
Dale L. Martin, Molecular Beam Epitaxy of IV–VI Compound Heterojunctions
Robert L. Gunshor, Leslie A. Kolodziejski, Arto V. Nurmikko, and Nobuo Otsuka, Molecular Beam Epitaxy of II–VI Semiconductor Microstructures

Volume 34 Hydrogen in Semiconductors

J. I. Pankove and N. M. Johnson, Introduction to Hydrogen in Semiconductors
C. H. Seager, Hydrogenation Methods
J. I. Pankove, Hydrogenation of Defects in Crystalline Silicon
J. W. Corbett, P. Deák, U. V. Desnica, and S. J. Pearton, Hydrogen Passivation of Damage Centers in Semiconductors
S. J. Pearton, Neutralization of Deep Levels in Silicon
J. I. Pankove, Neutralization of Shallow Acceptors in Silicon
N. M. Johnson, Neutralization of Donor Dopants and Formation of Hydrogen-Induced Defects in n-Type Silicon
M. Stavola and S. J. Pearton, Vibrational Spectroscopy of Hydrogen-Related Defects in Silicon
A. D. Marwick, Hydrogen in Semiconductors: Ion Beam Techniques
C. Herring and N. M. Johnson, Hydrogen Migration and Solubility in Silicon
E. E. Haller, Hydrogen-Related Phenomena in Crystalline Germanium
J. Kakalios, Hydrogen Diffusion in Amorphous Silicon
J. Chevallier, B. Clerjaud, and B. Pajot, Neutralization of Defects and Dopants in III–V Semiconductors
G. G. DeLeo and W. B. Fowler, Computational Studies of Hydrogen-Containing Complexes in Semiconductors
R. F. Kiefl and T. L. Estle, Muonium in Semiconductors
C. G. Van de Walle, Theory of Isolated Interstitial Hydrogen and Muonium in Crystalline Semiconductors

Volume 35 Nanostructured Systems

Mark Reed, Introduction
H. van Houten, C. W. J. Beenakker, and B. J. van Wees, Quantum Point Contacts
G. Timp, When Does a Wire Become an Electron Waveguide?
M. Büttiker, The Quantum Hall Effect in Open Conductors
W. Hansen, J. P. Kotthaus, and U. Merkt, Electrons in Laterally Periodic Nanostructures

Volume 36 The Spectroscopy of Bulk and Artificially Structured Semiconductors

D. Heiman, Laser Spectroscopy of Semiconductors at Low Temperatures and High Magnetic Fields
Arto V. Nurmikko, Transient Spectroscopy by Ultrashort Laser Pulse Techniques
A. K. Ramdas and S. Rodriguez, Piezospectroscopy of Semiconductors
Orest J. Glembockii and Benjamin V. Shanabrook, Photoreflectance Spectroscopy of Microstructures
David S. Seiler, Christopher L. Littler, and Margaret H. Weiler, One- and Two-Photon Magneto-Optical Spectroscopy of InSb and $Hg_{1-x}Cd_xTe$

Volume 37 The Mechanical Properties of Semiconductors

A.-B. Chen, Arden Sher and W. T. Yost, Elastic Constants and Related Properties of Semiconductor Compounds and Their Alloys
David R. Clarke, Fracture of Silicon and Other Semiconductors
Hans Siethoff, The Plasticity of Elemental and Compound Semiconductors
Sivaraman Guruswamy, Katherine T. Faber, and John P. Hirth, Mechanical Behavior of Compound Semiconductors
S. Mahajan, Deformation Behavior of Compound Semiconductors
John P. Hirth, Injection of Dislocations into Strained Multilayer Structures
Don L. Kendall, Charles B. Fleddermann, and Kevin J. Malloy, Critical Technologies for the Micromachining of Silicon
Ikuo Matsuba and Kinji Mokuya, Processing and Semiconductor Thermoelastic Behavior

ISBN 0-12-752137-2